Oleogels, Bigels, and Emulgels: Fabrication, Application and Research Trends

Oleogels, Bigels, and Emulgels: Fabrication, Application and Research Trends

Guest Editors

Cristina Ghinea
Ana Leahu

MDPI

Basel • Beijing • Wuhan • Barcelona • Belgrade • Novi Sad • Cluj • Manchester

Guest Editors
Cristina Ghinea
Faculty of Food Engineering
Stefan cel Mare University
of Suceava
Suceava
Romania

Ana Leahu
Faculty of Food Engineering
Stefan cel Mare University
of Suceava
Suceava
Romania

Editorial Office
MDPI AG
Grosspeteranlage 5
4052 Basel, Switzerland

This is a reprint of the Special Issue, published open access by the journal *Gels* (ISSN 2310-2861), freely accessible at: https://www.mdpi.com/journal/gels/special_issues/C72NR774CM.

For citation purposes, cite each article independently as indicated on the article page online and as indicated below:

Lastname, A.A.; Lastname, B.B. Article Title. *Journal Name* **Year**, *Volume Number*, Page Range.

ISBN 978-3-7258-6374-7 (Hbk)
ISBN 978-3-7258-6375-4 (PDF)
https://doi.org/10.3390/books978-3-7258-6375-4

Contents

About the Editors

Cristina Ghinea

Cristina Ghinea is a Lecturer at the Faculty of Food Engineering, "Stefan cel Mare" University of Suceava, and holds a PhD in Environmental Engineering (2012), co-supervised by "Gheorghe Asachi" Technical University of Iasi, Romania, and Twente University, the Netherlands. Her teaching experience includes courses in Organic Chemistry, Life Cycle Assessment of Products, and Techniques for Investigating Environmental Factors. Her research interests are waste generation, food waste, waste management and waste valorization processes, life cycle assessment, waste management modeling and optimization, and statistical analysis. She has authored or co-authored 11 books and chapters, 1 book as Editor, 2 editorials in ISI journals, and 81 articles, of which 42 articles were published in ISI journals, 25 in international database journals, 7 in ISI proceedings, and 7 in IEEE conference publications.

Ana Leahu

Ana Leahu, with a PhD in Science and Agroforestry (2005), is an Associate Professor PhD. Eng at the Faculty of Food Engineering, "Stefan cel Mare" University of Suceava. Her doctoral thesis was conceived and carried out during her work at the Office of Pedological and Agrochemical Studies in Suceava, where she worked as a research chemical engineer with responsibilities in soil and plant sample analysis, thus benefiting from access to experimental fields of pasture and forage plants and the agro-phytotechnical laboratories of the Agricultural Research Center in the same locality. Ana Leahu has the ability to analyze functional products and focuses on the research and development of functional foods fortified with food waste as raw materials. She is also good at analyzing bioactive ingredients, covering phenolic compounds, antioxidant activity, and anthocyanins. She focuses on food chemistry and food testing technology and is proficient in using spectroscopy (such as nuclear magnetic resonance, Fourier-transform infrared spectroscopy, and ultraviolet–visible spectroscopy) and chromatography (gas chromatography–mass spectrometry) to carry out related research. Additionally, she is interested in waste management and waste value-added treatment and committed to exploring innovative paths for resource recycling. Ana Leahu has fruitful academic achievements, publishing 3 books, 48 papers in ISI-indexed journals, 74 papers in Romanian and international peer-reviewed journals included in databases, and 8 papers in peer-reviewed conference proceedings (4 of which are included in ISI conference proceedings). She also has two patents.

gels

Editorial

Oleogels, Bigels, and Emulgels: Fabrication, Application and Research Trends

Cristina Ghinea * and Ana Leahu

Faculty of Food Engineering, Stefan cel Mare University of Suceava, 720229 Suceava, Romania; analeahu@fia.usv.ro
* Correspondence: cristina.ghinea@fia.usv.ro

1. Introduction

Gels are created by entrapping liquid oil (oleogels) or water (hydrogels) into the well-organized three-dimensional network of a gelling agent [1]. Mixing oleogels and hydrogels in different proportions forms hybrid gels or bigels [1,2], while emulsion and gel mixtures can form emulgels [3].

Oleogels are a healthier alternative to saturated and trans fats [4], with applications in many areas of the food industry, such as bakery [5,6], meat [7,8], confectionary [9,10], dairy [11,12], etc. Oleogels can also be used in the pharmaceutical industry as carriers for nutraceuticals and in cosmetics for the delivery of lipophilic compounds or in dermatological products targeting skin barrier problems [13], while hydrogels have demonstrated significant environmental benefits in wastewater treatment [14]. Bigels also have applications in the food industry, but these have been under-investigated in the literature, compared to pharmaceutical applications, where they are mainly used as a drug delivery vehicle, particularly in transdermal applications [15]. Emulgels have been investigated as fat replacers (e.g., a margarine alternative in sponge cakes) in the production of functional foods [16] or for the treatment of various skin disorders [17].

The primary aim of this Special Issue, "Oleogels, Bigels, and Emulgels: Fabrication, Application, and Research Trends", is to investigate recent advances and current knowledge in the study of different gel types in the food and cosmetic industries, dentistry (i.e., for treating infected root canals), and water management. This Special Issue comprises nine contributions (seven original research articles and two review articles) that address thematic investigations into the preparation and use of gels as follows: as inks for 3D food printing technology, for replacing solid fats in cookies and sausages, in skin care applications, in antibacterial therapies for root canal disinfection, and for water control under extreme reservoir conditions.

2. Overview of the Publications in This Special Issue

In "Design of Aerated Oleogel–Hydrogel Mixtures for 3D Printing of Personalized Cannabis Edibles", Andriotis et al. (contribution 3) developed and investigated 3D printable edible inks by mixing oleogel and hydrogel phases. They used oleogels from cannabis seed oil and xanthan gum hydrogels mixed with air (via a syringe coupling method) to prepare these inks. The visual appearance and flow rate of different ink formulations, as well as printability, were evaluated. The rheological properties of different ink formulations, the microstructure of the inks, and the thermal behavior during heating were also investigated.

The results of this study contribute to 3D food printing technology, posing a method to introduce new ingredients into the diet and a possible solution to food insecurity [18].

The second study, "Effect of Cooling Rate on Properties of Beeswax and Stearic Acid Oleogel Based on Rice Bran Oil and Sesame Oil" by Sivakanthan et al. (contribution 4), explored the effects of cooling rates on the rheological properties, microstructure, and oil-binding capacity of oleogels. Oleogels were obtained via direct oleogelation from sesame oil–rice bran oil and beeswax–stearic acid mixtures. Their results showed that the oil-binding capacity was higher for oleogels that cooled more slowly. Also, rheological properties showed that slow cooling yields more desirable properties in the beeswax–stearic acid oleogel compared to rapid cooling rates. The authors stated that the higher oil-binding capacity could be attributed to the larger fractal dimensions of oleogels cooled at a slower rate. The results obtained from this research present potential for the food industry, especially in manipulating and improving oleogel properties.

The third study, "Textural, Color, and Sensory Analysis of Cookies Prepared with Hemp Oil-Based Oleogels", by Leahu et al. (contribution 5), examined the rheological and textural profile of cookies formulated with hemp seed oil oleogels. The oleogelation agents used were beeswax, candelilla wax, carnauba wax, sitosterol, pea protein, and xanthan gum. In this work, the potential of oleogels as solid fat substitutes in cookies was investigated. The results showed increased hardness of cookie doughs with oleogels and of baked products. The colors of the cookie samples were different depending on the oleogel used; for example, the color of those with natural wax and hemp seed oil is darker, with a red and yellow hue. Investigating the use of different gelators for oleogels could have practical applications as a replacement for solid fats.

In "Injectable and Conductive Polyurethane Gel with Load-Responsive Antibiosis for Sustained Root Canal Disinfection", Mu et al. (contribution 6) developed injectable conductive polyurethane gels incorporating piezoelectric n-$BaTiO_3$ as a sustained bacteriostasis strategy for root canal therapy. The synthesis and characterization of conductive polyurethane-based prepolymers, physicochemical properties of injectable conductive polyurethane gels for root canal therapy, characterization of electrochemical properties of cured gels, bacteriostasis properties of cured gels, in vitro biofilm eradication and bacterial killing efficiency of cured gels, and others were discussed in this paper. This study provides a significant advance in active response therapeutic systems for prolonged infection control in root canal therapy.

The following article, "Development of Liposome-Based Hydrogel Patches Incorporating Essential Oils of African Plants and Deep Eutectic Solvents" by Jiang et al. (contribution 7), introduced liposome-based hydrogel patches that integrate deep eutectic solvent, argan oil, and passion fruit seed oil. The authors use liposomal encapsulation to stabilize the deep eutectic solvent (composed of betaine and phytic acid) against hydration-induced degradation and synergy with essential oils to enhance transdermal permeability in advanced skin care applications. Spectral and chromatographic analysis of betaine-phytic acid, analysis of the essential oil's properties (pH, density, apparent viscosity, percentage of free radical scavenging capacity), and characterization of nanoliposome structure and properties were performed. Nanoliposome-loaded hydrogel patches were also investigated and characterized. Their results showed that the hydrogel has exceptional hydration properties, which can improve skin texture and appearance. Also, colloidal stability and transdermal permeability were significantly improved by encapsulating bioactive oils in nanoliposomes.

In "Phytochemical Characterization, Bioactivities, and Nanoparticle-Based Topical Gel Formulation Development from Four Mitragyna speciosa Varieties", Anantaworasakul et al. (contribution 9) developed topical gels based on nanoparticles from ethanolic extracts of four varieties of kratom for cosmeceutical skincare applications. The kratom nanoparticle formulations were characterized based on particle size and polydispersity index, as well as zeta potential values. Moreover, phytochemical analysis, antioxidant activities (DPPH, ABTS, and FRAP), and collagenase and elastase inhibitory activities were investigated. The results of Anantaworasakul et al.'s study (contribution 9) revealed that kratom nanoparticle-loaded gels had a brownish appearance, with good clarity and the uniform dispersion of nanoparticles within the gel. The gels obtained in this work seem suitable for skin application, as they have an acceptable pH, constant viscosity, and stable appearance.

The article "In Situ-Prepared Nanocomposite for Water Management in High-Temperature Reservoirs", by Yang et al. (contribution 8), presented the development of a hydrogel system via in situ crosslinking of polyacrylamide with phenolic resin, reinforced with silica sol nanoparticles, for application in water control in high-temperature reservoirs. The authors investigated improvements in gelation performance, thermal stability, water retention, strengthening mechanism, and relevant field performance. They determined that a gelation period of 72 h at 130 °C was optimal for the gelation system. Thus, under these conditions, a plugging efficiency of 92.4% was achieved. Finally, their results outline a practical solution for water control under extreme reservoir conditions that require delayed gelation and sustained mechanical performance.

This Special Issue also includes two comprehensive review articles that synthesize recent advances in the fabrication, application, and research trends of oleogels, bigels, and emulgels. The first review, "Food-Grade Bigel Systems: Formulation, Characterization, and Applications for Novel Food Product Development" by Zampouni et al. (contribution 1), described recent advances in the development of the two structured phases (oleogels and hydrogels) of bigels and their use in various food applications. The authors discussed the types and characteristics of oleogelators, oils, and hydrogels. They also presented the bigel preparation techniques (cold and hot homogenization) and the most used characterization techniques for bigel systems (physical properties, microstructure, textural and rheological properties, thermal properties, swelling capacity, and stability). Bigels can be used as delivery systems for bioactive compounds (lycopene, lutein, β-carotene, quercetin, fatty acids from vegetable oils), as fat replacers in meat and bakery products, or in 3D food printing and other applications, such as edible coatings.

The second review, "Exploring the Feasibility of Direct-Dispersion Oleogels in Healthier Sausage Formulations" by Mahmud et al. (contribution 2), evaluated the potential use of oleogels in healthier sausage formulations. The authors presented a brief history and current size of the sausage market, the health risks of sausage consumption, and market trends in the production of healthier sausages. Incorporating oleogels into sausages may be a solution to the challenges of low-fat and low-saturated fat sausages. This review presents and outlines various direct dispersion oleogels for sausage manufacturing applications, the incorporation of oleogels in the mixing step of sausage processing, oleogel mechanisms, emulsion stability and slice-ability, cooking yields, and the effects of oleogels on nutritional, textural, and sensory properties of sausages. By carefully optimizing formulation, processing, and equipment to maintain consistent quality and functionality, oleogels could be industrially applied in sausage manufacturing to meet public health objectives. But rigorous quality control measures, optimization of formulation, processing, and equipment, cost efficiency, regulatory compliance, and appropriate labeling will be required.

3. Conclusions

The research and reviews presented in this Special Issue highlight the methods of preparation, analysis, and use of different types of gels in the food industry and beyond. Innovation starts from gel compositions (kratom nanoparticle-loaded gels; cannabis seed oil, hemp seed oil, sesame oil, and rice bran oil oleogels; xanthan gum hydrogel; hydrogel system with polyacrylamide, phenolic resin, reinforced by silica sol nanoparticles; encapsulation of *Argania spinosa* and *Passiflora edulis* essential oils in hydrogel; polyurethane gels incorporating piezoelectric n-$BaTiO_3$) and selection of preparation methods (like cold and hot homogenization techniques for bigel preparation), and continues with the characterization of gels to understand and optimize their properties (gelation and application performances; antioxidant activities; appearance, microstructure, swelling, and mechanical properties of hydrogel patches; antibacterial properties of gels; oil-binding capacity and rheological properties of oleogels, etc.). The microstructure, mechanical, and thermal properties and stability of gels should be determined before their use in food applications to ensure desired functionality and consumer acceptance. The uniform distribution of oleogels in the food matrix, with the preservation of the gel structure under different conditions, as well as the replication of the sensory and textural characteristics of animal fats by oleogels, represents a key challenge in the food industry. Furthermore, long-lasting skin hydration by hydrogels is crucial for skin care applications. Investigating the antibacterial properties of gels under pressure-induced activation for root canal therapy is particularly important in dentistry for more effective and dynamic long-term disinfection compared to traditional static methods. Extended gelation time, high thermal stability, effective operational efficiency, and gel degradability are critical for enhanced oil recovery in high-temperature reservoirs, particularly for water control.

The future of gels is very promising in the food, cosmetic, pharmaceutical, dental, and environmental sectors, due to their unique properties. However, much more detailed research is needed. The full potential of gels can be unlocked through collaborative interdisciplinary research, essential for the transition from laboratory experiments to industry; therefore, engineers, chemists, biologists, and clinicians should work together.

Author Contributions: Conceptualization, C.G. and A.L.; writing—original draft preparation, C.G. and A.L.; review and editing, C.G. and A.L. All authors have read and agreed to the published version of the manuscript.

Acknowledgments: As the Guest Editors, we would like to thank all the authors who submitted their work to this Special Issue. We also extend our special thanks to all anonymous reviewers involved in this Special Issue who helped the authors to improve their manuscripts. We would also like to express our many thanks to the editorial staff of *Gels*—particularly the Managing Editors and Assistant Editors—for their invaluable support and careful handling of the publication process. Given the success and strong engagement of this Special Issue, we are pleased to announce that we have initiated the second edition of "Oleogels, Bigels, and Emulgels: Fabrication, Application, and Research Trends" to continue highlighting research in this field. We invite colleagues from academia and industry to contribute to this ongoing effort.

Conflicts of Interest: The authors declare no conflicts of interest.

List of Contributions:

1. Zampouni, K.; Dimakopoulou-Papazoglou, D.; Katsanidis, E. Food-Grade Bigel Systems: Formulation, Characterization, and Applications for Novel Food Product Development. *Gels* **2024**, *10*, 712. https://doi.org/10.3390/gels10110712.
2. Mahmud, N.; Ferdaus, M.J.; Silva, R.C.d. Exploring the Feasibility of Direct-Dispersion Oleogels in Healthier Sausage Formulations. *Gels* **2024**, *10*, 819. https://doi.org/10.3390/gels10120819.
3. Andriotis, E.G.; Paraskevopoulou, A.; Fatouros, D.G.; Zhang, H.; Ritzoulis, C. Design of Aerated Oleogel-Hydrogel Mixtures for 3D Printing of Personalized Cannabis Edibles. *Gels* **2024**, *10*, 654. https://doi.org/10.3390/gels10100654.

4. Sivakanthan, S.; Fawzia, S.; Mundree, S.; Madhujith, T.; Karim, A. Effect of Cooling Rate on Properties of Beeswax and Stearic Acid Oleogel Based on Rice Bran Oil and Sesame Oil. *Gels* **2024**, *10*, 697. https://doi.org/10.3390/gels10110697.
5. Leahu, A.; Ghinea, C.; Ropciuc, S.; Damian, C. Textural, Color, and Sensory Analysis of Cookies Prepared with Hemp Oil-Based Oleogels. *Gels* **2025**, *11*, 46. https://doi.org/10.3390/gels11010046.
6. Mu, B.; Lei, X.; Zhang, Y.; Zhang, J.; Du, Q.; Li, Y.; Huang, D.; Wang, L.; Li, J.; Li, Y.; et al. Injectable and Conductive Polyurethane Gel with Load-Responsive Antibiosis for Sustained Root Canal Disinfection. *Gels* **2025**, *11*, 346. https://doi.org/10.3390/gels11050346.
7. Jiang, W.; Toufouki, S.; Mahmood, S.; Ahmad, A.; Yohannes, A.; Xiang, Y.; Yao, S. Development of Liposome-Based Hydrogel Patches Incorporating Essential Oils of African Plants and Deep Eutectic Solvents. *Gels* **2025**, *11*, 364. https://doi.org/10.3390/gels11050364.
8. Yang, H.; Zhang, J.; Wang, Z.; Li, S.; Wei, Q.; He, Y.; Li, L.; Zhao, J.; Xu, C.; Zhang, Z. In Situ-Prepared Nanocomposite for Water Management in High-Temperature Reservoirs. *Gels* **2025**, *11*, 405. https://doi.org/10.3390/gels11060405.
9. Anantaworasakul, P.; Preedalikit, W.; Anantaworasakul, P.; Singh, S.; Intharuksa, A.; Arunotayanun, W.; Na Takuathung, M.; Yotsawimonwat, S.; Chittasupho, C. Phytochemical Characterization, Bioactivities, and Nanoparticle-Based Topical Gel Formulation Development from Four *Mitragyna speciosa* Varieties. *Gels* **2025**, *11*, 494. https://doi.org/10.3390/gels11070494.

References

1. Zampouni, K.; Mouzakitis, C.K.; Lazaridou, A.; Moschakis, T.; Katsanidis, E. Physicochemical properties and microstructure of bigels formed with gelatin and κ-carrageenan hydrogels and monoglycerides in olive oil oleogels. *Food Hydrocoll.* **2023**, *140*, 108636. [CrossRef]
2. Łętocha, A.; Miastkowska, M.; Sikora, E.; Michalczyk, A.; Liszka-Skoczylas, M.; Witczak, M. Hybrid Systems of Oleogels and Probiotic-Loaded Alginate Carriers for Potential Application in Cosmetics. *Molecules* **2024**, *29*, 5984. [CrossRef] [PubMed]
3. Milutinov, J.; Krstonošić, V.; Ćirin, D.; Pavlović, N. Emulgels: Promising Carrier Systems for Food Ingredients and Drugs. *Polymers* **2023**, *15*, 2302. [CrossRef] [PubMed]
4. Manzoor, S.; Masoodi, F.A.; Naqash, F.; Rashid, R. Oleogels: Promising alternatives to solid fats for food applications. *Food Hydrocoll. Health* **2022**, *2*, 100058. [CrossRef]
5. Mert, B.; Demirkesen, I. Reducing saturated fat with oleogel/shortening blends in a baked product. *Food Chem.* **2016**, *199*, 809–816. [CrossRef] [PubMed]
6. Kim, J.Y.; Lim, J.; Lee, J.; Hwang, H.S.; Lee, S. Utilization of oleogels as a replacement for solid fat in aerated baked goods: Physicochemical, rheological, and tomographic characterization. *J. Food Sci.* **2017**, *82*, 445–452. [CrossRef] [PubMed]
7. Zetzl, A.K.; Marangoni, A.G.; Barbut, S. Mechanical properties of ethylcellulose oleogels and their potential for saturated fat reduction in frankfurters. *Food Funct.* **2012**, *3*, 327–337. [CrossRef] [PubMed]
8. Oh, I.; Lee, J.; Lee, H.G.; Lee, S. Feasibility of hydroxypropyl methylcellulose oleogel as an animal fat replacer for meat patties. *Food Res. Int.* **2019**, *122*, 566–572. [CrossRef] [PubMed]
9. Patel, A.R.; Rajarethinem, P.S.; Gredowska, A.; Turhan, O.; Lesaffer, A.; De Vos, W.H.; Van de Walle, D.; Dewettinck, K. Edible applications of shellac oleogels: Spreads, chocolate paste and cakes. *Food Funct.* **2014**, *5*, 645–652. [CrossRef] [PubMed]
10. Si, H.; Cheong, L.-Z.; Huang, J.; Wang, X.; Zhang, H. Physical properties of soybean oleogels and oil migration evaluation in model praline system. *J. Am. Oil Chem. Soc.* **2016**, *93*, 1075–1084. [CrossRef]
11. Park, C.; Bemer, H.L.; Maleky, F. Oxidative Stability of Rice Bran Wax Oleogels and an Oleogel Cream Cheese Product. *J. Am. Oil Chem. Soc.* **2018**, *95*, 1267–1275. [CrossRef]
12. Ropciuc, S.; Ghinea, C.; Leahu, A.; Prisacaru, A.E.; Oroian, M.A.; Apostol, L.C.; Dranca, F. Development and Characterization of New Plant-Based Ice Cream Assortments Using Oleogels as Fat Source. *Gels* **2024**, *10*, 397. [CrossRef] [PubMed]
13. Pawar, V.U.; Dessai, A.D.; Nayak, U.Y. Oleogels: Versatile novel semi-solid system for pharmaceuticals. *AAPS PharmSciTech* **2024**, *25*, 146. [CrossRef] [PubMed]
14. Yang, G.; Zhang, J.; Abdullah, R.; Cheah, W.Y.; Zhao, D.; Ling, T.C. Comprehensive Advancements in Hydrogel, and Its Application in Microalgae Cultivation and Wastewater Treatment. *J. Microbiol. Biotechnol.* **2024**, *35*, e2407038. [CrossRef] [PubMed]
15. Francavilla, A.; Corradini, M.G.; Joye, I.J. Bigels as Delivery Systems: Potential Uses and Applicability in Food. *Gels* **2023**, *9*, 648. [CrossRef] [PubMed]

16. Bruno, E.; Lupi, F.R.; Mammolenti, D.; Mileti, O.; Baldino, N.; Gabriele, D. Emulgels Structured with Dietary Fiber for Food Uses: A Rheological Model. *Foods* **2022**, *11*, 3866. [CrossRef]
17. Alexander, A.; Khichariya, A.; Gupta, S.; Patel, R.J.; Giri, T.K.; Tripathi, D.K. Recent expansions in an emergent novel drug delivery technology: Emulgel. *J. Control. Release* **2013**, *171*, 122–132. [CrossRef] [PubMed]
18. Zhu, W.; Iskandar, M.M.; Baeghbali, V.; Kubow, S. Three-Dimensional Printing of Foods: A Critical Review of the Present State in Healthcare Applications, and Potential Risks and Benefits. *Foods* **2023**, *12*, 3287. [CrossRef] [PubMed]

gels

MDPI

Article

Design of Aerated Oleogel–Hydrogel Mixtures for 3D Printing of Personalized Cannabis Edibles

Eleftherios G. Andriotis [1,*], Adamantini Paraskevopoulou [2], Dimitrios G. Fatouros [3], Hui Zhang [4] and Christos Ritzoulis [1]

1 Department of Food Science and Technology, International Hellenic University, 57400 Thessaloniki, Greece
2 Laboratory of Food Chemistry and Technology, School of Chemistry, Aristotle University of Thessaloniki, 54124 Thessaloniki, Greece
3 Laboratory of Pharmaceutical Technology, Department of Pharmacy, Faculty of Health Sciences, Aristotle University of Thessaloniki, 54124 Thessaloniki, Greece
4 College of Biosystems Engineering and Food Science, Zhejiang University, Hangzhou 310058, China
* Correspondence: e.andriotis@ihu.gr

Abstract: Cannabis seed oil oleogel structured with Glycerol Monostearate (20% *w*/*w*) was mixed with xanthan gum hydrogel (2% *w*/*w*) at different ratios ranging from 0% *w*/*w* hydrogel to 75% *w*/*w* hydrogel, using a syringe-to-syringe apparatus, for the preparation of 3D-printable food inks. This process enabled the simultaneous blend of oleogel and hydrogel phases and the incorporation of air in a reproducible and accurate manner. The printability of bigel inks with different mass ratios was evaluated by using a conventional benchtop food 3D printer. The printability of the inks was found to be negatively affected by the presence of higher portions of the hydrogel phase, while the printing performance of pure cannabis seed oil oleogel was superior compared to the printing performance of the bigel inks. The physicochemical properties of hybrid gels were investigated with rheological studies, thermophysical studies (Differential Scanning Calorimetry), Polarized Light Microscopy, and Confocal Laser Scanning Microscopy. The microstructure of the aerated inks was affected by the presence of a higher oleogel fraction, in terms of air bubble shape and distribution. The addition of hydrogel at concentrations higher than 50% *w*/*w* had a strong negative effect on the mechanical properties of the inks leading to a partial collapse of the printed structures and subsequently to poor printing performance.

Keywords: oleogels; bigels; cannabis seed oil; 3D printing; personalized dosage forms

1. Introduction

Cannabis Sativa L. is one of the most recognizable and the oldest plants used for medical purposes, as it is known to contain a vast number of pharmaceutical compounds. There are more than 1200 different chemical compounds in the cannabis plant, while there are more than 100 known cannabinoids [1]. The hemp plant is also known for production of other products, like hemp or cannabis seed oil (CSO). CSO is very valuable in nutritional terms due to its composition (unsaturated fatty acids), and it is considered one of the best carriers for hosting active cannabinoids, like cannabidiol (CBD) or cannabigerol (CBG) [2]. Hempseeds are commonly consumed due to their pleasant nutty flavor and rich nutritional content [3]. They serve as an excellent source of essential fatty acids, minerals, vitamins, dietary fibers, and proteins such as edestin and albumin [3]. On the other hand, hempseed oil consists of polyunsaturated fatty acids, which are recognized for their positive impact on cardiovascular health, cancer prevention, and inflammatory conditions [3,4]. Consuming small quantities of hempseed oil (2 g/day) did not impact plasma levels of total cholesterol, high-density cholesterol (HDL), low-density cholesterol (LDL), or triglycerides. Additionally, it did not influence platelet aggregation or circulating inflammatory markers. While the evidence is limited, these findings imply that incorporating hempseed oil into one's diet may offer potential health benefits [3,5]. On the other hand, the presence of polyunsaturated fatty acids in cannabis seed oil make it prone to oxidation during storage and exposure to heat. As a result, it is not recommended for food preparations that involve prolonged high-temperature cooking. To mitigate this issue, consumers can use smaller containers to manage consumption and ensure freshness [5]. The oxidative stability of cannabis seed oil is one of the reasons why this oil is not widely applied for the preparation of cannabinoids infused oil, suitable for the preparation of food products containing cannabis extract, known as cannabis edibles [6]. The concentration of active cannabinoids in edibles can vary across a single product and across batches formulated at different times, making it difficult for users to estimate the consumption amount. The lack of consistency may cause users of cannabis edibles to consume higher or lower than intended amounts of the active ingredients. Additive manufacturing of food systems can also be applied to address those issues. Food 3D printing (F3DP) is an emerging technology in the food research field, as it can provide novel routes for preparation of the food materials towards the fabrication of accurate 3D structures with defined textural properties [7,8].

The most widely applied method of food 3D printing is extrusion-based printing, which involves the extrusion of the food-stuff through a nozzle in the form of a paste [8,9]. Food-grade biopolymers such as xanthan, gellan gums, agar, carrageenan, alginate, gelatine, collagen, and chitosan are extensively studied as potential substrates for extrusion-based food 3D printing [10]. The final properties of the 3D printed specimen are highly dependent on the initial 3D printable ink properties, and subsequently through this dependent accuracy, personalization can be realized [11]. At this moment, the most extensively studied class of materials as 3D printable inks turns out to be hydrogels. Even though hydrogels can cover a wide range of structural and textural properties, with applications ranging from food products to pharmaceutical products for controlled drug release [12,13], they also have some limitations mainly in terms of printing precision and overall structural stability [8,14–16]. One suggested way to overcome these limitations is introducing solid lipids in the form of an oleogel, which could inhibit the network structure and alter the viscosity of the system, making it more stable [8,17]. Additionally, one of the most advantageous features of lipid-based food 3D printing is the broadening of the nutrient sources and the possibility of introducing poorly soluble substances to the system.

Bigels are biphasic systems consisting of phases with varying polarities. These systems exhibit stable structures characterized as viscoelastic semisolids [8,18,19], owing to the advantages of both hydrogel and organogel [8,20]. Additionally, bigels can be applied for the delivery of both hydrophilic and hydrophobic actives. They are easy to prepare and can be easily modified by simply adjusting the proportion of each phase in the system [8]. According to the existing literature, the application of bigels in 3D printing technology was justified by the fact that bigels could combine two phases and, subsequently, they could be a carrier system for more nutritional or pharmaceutical agents to enhance nutritional diversity [8] or personalized dosage forms for the accurate dosing of active pharmaceutical ingredients [1,21]. The 3D printing of such systems is already demonstrated in the literature where beeswax oleogel mixtures with carrageenan and xanthan gum [8] or candelilla wax oleogel mixed with gelatin hydrogel [22] were successfully printed using a 3D printer.

To ensure personalization and to avoid cannabis seed oil storage, processes that could be applied to small volumes of oil should be adopted. One such process is the syringe coupler method, inspired by the modified Tessari method for producing sclerosing foam [23]. This method is suitable for mixing small volumes ($\leq$5 mL) and can be applied under fully aseptic conditions and even in the presence of inert gases like nitrogen [23]. The necessary equipment for the syringe coupler method is considered as low cost and easily accessible [23]. The method exploits the shear forces that are developed during compression and expansion of a mixture through a syringe nozzle [23]. The application of this method is always followed by the incorporation of a small amount of gas into the mixture, which is very difficult to avoid [23]. Nevertheless, the syringe coupler method can be applied for the precise incorporation of air into the system, towards the formation of foams with predetermined gas concentration [23].

To the best of our knowledge, there is a limited number of studies concerning the storage stability of cannabis seed oil, while there is no available data about the use of cannabis seed oil as a raw material for the preparation of 3D-printed cannabis edibles. In this study, cannabis edibles were considered as both food and dosage forms for the administration of cannabinoids. This approximation required the preparation of novel food systems in such a way that would be ready to host accurate doses of the active ingredients of cannabis. For this purpose, the technology of food additive manufacturing (food 3D printing) was applied to prepare precise and personalized cannabis edibles. To achieve this goal, 3D-printable food inks must be prepared with enhanced printability, to be suitable to serve as carriers for both hydrophilic and hydrophobic substances. The present study was focused on the development and evaluation of 3D-printable food systems that could be used as platform formulations (food printing inks) for printing food specimens with predetermined geometrical and physical characteristics. Thus, different formulations were characterized and evaluated for their printing performance by printing different model shapes like 1D, 2D, and 3D shapes.

2. Results and Discussion

2.1. Inks Appearance and Overrun Evaluation

The visual appearance of different formulations after the air-mixing process are exhibited in Figure 1A. In all cases, thick foam-like inks with excellent self-supporting characteristics were achieved. Additionally, the maximum overrun values for the different formulations are depicted in Figure 1B. All formulations were off-white, self-supporting semi-solid soft materials. The increase in the hydrogel portion in the formulation had a negative effect on the shape and uniformity of the foams. On the other hand, apart from the C3X1 sample, the maximum overrun was process-dependent, as there was no significant difference between the different samples. The C3X1 sample exhibited a higher overrun, suggesting a synergistic effect of the hydrogel on the stabilization of the air bubbles when used in smaller concentrations. The overall foaming properties of bigels with a higher concentration of oleogel (fatty acid crystals) agreed with the existing literature where it was reported that the foamability of bigels was affected by the presence of oleogel indicating that fatty acid crystals in the oil stabilized the air bubbles in whipped oleogels [24]. Conclusively, the as-prepared aerated bigels were considered visually homogeneous and were suitable for extrusion-based printing.

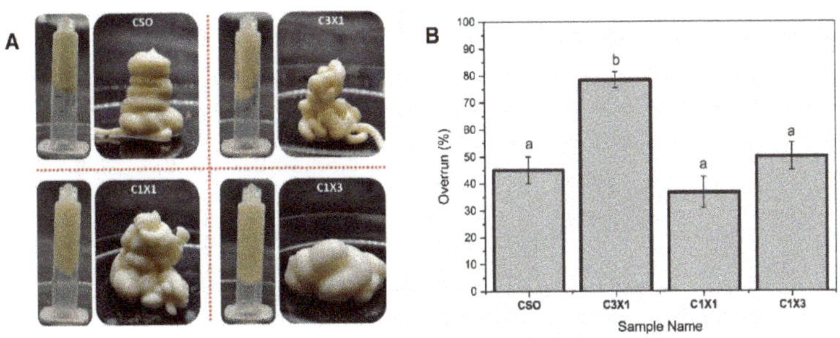

Figure 1. (**A**) Appearance evaluation and (**B**) overrun of the different ink formulations. Different lowercase letters indicate statistically significant differences.

2.2. Printability Evaluation

This study aimed to develop a 3D-printable food ink designed for the accurate dosing of cannabinoids. To ensure that the printed dosage form could be accurately printed, the printing process must be evaluated during the one-dimensional printing step (single-line filament extrusion), the two-dimensional printing step (single-layer printing), and finally the three-dimensional printing step (3D-printed specimen). The printing of the different model shapes are presented in Figure 2A. Specifically, the different printing steps that were in the printing process are separately discussed (namely 1D printing, 2D printing (2 mm, 3 mm, and 10 mm × 20 mm single layer), and 3D printing, from right to left, respectively). According to Figure 2A, all the different inks can be accurately printed in one-dimension, with minor defects, as in the case of C1X3, where there was minor material dragging by the nozzle. Nevertheless, the accuracy of the shape was followed by a discrepancy in the accuracy of the dimensions, compared to the given design, which stands for all the formulations. The deviation from the initial dimensions based on layer area is summarized in Figure 2C. When it came to the printing of a single layer, samples of CSO and C3X1 performed more accurately when compared to samples of C1X1 and C1X3. More specifically, in the case of the C1X1 formulation, there were structural failures due to extrusion discontinuity caused by partial nozzle clogging that could momentarily change the nozzle diameter (under-extrusion conditions). The partial clogging of the nozzle is attributed to possible inhomogeneity of the ink due to the manual mixing process. Total blockage of the nozzle was not observed as the printing was continued without over-extrusion signs. In the case of the C1X3 formulation, the under-extrusion conditions were more intense, resulting in a very round-shaped printing that deviated from the sharp edges of the initial design. The three-dimensional printing performance of the different formulations was evaluated by qualitatively comparing the 3D printed specimens (Figure 2B,D). As previously mentioned, in the case of samples of C1X1 and C1X3, there was a build-up of discontinuities caused by the continuous under-extrusion conditions. Additionally, C1X3 showed a partial layer collapse and failed to achieve self-support. The lack of self-support resulted in the specimen exhibiting a trapezoid shape instead of the intended rectangular shape of the initial design. The same deviation due to partial layer collapse was also observed for the samples of C3X1 and C1X1 but in a different way. Specifically, due to the differences in layer heights, the distance of the nozzle from the last layer was higher than 1 mm, causing the extruded material to drop from a distance. This behavior is directly linked to the rheological properties of the ink formulations, and it is furthered discussed in the respective section. These conditions caused the material to form curls that resulted in a less sharp shape. Finally, the printability of the CSO was the most optimum in terms of sharpness, shape accuracy, and homogeneity. It must be noted that the printing conditions could be adjusted to fit every formulation property and achieve a satisfactory printing performance.

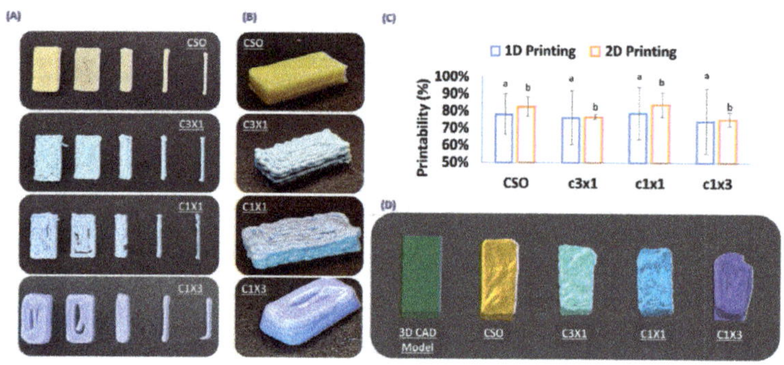

Figure 2. (**A**) Printability test of different formulations during 1D (single line), 2D (single layer), and 3D printing. (**B**) Printing performance of the different formulations (printing of a 3D rectangular shape; 10 mm × 20 mm × 5 mm). (**C**) Quantification of the printing performance for 1D and 2D printing. (**D**) 3D scanned printed specimens and comparison with initial 3D CAD model. Nile blue was used as a water-soluble colorant for the hydrogel phase. Different lowercase letters indicate statistically significant differences.

2.3. Rheological Studies

The three pillars of extrusion-based three-dimensional printing could be described as (i) extrusion, (ii) recovery, and (iii) self-support. These distinctive pillars were directly related to the rheological properties of the extruded material and could be described by the viscosity, thixotropic characteristics, elastic modulus (G′), viscous modulus (G″), and complex modulus (G*) [8,25].

In Figure 3, the viscosity of the different formulations was plotted against the shear rate, to evaluate the rheological properties of the materials during the extrusion process. It was observed that the increase in shear rate was followed by a decrease in the apparent viscosity, for all samples, indicating that the inks were non-Newtonian fluids. Shear thinning is essential for 3D-printable inks to be able to be extruded through the print head nozzle smoothly. Both C1X1 and C1X3 samples had similar apparent viscosity values that were lower than those of the samples CSO and C3X1 at lower shear rates. On the other hand, CSO and C3X1 samples also possessed similar apparent viscosity values (between them) that were higher than those of the other two samples. For higher values of shear rates, all samples displayed similar apparent viscosity values. This phenomenon was also observed in previously reported studies for bigels when the fraction of hydrogel was higher than 50% *w*/*w*, indicating a system transition from a hydrogel-dominated system to an oleogel-dominated system [8]. All previous data were fitted using the Ostwald-de Waele power law model (Table 1). The power law index n could be used to evaluate the extrusion ability of the samples through the nozzle. For all samples, n values were less than 1, indicating that the inks exhibited a shear-thinning character. The negative n values of samples CSO and C3X1 indicate stronger shear-thinning properties. The power law index values were also divided into two groups, as was mentioned earlier. The shear thinning properties of all samples directly affect the extrudability and printability of the inks. Materials that exhibit shear-thickening properties induce over-extrusion conditions during printing, attributed to pressure built-up within the syringe barrel caused by nozzle clogging. The K value was indicative of the different apparent viscosity values of the samples. When aerated oleogel (CSO) was used, the K value was the maximum (Table 1), though without significant statistical difference from the next sample, which also had a higher oleogel fraction. These two samples differed significantly from the C1X1 sample which had almost half of the K value. Finally, the significantly lowest K value was observed for the sample C1X3, indicating that when the hydrogel fraction exceeded 50%, there was a decrease in viscosity, due to the domination of the hydrogel phase caused by hydrophobic interactions [8,25–29]. The higher K values indicated that the system showed solid-like characteristics during extrusion, meaning that the risk of clotting is higher [25]. As was previously discussed, there were no clogging signs (instantaneous over-extrusion) during the printing of all the under-study materials. This analysis supported the already discussed printability of the systems, where the printing of the CSO sample resulted in the sharpest and most accurate printed shape, and it can be correlated to the samples' higher K and lower n values. Following the extrusion process during printing, the inks must recover their initial solid-like state (recovery). The ability of the inks to recover was evaluated by the application of the three intervals thixotropy test (Figure 4A).

Table 1. Consistency index (K) and flow index (n) of the different ink formulations.

Sample Name	K (10^5 Pa*s^n)	n	R^2
CSO	215.1 ± 16.6 [a]	-0.22 ± 0.02 [a]	0.99
C3X1	206.2 ± 10.6 [a]	-0.29 ± 0.01 [a]	0.99
C1X1	100.9 ± 1.6 [b]	0.14 ± 0.09 [b]	0.99
C1X3	63.7 ± 3.2 [c]	0.13 ± 0.07 [b]	1

Different letters in the same column state significant differences ($p < 0.05$).

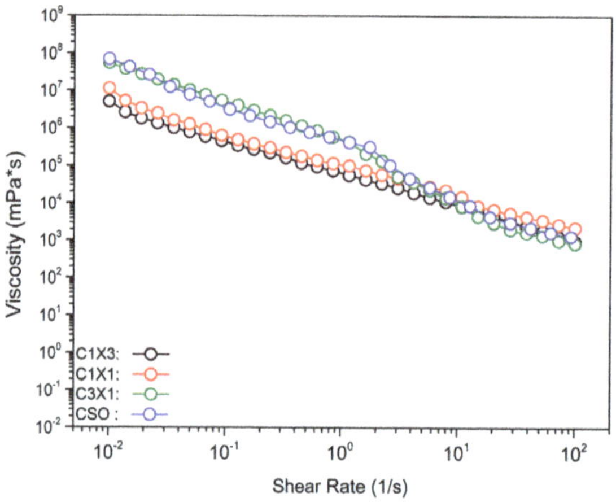

Figure 3. Viscosity vs. shear rate plot for the different ink formulations.

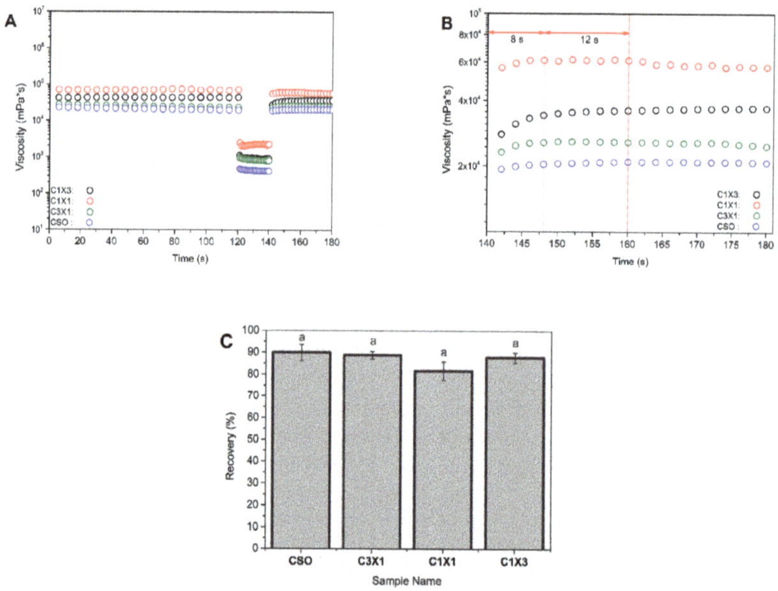

Figure 4. Viscosity recovery test for different ink formulations. (**A**) Three intervals thixotropy test, (**B**) recovery time, and (**C**) recovery percentage (the same letters indicate no statistically significant difference).

All samples were subjected to the thixotropy test and exhibited a similar ability to recover their initial viscosity up to 89.93% (CSO sample, Figure 4C) following the termination of the high-speed shearing rate. The recovery time of the samples was defined as the time required for the samples to reach a plateau. In all cases, the viscosity reached a plateau within the first 8 s, and the viscosity was kept constant, indicating that the samples can self-support after the extrusion through the nozzle. It should be noted that samples of C1X1 and C3X1 showed a mild decrease in the viscosity after 20 s. This decrease could be explained by the partial collapse of the printed layers resulting to the formation of the curls and coils during printing (Figure 4B). This discrepancy could be considered as a longer recovery time for these samples and was consistent with the existing literature where it was reported that the recovery time of the printed gels was directly related to the ability of the inks to maintain their structure and their mechanical properties after 3D printing. Even though oleogels were known to lack strong connection points between crystals [8,30], in the present study, there was an opposite effect. Both CSO and C1X3 had a very fast recovery time, while the samples with intermediate concentrations (C3X1 and C1X1) exhibited a decrease in viscosity after the first 20 s, indicating that the recovery time was (in this case) oleogel-fraction-independent.

The elastic modulus (G′), viscous modulus (G″), and complex modulus (G*) could be used to study the mechanical properties of the samples and provide us with information concerning their ability to self-support after 3D printing [8,31]. Figure 5A showed that the G′ of all samples was higher than the G″, indicating that the elastic properties of the inks were stronger than plastic properties (liquid-like properties) [8]. Additionally, the G* values increased as the oleogel fraction increased (Figure 5B), meaning that the inks with higher oleogel fractions tended to be mechanically stronger. This observation agreed with the printability evaluation that was previously discussed where specimens printed with inks with higher oleogel fractions performed in a better way. The smallest amount of oleogel fraction used (sample C1X3) was the least stiff among the samples that were studied. This difference in the stiffness was considered the main reason for the partial layer collapse that caused the formation of a trapezoid-shaped specimen instead of a rectangular shape, as is presented in Figure 2B, which was attributed to partial collapse of the printed layers. According to the weak gel model, the increased values of z indicated an increase in the interaction numbers of the bigel inks [8]. On the other hand, the same model dictated that the increased values of A were an indication of stronger interactions within the gel network [8,28]. Figure 5C revealed that the number of interactions (z values) for all samples were similar, except for CSO samples where the z value was considerably higher. The higher z value of CSO sample is directly linked with the better printing performance of this ink (Figure 2B). Figure 5D reveals a stronger dependency of the oleogel fraction on the overall strength of interactions within the gel network, with the CSO sample owning a considerably higher value of A. For the sample of CSO, there was no interaction between the oleogel and hydrogel while an interaction between the air and oleogel occurred. For samples containing hydrogels, the higher A values indicated the increased interactions between the two phases, and according to Figure 5D, this was related to the amount of oleogel present, resulting in samples with enhanced structural stability and mechanical properties [8], due to the interpenetration of the oleogel and hydrogel network. These properties could be related to the ability of the inks to maintain layer-by-layer adhesion after printing [8].

2.4. *Inks' Microstructure Analysis*

PLM micrographs of different ink formulations were used to interpret the distribution of the air and crystalline area (oleogel phase) within the bulk of the inks. As presented in Figure 6, the CSO sample exhibited a relatively uniform distribution of air bubbles of various sizes within its bulk. Most of the air bubbles were round shaped, indicating that there were no intense disproportionation phenomena (air bubbles merging or volume changing following the mixing or printing process). Additionally, there was a visible bright area (no crystallinity) that was indicative of a mild phase separation, demonstrated by the formation of oil islets within the ink. The microstructures of C3X1 and C1X1 samples showed distanced and elongated larger air bubbles, suggesting the presence of disproportionation phenomena. The homogeneity of the crystalline phase was also observed in the CLSM micrographs. There was an intermediate color presented throughout the image that could be compared to a bicontinuous structure. This observation was more obvious for the C1X1 sample for which both the conventional and split CLSM images showed the evenly distributed phases. Finally, the PLM micrograph of the C1X3 sample was indicative of more intense disproportionation phenomenon along with phase separation which was the reason for lower performance of the ink during printing (trapezoid shape due to partial collapse). CLSM images of the C1X1 and C1X3 samples revealed that the air was mostly presented in the form of cracks rather than bubbles, leading to poor mechanical properties. The microstructure of these samples could also be compared to a bicontinuous system as there was an even distribution of the two phases (hydrogel and oleogel) within the bulk of the inks.

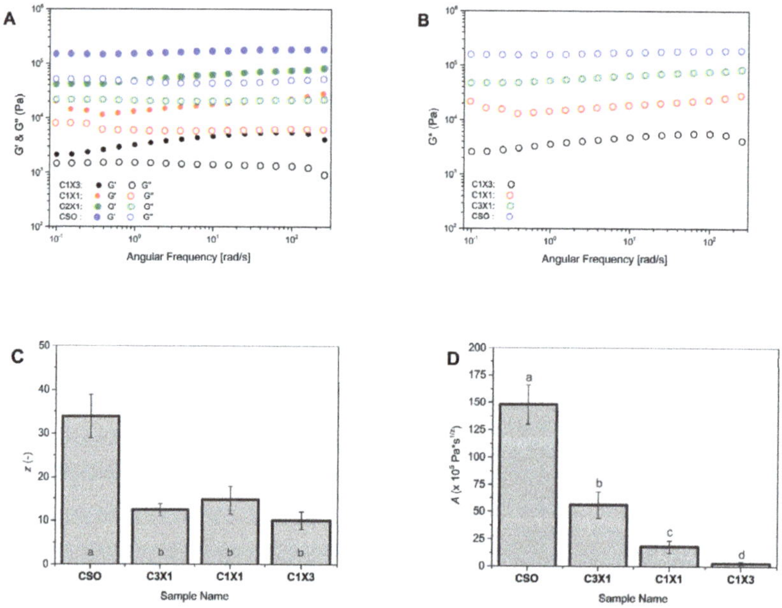

Figure 5. Rheological properties of the different ink formulations. (**A**) Elastic modulus (G′) and viscous modulus (G′′) vs. frequency plots, (**B**) complex modulus (G*) vs. frequency plots, and weak gel model parameter in terms of (**C**) network strength; (**A**,**D**) network strength of network extension (z). Different lowercase letters indicate statistically significant differences.

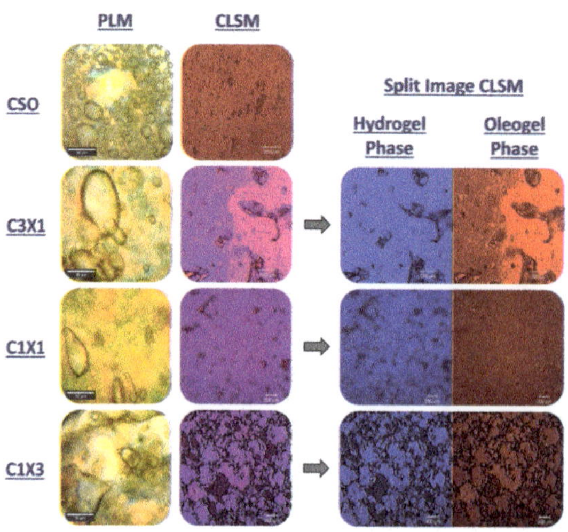

Figure 6. Polarized Light Microscopy (scale bar set at 50 µm) and Confocal Laser Scanning Microscopy (scale bar set at 200 µm) micrographs, along with CLSM split images (scale bar set at 200 µm) showing the two different phases (oleogel and hydrogel) present in the samples.

2.5. Thermal Behavior during Heating

Differential Scanning Calorimetry was applied to monitor the melting behavior of the formulations during heating at temperatures that mimic the expected working temperatures of the system. DSC analysis is considered complementary to the present study as the melting behavior of the inks is indicative of the structural characteristics of the materials. Bigel formulations are expected to be mainly stabilized by the fat crystals present in the organogel phase [24]. In Figure 7, all samples exhibited a broad transition peak at around 45 °C. Only the thermograph of the CSO sample showed a distinct double peak, which was indicative of the melting of the inverse lamellar phase following the melting of the sub-α crystals [32]. The melting behavior of all the bigel formulations was characterized by the presence of a single peak with a distinct shoulder at lower temperatures, which was attributed to the presence of the hydrogel (Figure 7a–c). Additionally, the overall lower peak melting temperatures of the bigels, compared to the CSO sample (Table 2), could be attributed to a possible plasticizing effect of the hydrogel phase to the oleogel phase [32]. Finally, according to Table 2, the observed melting enthalpies of the samples were divided into two statistically similar groups: the CSO—C3X1 group and the C1X1—C1X3 group. The melting enthalpy of the CSO sample was indicative of the disruption of van der Waals forces developed by the crystal matrix of the oleogel phase. The value of the observed enthalpy was expected to be lower by reducing the fraction of the oleogel phase in the C3X1 sample, but it appears to be unaffected, due to the presence of interfacial phenomena that were induced by the hydrogel phase, increasing to some extent the observed enthalpy value, counteracting the expected lowering due to the lower oleogel fraction. This phenomenon was not observed for the second group of samples, where the observed melting enthalpy values were further decreased. This observation was attributed to the decrease in the participation of interfacial phenomena, as the system seems to transit to a bicontinuous state, as is observed in Figure 6. The results are indicative of the effect of the hydrogel concentration on the properties of the bigels and they provide useful insights for understanding the under-study materials. The transition to a bicontinuous state that was observed for C1X1 and C1X3 samples could possibly explain their inferior performance as the non-bicontinuous samples exhibit a distinct dispersion of the hydrogel phase within the oleogel phase that results in a more stable structure.

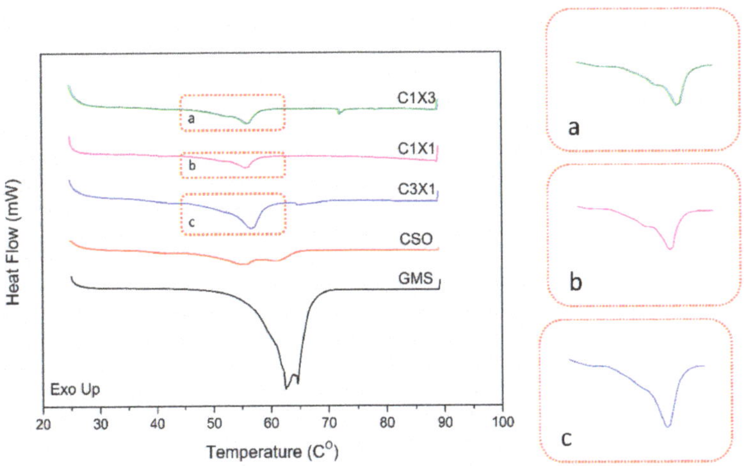

Figure 7. DSC thermograms of heating curves for the different ink formulations. The embedded pics a, b, and c are the zoomed areas of the thermographs that highlight the presence of the respective peaks and shoulders.

Table 2. Peak melting temperatures T_m and melting enthalpies ΔHm of bigel inks with different formulations.

Sample Name	$T_{m\ onset}$ (°C)	T_{m1} (°C)	T_{m2} (°C)	$\Delta H_{m\ total}$ (J/g)
CSO	45.2 ± 1.2	55.5 ± 1.6	61.2 ± 0.9	18.6 ± 3.9 [a]
C3X1	45.4 ± 1.3	51.4 ± 1.1 *	56.9± 1.3	23.0 ± 2.9 [a]
C1X1	45.2 ± 0.9	51.1 ± 0.8 *	56.1 ± 1.7	9.9 ± 2.6 [b]
C1X3	45.1 ± 1.0	51.7 ± 1.2 *	56.3 ± 1.7	11.2 ± 2.7 [b]

* Temperature refers to a peak shoulder; different letters in the same column state significant differences ($p <$ 0.05).

3. Conclusions

In this study, aerated bigel inks with varying hydrogel-to-oleogel mass ratios were used as printing materials for an extrusion-based additive manufacturing process (food 3D printing). The printability of the inks was found to be negatively affected by the presence of higher portions of the hydrogel phase, even though the printing of the bigels had been characterized as successful. The CSO ink samples exhibited superior printing performance compared to the bigel inks, while the C1X3 ink sample had inferior printing performance due to intense disproportionation along with phase separation phenomena. The presence of hydrogel had a strong effect on the mechanical properties of the inks, leading to a partial collapse of the printed structures and poor printability. The microstructure of the aerated inks was affected by the presence of a higher oleogel fraction, in terms of air bubble shape and distribution, and the results of this study also indicate the presence of intense disproportionation along with phase separation phenomena, especially for C1X3 ink. The same phenomena are also supported by the rheological and DSC study, rendering the CSO ink the best performing 3D-printable ink and the C1X3 ink sample the least favorable, in terms of their possible application as a host system for the delivery of cannabinoids, as the poor printing performance could lead to inadequate predictability of the dosing and subsequently to the delivery of the active ingredients. As an overall conclusion, the three-dimensional printing of cannabis seed oil oleogel and its bigel with xanthan gum hydrogel was plausible and promising for the development of personalized dosage forms for hosting and delivering cannabinoids. These systems are ready to be used alone or in combination with other printable food systems towards the development of a fully customizable process for the preparation of personalized cannabis edibles with tunable properties.

4. Materials and Methods

4.1. Materials

Cannabis seed oil was kindly donated from Hempoil®, Athens, Greece. Xanthan gum (G1253-Xanthan gum from *Xanthomonas campestris*) was purchased from Sigma-Aldrich (Zwijndrecht, The Netherlands). Glycerol Monostearate (40–55 (Type I, EP), GELEOL) was purchased from Gattefossé (Saint-Priest, France). All substances were used as received without any additional purification.

4.2. Sample Preparation

4.2.1. Hydrogel and Oleogel Preparation

Xanthan gum (XG) hydrogels were prepared according to the literature, with modifications [2]. Briefly, XG powder was dispersed in distilled water (2% *w*/*w*) at room temperature, under continuous magnetic stirring for 24 h. Cannabis seed oil oleogels (CSOs) were prepared according to previously reported studies [32]. Glyceryl Monostearate (GMS) (20% *w*/*w*) was mixed with cannabis seed oil (80% *w*/*w*), under continuous magnetic stirring, and heated at 80 °C. When GMS was completely dissolved in CSO (clear solution), the system was kept for 5 additional minutes at 80 °C, and then it was cooled to room temperature at a cooling rate of ≤1 °C/min. The prepared oleogels were stored at 4 °C (overnight) before any further use.

4.2.2. 3D-Printable Inks Preparation

The 3D-printable inks were prepared by mixing oleogels and hydrogels with different mass ratios (detailed formulations in Table 3). A syringe coupler method, inspired by the modified Tessari method for producing sclerosing foam [23], was applied to mix the two phases.

Table 3. Different formulations of bigel inks.

Sample Name	Cannabis Seed Oil Oleogel (% *w*/*w*)	Xanthan Gum Hydrogel (% *w*/*w*)
CSO	100	0
C3X1	75	25
C1X1	50	50
C1X3	25	75

In a typical experiment, predetermined amounts of hydrogel and oleogel were transferred to two separated 5 mL syringes that were connected to a three-way tap syringe connector (Figure 8). Following the secure lock of the syringes, the plungers were pressed back and forth (20 times or 10 cycles) to ensure thorough mixing. Afterwards, the mixture was further mixed with air using the same apparatus. For this process, one of the two syringes connected to the three-way tap connector was filled with air and subsequently mixed with the previously prepared mixture by pressing the plungers back and forth (20 times or 10 cycles). The prepared bigel foam was transferred to a new 5 mL syringe and used as a 3D-printing ink. The foamability and printability of the formulations were evaluated immediately after the air-mixing process. All samples were mixed with air at a final concentration of 20% v/v, as a benchmark concentration that was less than the higher possible amount of the incorporated air (according to Figure 1B), ensuring the accurate, precise, and reproducible incorporation of the same amount of air in all samples.

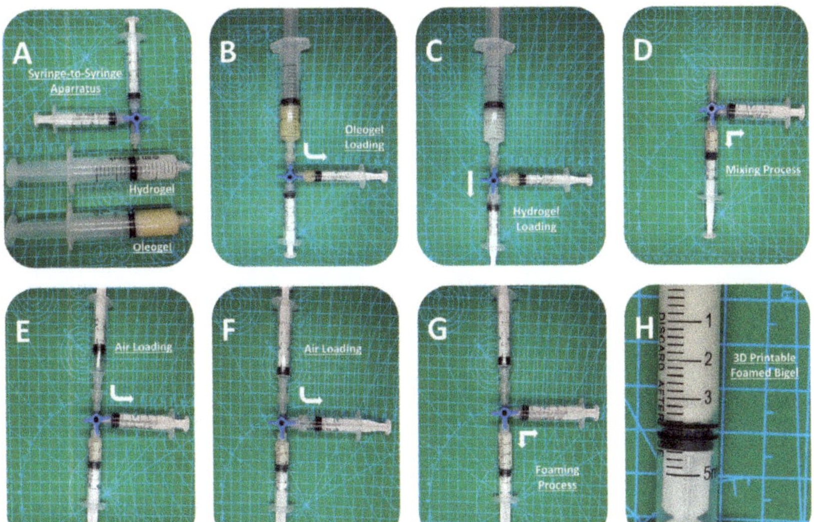

Figure 8. Syringe coupler mixing and foaming process. (**A**) Raw materials and mixing apparatus; (**B**) oleogel loading; (**C**) hydrogel loading; (**D**) apparatus before mixing; (**E**) apparatus after mixing (bigel formation); (**F**) air loading apparatus before foaming; (**G**) apparatus after foaming; and (**H**) foamed bigel (close-up photo).The arrows indicate movement direction.

4.3. Overrun Measurements

The quantification of the samples' overrun was performed to determine the overall amount of air that was capable of being incorporated within the bulk of the bigels. Overrun was determined by applying the following formula [24]:

$$Overrun\ (\%) = 100 \times (V_1 - V_0)/V_0, \quad (1)$$

where V_1 is the volume of the foam after mixing the air with the bigels, while V_0 is the initial volume of the bigel before mixing with the air.

4.4. Rheological Studies

Rheological measurements of the ink formulations were conducted using an Anton Paar MCR 92 (Anton Paar GmbH, Graz, Austria) strain rate-controlled rheometer equipped with a Peltier module. A 25 mm diameter parallel-plate geometry with a 1 mm gap was selected for all rheological experiments [8,32]. To evaluate the rheological properties of the samples during the extrusion process (during printing), the shear-thinning properties were evaluated and the Ostwald-de Waele power law model (Equation (2)) was applied to determine the correlation between apparent viscosity and shear rate [8,27]:

$$\left(\eta = K \cdot \dot{\gamma}^{(n-1)}\right), \quad (2)$$

where η, $\dot{\gamma}$, K, and n are the apparent viscosity, shear rate, consistency coefficient, and power law index, respectively.

To evaluate the recovery ability of the 3D-printable inks, a thixotropy experiment with three different shear rates (1/s, 100/s, and 11/s) was carried out. A frequency sweep test was performed at the range of 10 to 100 rad/s within the linear viscoelastic range (LVR) of 0.01, and the weak gel model (Equation (3)) was applied, according to the literature [8,28]:

$$G^* = (G'^2 + G''^2)^{0.5} = A \cdot \omega^{1/z}, \quad (3)$$

where z represents the interaction number of the bigels, and A represents the interaction strength.

4.5. Printability Evaluation

4.5.1. Food 3D Printing Process

A conventional 3D printer (Ender 3 S1Pro 3D printer, Creality, Shenzhen, China) equipped with a syringe extrusion print head module suitable for food extrusion (LuckyBot Food Extruder, Wiibox Group Limited, Wan Chai, HongKong) was used to print the different specimens. The nozzle diameter and the layer height were 1 mm, while the printing speed was 10 mm/s.

4.5.2. Printing Quality Evaluation

To evaluate the printing quality, different model shapes were printed, including 1D printing (1 mm × 20 mm single-filament/line printing), 2D printing (2, 5, and 10 mm × 20 mm single-layer printing), and 3D printing (5 mm × 10 mm × 20 mm rectangular 3D model). All layers were of the same thickness, and no additional walls were used. A concentric infill pattern (100% infill) was selected as a typical infill pattern for food 3D printing. Room temperature (25 °C) was selected as the print head and print-bed temperature. The overall printability was quantitively evaluated based on the single-line and single-layer printing performance (compared to the initial CAD model dimensions), while the 3D-printed specimens were evaluated qualitatively by comparing the 3D-scanned specimens (CR-Scan Lizard 3D scanner, Creality, Shenzhen, China) to the CAD model.

4.6. Confocal Laser Scanning Microscopy Observation (CLSM)

The inks' structure was observed with an inverted Zeiss LSM 700 confocal microscope (Carl Zeiss, CZ Microscopy GmbH, Jena, Germany) in optical mode with a 20× lens. Before the examination, 10 μL of 0.1 mg mL^{-1} Nile Red and 10 μL of 0.1 mg mL^{-1} Nile Blue were added into the oil and aqueous phase, respectively [32,33]. A small amount of each ink was placed on a glass slide and was covered with a coverslip before imaging.

4.7. Polarized Light Microscopy Observation (PLM)

The microstructure of the prepared inks was captured using an Olympus BX 41 polarized light microscope (Olympus Corporation, Tokyo, Japan). All measurements were performed at 25 °C. The ink samples were placed on a microscope glass slide and covered by a coverslip to form a thin uniform film [32].

4.8. Differential Scanning Calorimetry (DSC)

Thermal analyses were performed by using a TA Instruments temperature-modulated DSC (TA Q2000). The instrument was calibrated with the indium standard for the heat flow and temperature, while heat capacity was evaluated using the sapphire standard. A nitrogen gas flow of 50 mL/min was purged into the DSC cell for all measurements. A total of 5 mg of sample was weighted and placed into hermetically sealed aluminum pans and heated from 25 to 90 °C at a heating rate of 10 °C/min. An empty DSC aluminum pan was used as a reference sample. Peak melting temperature (Tp_m) and apparent melting enthalpy (ΔHm, J/g) of the different samples were obtained from the endothermic peaks of the thermograms [32].

4.9. Statistical Analysis

All measurements were performed in triplicates and mean average values were presented. Student's *t*-test (Microsoft® Excel®, Microsoft 365 (version 2205 Build 16.0.15225.20028), Redmond, WA, USA) was used for the analysis of the acquired data. The significance level was set at $p < 0.05$.

Author Contributions: E.G.A.: Conceptualization, experiment design, data collection, data analysis, writing—original draft, funding acquisition, project administration, supervision; A.P.: Conceptualization, review and editing; D.G.F.: Conceptualization, review and editing; H.Z.: Data analysis, review and editing; C.R.: Conceptualization, experiment design, review and editing, supervision. All authors have read and agreed to the published version of the manuscript.

Funding: The work was supported by the Hellenic Foundation for Research and Innovation (H.F.R.I.) under the "3rd Call for H.F.R.I. Research Projects to support Post-Doctoral Researchers" (Project Number: 7591).

Institutional Review Board Statement: Not applicable.

Informed Consent Statement: Not applicable.

Data Availability Statement: The original contributions presented in the study are included in the article, further inquiries can be directed to the corresponding author.

Conflicts of Interest: The authors declare no conflicts of interest.

References

1. Monou, P.K.; Mamaligka, A.M.; Tzimtzimis, E.K.; Tzetzis, D.; Vergkizi-Nikolakaki, S.; Vizirianakis, I.S.; Andriotis, E.G.; Eleftheriadis, G.K.; Fatouros, D.G. Fabrication and Preliminary In Vitro Evaluation of 3D-Printed Alginate Films with Cannabidiol (CBD) and Cannabigerol (CBG) Nanoparticles for Potential Wound-Healing Applications. *Pharmaceutics* **2022**, *14*, 1637. [CrossRef] [PubMed]
2. Baniasadi, H.; Kimiaei, E.; Polez, R.T.; Ajdary, R.; Rojas, O.J.; Österberg, M.; Seppälä, J. High-Resolution 3D Printing of Xanthan Gum/Nanocellulose Bio-Inks. *Int. J. Biol. Macromol.* **2022**, *209*, 2020–2031. [CrossRef]
3. Cerino, P.; Buonerba, C.; Cannazza, G.; D'Auria, J.; Ottoni, E.; Fulgione, A.; Di Stasio, A.; Pierri, B.; Gallo, A. A Review of Hemp as Food and Nutritional Supplement. *Cannabis Cannabinoid Res.* **2021**, *6*, 19–27. [CrossRef] [PubMed]

4. Mikulcová, V.; Kašpárková, V.; Humpolíček, P.; Buňková, L. Formulation, Characterization and Properties of Hemp Seed Oil and Its Emulsions. *Molecules* **2017**, *22*, 700. [CrossRef]
5. Matthäus, B.; Brühl, L. Virgin Hemp Seed Oil: An Interesting Niche Product. *Eur. J. Lipid Sci. Technol.* **2008**, *110*, 655–661. [CrossRef]
6. RTI International; Barrus, D.; Capogrossi, K.; Cates, S.; Gourdet, C.; Peiper, N.; Novak, S.; Lefever, T.; Wiley, J. *asty THC: Promises and Challenges of Cannabis Edibles*; RTI Press: Research Triangle Park, NC, USA, 2016.
7. Gholamipour-Shirazi, A.; Norton, I.T.; Mills, T. Designing Hydrocolloid Based Food-Ink Formulations for Extrusion 3D Printing. *Food Hydrocoll.* **2019**, *95*, 161–167. [CrossRef]
8. Qiu, R.; Wang, K.; Tian, H.; Liu, X.; Liu, G.; Hu, Z.; Zhao, L. Analysis on the Printability and Rheological Characteristics of Bigel Inks: Potential in 3D Food Printing. *Food Hydrocoll.* **2022**, *129*, 107675. [CrossRef]
9. Lille, M.; Nurmela, A.; Nordlund, E.; Metsä-Kortelainen, S.; Sozer, N. Applicability of Protein and Fiber-Rich Food Materials in Extrusion-Based 3D Printing. *J. Food Eng.* **2018**, *220*, 20–27. [CrossRef]
10. Sharma, R.; Chandra Nath, P.; Kumar Hazarika, T.; Ojha, A.; Kumar Nayak, P.; Sridhar, K. Recent Advances in 3D Printing Properties of Natural Food Gels: Application of Innovative Food Additives. *Food Chem.* **2024**, *432*, 137196. [CrossRef]
11. Pulatsu, E.; Lin, M. A Review on Customizing Edible Food Materials into 3D Printable Inks: Approaches and Strategies. *Trends Food Sci. Technol.* **2021**, *107*, 68–77. [CrossRef]
12. Ghosal, K.; Nanda, A. Development of Diclofenac Potassium Gel from Hydrophobically Modified HPMC. *Iran. Polym. J.* **2013**, *22*, 457–464. [CrossRef]
13. Batheja, P.; Sheihet, L.; Kohn, J.; Singer, A.J.; Michniak-Kohn, B. Topical Drug Delivery by a Polymeric Nanosphere Gel: Formulation Optimization and In Vitro and In Vivo Skin Distribution Studies. *J. Control. Release* **2011**, *149*, 159–167. [CrossRef] [PubMed]
14. Dick, A.; Dong, X.; Bhandari, B.; Prakash, S. The Role of Hydrocolloids on the 3D Printability of Meat Products. *Food Hydrocoll.* **2021**, *119*, 106879. [CrossRef]
15. Tian, H.; Wang, K.; Lan, H.; Wang, Y.; Hu, Z.; Zhao, L. Effect of Hybrid Gelator Systems of Beeswax-Carrageenan-Xanthan on Rheological Properties and Printability of Litchi Inks for 3D Food Printing. *Food Hydrocoll.* **2021**, *113*, 106482. [CrossRef]
16. Uribe-Alvarez, R.; O'Shea, N.; Murphy, C.P.; Coleman-Vaughan, C.; Guinee, T.P. Evaluation of Rennet-Induced Gelation under Different Conditions as a Potential Method for 3D Food Printing of Dairy-Based High-Protein Formulations. *Food Hydrocoll.* **2021**, *114*, 106542. [CrossRef]
17. Shi, Y.; Zhang, M.; Bhandari, B. Effect of Addition of Beeswax Based Oleogel on 3D Printing of Potato Starch-Protein System. *Food Struct.* **2021**, *27*, 100176. [CrossRef]
18. Singh, V.K.; Anis, A.; Banerjee, I.; Pramanik, K.; Bhattacharya, M.K.; Pal, K. Preparation and Characterization of Novel Carbopol Based Bigels for Topical Delivery of Metronidazole for the Treatment of Bacterial Vaginosis. *Mater. Sci. Eng. C* **2014**, *44*, 151–158. [CrossRef]
19. Shakeel, A.; Farooq, U.; Iqbal, T.; Yasin, S.; Lupi, F.R.; Gabriele, D. Key Characteristics and Modelling of Bigels Systems: A Review. *Mater. Sci. Eng. C* **2019**, *97*, 932–953. [CrossRef]
20. Rehman, K.; Zulfakar, M.H. Recent Advances in Gel Technologies for Topical and Transdermal Drug Delivery. *Drug Dev. Ind. Pharm.* **2014**, *40*, 433–440. [CrossRef]
21. Andriotis, E.G.; Chachlioutaki, K.; Monou, P.K.; Bouropoulos, N.; Tzetzis, D.; Barmpalexis, P.; Chang, M.-W.; Ahmad, Z.; Fatouros, D.G. Development of Water-Soluble Electrospun Fibers for the Oral Delivery of Cannabinoids. *AAPS PharmSciTech* **2021**, *22*, 23. [CrossRef]
22. Xie, D.; Hu, H.; Huang, Q.; Lu, X. Development and Characterization of Food-Grade Bigel System for 3D Printing Applications: Role of Oleogel/Hydrogel Ratios and Emulsifiers. *Food Hydrocoll.* **2023**, *139*, 108565. [CrossRef]
23. Xu, J.; Wang, Y.; Chen, A.; Wang, T.; Liu, S. A Modified Tessari Method for Producing More Foam. *SpringerPlus* **2016**, *5*, 129. [CrossRef]
24. Zheng, R.; Chen, Y.; Wang, Y.; Rogers, M.A.; Cao, Y.; Lan, Y. Microstructure and Physical Properties of Novel Bigel-Based Foamed Emulsions. *Food Hydrocoll.* **2023**, *134*, 108097. [CrossRef]
25. Liu, Z.; Bhandari, B.; Prakash, S.; Mantihal, S.; Zhang, M. Linking Rheology and Printability of a Multicomponent Gel System of Carrageenan-Xanthan-Starch in Extrusion Based Additive Manufacturing. *Food Hydrocoll.* **2019**, *87*, 413–424. [CrossRef]
26. Bollom, M.A.; Clark, S.; Acevedo, N.C. Development and Characterization of a Novel Soy Lecithin-Stearic Acid and Whey Protein Concentrate Bigel System for Potential Edible Applications. *Food Hydrocoll.* **2020**, *101*, 105570. [CrossRef]
27. Jagadiswaran, B.; Alagarasan, V.; Palanivelu, P.; Theagarajan, R.; Moses, J.A.; Anandharamakrishnan, C. Valorization of Food Industry Waste and By-Products Using 3D Printing: A Study on the Development of Value-Added Functional Cookies. *Future Foods* **2021**, *4*, 100036. [CrossRef]
28. Lupi, F.R.; Gentile, L.; Gabriele, D.; Mazzulla, S.; Baldino, N.; De Cindio, B. Olive Oil and Hyperthermal Water Bigels for Cosmetic Uses. *J. Colloid. Interface Sci.* **2015**, *459*, 70–78. [CrossRef] [PubMed]
29. Wilson, S.A.; Cross, L.M.; Peak, C.W.; Gaharwar, A.K. Shear-Thinning and Thermo-Reversible Nanoengineered Inks for 3D Bioprinting. *ACS Appl. Mater. Interfaces* **2017**, *9*, 43449–43458. [CrossRef]
30. Doan, C.D.; Van De Walle, D.; Dewettinck, K.; Patel, A.R. Evaluating the Oil-Gelling Properties of Natural Waxes in Rice Bran Oil: Rheological, Thermal, and Microstructural Study. *J. Am. Oil Chem. Soc.* **2015**, *92*, 801–811. [CrossRef]

31. Montoya, J.; Medina, J.; Molina, A.; Gutiérrez, J.; Rodríguez, B.; Marín, R. Impact of Viscoelastic and Structural Properties from Starch-Mango and Starch-Arabinoxylans Hydrocolloids in 3D Food Printing. *Addit. Manuf.* **2021**, *39*, 101891. [CrossRef]
32. Andriotis, E.G.; Monou, P.-K.; Komis, G.; Bouropoulos, N.; Ritzoulis, C.; Delis, G.; Kiosis, E.; Arsenos, G.; Fatouros, D.G. Effect of Glyceryl Monoolein Addition on the Foaming Properties and Stability of Whipped Oleogels. *Gels* **2022**, *8*, 705. [CrossRef] [PubMed]
33. Koliastasi, A.; Kompothekra, V.; Giotis, C.; Moustakas, A.K.; Skotti, E.P.; Gerakis, A.; Kalogianni, E.; Ritzoulis, C. Emulsifiers from Partially Composted Olive Waste. *Foods* **2019**, *8*, 271. [CrossRef] [PubMed]

gels

MDPI

Article

Effect of Cooling Rate on Properties of Beeswax and Stearic Acid Oleogel Based on Rice Bran Oil and Sesame Oil

Subajiny Sivakanthan [1,2,3], Sabrina Fawzia [4], Sagadevan Mundree [5], Terrence Madhujith [6] and Azharul Karim [1,*]

1 School of Mechanical, Medical and Process Engineering, Faculty of Engineering, Queensland University of Technology, Brisbane, QLD 4000, Australia; ssubajiny@univ.jfn.ac.lk
2 Department of Agricultural Chemistry, Faculty of Agriculture, University of Jaffna, Kilinochchi 44000, Sri Lanka
3 Postgraduate Institute of Agriculture, University of Peradeniya, Peradeniya 20400, Sri Lanka
4 School of Civil and Environmental Engineering, Faculty of Engineering, Queensland University of Technology, Brisbane, QLD 4000, Australia; sabrina.fawzia@qut.edu.au
5 School of Agriculture and Food Sustainability, Faculty of Science, University of Queensland, St Lucia, QLD 4067, Australia; s.mundree@uq.edu.au
6 Department of Food Science and Technology, Faculty of Agriculture, University of Peradeniya, Peradeniya 20400, Sri Lanka; tmadhujith@agri.pdn.ac.lk
* Correspondence: azharul.karim@qut.edu.au

Abstract: This study aimed to investigate how varying cooling rate impacts the characteristics of oleogels prepared using a sesame oil and rice bran oil blend (5:6, *w/w*) using a combination of beeswax and stearic acid (3:1, *w/w* at 12%, *w/w*) as the oleogelators. The study assessed three different cooling rates—0.5, 1.5, and 5 °C/min—with a focus on the attributes of the oleogels. The study revealed that the cooling rate had a substantial impact on the strength of the gel network. The cooling rate of 0.5 °C/min resulted in a higher oil-binding capacity and a stronger gel structure than fast cooling. The thermal properties and molecular interactions of the oleogels were not influenced by the cooling rate. The findings of this study indicated that the characteristics of beeswax and stearic acid oleogel prepared using sesame oil and rice bran oil blend could be tailored by manipulating the cooling rate.

Keywords: oil binding; oleogelation; rheology; saturated fat; structured lipid; trans fat

1. Introduction

Oleogels are structured lipid systems consisting of a liquid oil phase immobilized within a network of structuring agents, such as polymers, and low molecular weight compounds. In recent years, oleogels have garnered attention as a potential alternative to solid fats, such as margarine, which are rich in trans and saturated fats. Trans fats, which are formed during the hydrogenation process, are strongly linked to health implications, including increased risk of cardiovascular diseases [1]. Solid fats containing high trans and saturated fats have traditionally been used in food products to provide stability, texture, and functionality. However, the adverse health effects linked to the excessive intake of saturated and trans fats have driven scientists to explore alternatives. Oleogels have emerged as a focal point of academic interest due to their capability to offer a trans-free and lower saturated fat option while maintaining comparable functional attributes to conventional solid fats.

It is imperative to improve oleogel characteristics in order to mimic those of conventional solid fats, ensuring the desired texture and functionality [2]. Critical factors in this endeavor include processing parameters like heating temperature, duration of heating, and cooling rate. These variables significantly influence the properties of oleogels, such as their structure, rheological behavior, stability, and functionality [3]. In the process of developing oleogels by dispersing the oleogelators directly into the oil, the oil and oleogelator mixture is heated to make a homogeneous dispersion. This is followed by cooling, which promotes microstructure formation through crystallization, thereby trapping the oil within the 3D network formed by the gelators. The process of cooling significantly influences the formation and arrangement of this microstructure, ultimately shaping the properties of the oleogel [4,5]. The elasticity, strength, and oil-holding capacity of the oleogels depend on the size, morphology, and spatial arrangement of crystals. Moreover, the cooling process exerts influence over nucleation and crystallization kinetics [5].

The correlation between cooling rate and properties of oleogels has been reported for different oleogel systems, including rice bran wax-based [6], rice bran wax, candelilla wax, and carnauba wax-based [5], as well as monoacylglycerol-based oleogels [7]. These studies have emphasized the notable influence of cooling rate on oleogel properties. This emphasizes the importance of assessing the effect of cooling rate on oleogel characteristics when establishing a new oleogel system. Even though few studies have reported on the properties of oleogels as affected by cooling rates, the correlation between cooling rate and oleogel properties has been conflicting for different oleogel systems. For instance, high cooling rates were associated with desirable oleogel properties of rice candelilla wax, bran wax, and carnauba wax-based oleogels [5], as well as monoacylglycerol oleogels [7], whereas slower cooling rates were associated with improved oleogel properties in terms of gel hardness for rice bran wax oleogel [8] and fatty alcohol-based oleogels made from peanut oil [9]. The same study by Scharfe et al. [8] reported that slow cooling resulted in softer gels in sunflower wax oleogels. In contrast to the above studies, Valoppi et al. [10] reported that the possible differences in microcrystal dimensions resulting from different cooling rates did not influence the oleogel properties. Therefore, it is worth noting that the influence of cooling rate on oleogel characteristics depends on the type of oleogelator and the resulting mechanism of formation of the oleogel [4]. Given the absence of prior research on assessing the impact of cooling rate on oleogels based on beeswax and stearic acid, it is imperative to determine the optimal cooling rate for developing an oleogel using these two oleogelators. Therefore, this study has been conducted to advance the findings of our previous study [11] to develop a healthy oleogel based on beeswax and stearic acid using a blend of sesame oil and rice bran oil. Our earlier research focused on optimizing oleogel formulations with properties similar to commercial margarines. This was accomplished using a binary mixture of sesame oil and rice bran oil, with beeswax and stearic acid as the oleogelators. We found that beeswax and stearic acid displayed synergistic effects at a 3:1 ratio. The optimized oleogel formulation consisted of 0.40 g of sesame oil, 0.48 g of rice bran oil, 0.09 g of beeswax, and 0.03 g of stearic acid (with a sesame oil to rice bran oil ratio of 5:6, and a beeswax to stearic acid ratio of 3:1, resulting in a total oleogelator concentration of 12%). This study further explored how varying cooling rates affect the oil-binding capacity, as well as the rheological, microstructural, thermal, and molecular properties of the oleogel made with a combination of beeswax and stearic acid in a sesame oil and rice bran oil blend.

2. Results and Discussion

2.1. Cooling Temperature Profile

Figure 1 illustrates the cooling temperature profiles of oleogels cooled at different rates. All samples recorded a decrease in the cooling rate with time, and the cooling rate was calculated as the average cooling rate.

2.2. Oil-Binding Capacity

The cooling rates were expressed as the average cooling rate, calculated by dividing the change in temperature by the time. The results demonstrated that all the samples exhibited a decline in cooling rate with time. The oil-binding capacity of the oleogels is the capability of the gel network to hold the oil, which is expressed as the fraction of oil remaining in the structure after being subjected to a centrifugal force [12]. Table 1 presents the oil-binding capacities of the oleogels. All samples had oil-binding capacities higher than 99%, indicating that more than 99% of the oil was retained within the network formed by the oleogelators after applying a centrifugal force. This exhibits the excellent oil-binding capacity of the gel structure formed by the formula evaluated. These values align with the oil-binding capacity of the same formulation, as documented in our earlier study [11]. However, a notable discrepancy was observed in the oil-binding capacities of samples prepared at varying cooling rates. The oleogel cooled at a faster rate (5 °C/min) showed significantly lower oil-binding capacity than the other two oleogels prepared by cooling at slower rates. Natural wax-based oleogels have been reported to have very high oil-binding capacity. For instance, Shi et al. [13] also demonstrated an oil-binding capacity exceeding 99% for the beeswax-based oleogel made from fish oil and sunflower oil. Even though oleogels prepared by cooling at different rates had more than 99% of the oil-binding capacity, significant differences among them indicate that there is an influence of cooling rate on the structure of the gel network. Processing conditions that influence the structure, size, and shape of the crystals, as well as their spatial distribution in the network, have an influence on the oil-binding capacity of the oleogels [14].

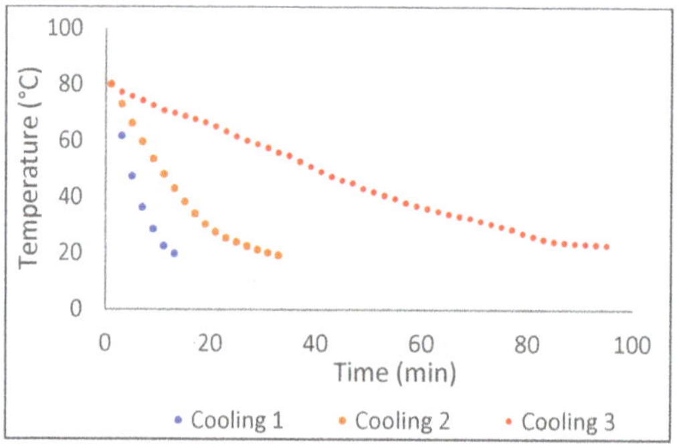

Figure 1. Temperature profiles recorded for the oleogels cooled at different rates. The calculated average cooling rates for cooling 1, cooling 2, and cooling 3 were 5 °C/min, 1.5 °C/min, and 0.5 °C/min, respectively.

Table 1. Oil-binding capacity and rheological parameters of oleogels produced at different cooling rates.

Average Cooling Rate	Oil-Binding Capacity (%)	Structural Recovery (%)	LVR (%)	G′ at LVR (Pa)
0.5 °C/min	99.92 ± 0.03 [a]	35.13 ± 3.42 [a]	0.054 ± 0.003 [a]	107,414 ± 869 [a]
1.5 °C/min	99.89 ± 0.04 [a]	25.75 ± 1.43 [b]	0.045 ± 0.001 [b]	90,261 ± 3700 [b]
5.0 °C/min	99.61 ± 0.02 [b]	22.93 ± 1.87 [b]	0.042 ± 0.001 [b]	88,912 ± 2244 [b]

Distinct superscript letters (a–b) within the same column indicate a significant difference ($p < 0.05$).

In agreement with the results of the present study, Scharfe et al. [8] also reported that rice bran wax-based gels exhibited increased hardness when cooled at slower rates, as slower cooling favored the development of more highly ordered wax crystal structures. The higher oil-binding capacity of the oleogels cooled at a slower rate in this study could be due to the fact that when an oleogel is cooled slowly, the structuring agents have more time to organize and form a well-defined three-dimensional network throughout the oil phase. It can be understood from the microstructure (Section 2.3) of the oleogels. Oleogel prepared at a faster cooling rate (5 °C/min) had smaller size crystals; however, they were less uniformly distributed compared to the other two oleogels. Blake and Marangoni [5] reported that enhancing homogeneity in crystal distribution can increase the oil-binding capacity of the oleogel. In this study, the reason for the less-ordered arrangement of the crystals at a faster cooling rate could be that there was not enough time for the organization of crystals at a faster cooling rate. Further, the impact of cooling rates on the oil-binding capacity of different oleogel systems may be different due to the differences in conformational arrangement and crystal type specific to the gelator, even though they are subjected to similar preparation methods [15]. Further explanation related to the higher oil-binding capacity at a slower cooling rate, in relation to microstructure (fractal dimension), is provided in Section 2.3.

The findings of this study regarding oil-binding capacity differ from those reported by Blake and Marangoni [5] for rice bran wax, sunflower wax, and candelilla wax-based oleogels, as well as from the results presented by Giacomozzi et al. [7] for monoglyceride oleogels. The researchers in these studies have documented that oleogels cooled at faster rates exhibit a greater oil-binding capacity compared to oleogels cooled at slower rates, and they interpreted that the smaller crystals formed by cooling at faster rates facilitate more surface area to entrap oil than those larger crystals generated at the slower cooling rates.

2.3. Rheological Properties

Oleogel exhibits viscoelastic behavior, displaying properties of both viscosity and elasticity. Rheological analysis of oleogels is essential for gaining a deeper understanding of the gel strength, elasticity, and viscosity of the system, which enables researchers to design the formulation and process parameters to ensure the desired functionality. In terms of rheology, oleogels can be described by storage modulus (elastic portion), G', with an apparent plateau, and by a loss modulus, G''. In the plateau region, G'' is less than the G' [16]. Analyzing the variations in G' and G'' with respect to frequency, as well as stress or strain, can offer valuable insights into the microstructural integrity and overall performance of oleogels across diverse applications [17].

In this study, oleogels produced by cooling at different rates were characterized by amplitude sweep, frequency sweep, and thixotropic experiment. Amplitude sweep is used to investigate how rheological properties of oleogels change with varying strain amplitudes. The region amplitude sweeps where both G' and G'' remain steady as strain increases is referred to as the LVR, indicating that the deformation of the structure is reversible. That is, LVR is a range of strain within which a material exhibits a linear response. After the LVR, there is a non-linear viscoelastic region where G' and G'' start to decrease with a rising strain [18]. During a frequency sweep, the performance of the oleogel at both low and high frequencies is assessed, revealing their behavior over the long term and short term, respectively, while maintaining a consistent strain value within the LVR. Amplitude sweep experiments were performed to evaluate the LVR and G' at the LVR, and the oleogels were further characterized through frequency sweep and thixotropy experiments, all conducted at a consistent strain value within the LVR.

Rheological parameters are shown in Table 1 and amplitude sweeps, complex viscosity, frequency sweeps, and thixotropic behavior are shown in Figure 2A–D. Within the LVR, it was observed that all samples displayed higher G' values than G'', confirming the gel-like structure of the oleogels. Specifically, the oleogel cooled at a rate of 0.5 °C/min demonstrated an elevated LVR limit and G' at LVR in comparison to the other two oleogels that were cooled at faster rates. The plateau value of G' within the LVR indicates the strength of the sample, and a more prolonged LVR suggests the highest resistance to structural deterioration under applied stress [19]. This is more evident from the higher structural recovery ability of the oleogels cooled at 0.5 °C/min than from the other two (Table 1 and Figure 2D). Scharfe et al. [8] have reported that cooling rates significantly altered the oleogel firmness in sunflower wax and rice bran wax oleogels; however, no significant changes were reported in the firmness of beeswax and candelilla wax oleogels. The same authors further reported that slow cooling resulted in softer gels in sunflower wax oleogels, whereas the same cooling rate resulted in harder gels for rice bran wax-based oleogels. Similarly, Trujillo-Ramírez et al. [4] also reported conflicting results for the influence of cooling rate on oleogel hardness. A faster cooling rate resulted in increased hardness of chia oil oleogels based on glycerol monostearate, whereas the same cooling rate caused a decrease in the hardness of the chia oleogel with sorbitan monostearate. Interestingly, despite these differences, both oleogels exhibited the formation of smaller crystals and a more condensed network structure. Palla et al. [20] also mentioned that a low cooling rate resulted in a greater hardness value (up to 39%) compared to fast cooling rates for the monoglyceride oleogels. Based on the existing literature and the outcomes of this study, it can be inferred that the impact of cooling rates on the mechanical strength of the oleogel depends on various factors, including the composition and type of oleogelator, and the composition and type of oil.

According to Patel et al. [21], an optimal solid fat should possess a G′ within the range of 1×10^5 to 5×10^6 Pa, which covers the spectrum from 'soft' to 'hard'. As per the G′ in the LVR shown in Table 1, only the oleogel cooled at 0.5 °C/min had the values within this range and the other two oleogels cooled at a faster rate had G′ less than these values. That is, even though they had a gel structure, they were too soft to be considered ideal for solid fat.

The complex viscosity of the samples obtained from the amplitude sweep experiment is shown in Figure 2B. Complex viscosity is a rheological parameter that describes the resistance of a material to flow under applied stress. All samples exhibited a shear thinning (non-Newtonian) behavior; that is, the complex viscosity decreased with increasing strain. The oleogel becomes less viscous and flows more easily as the applied stress increases. With strain, the structures deform, break, or realign, leading to a decrease in the complex viscosity. Shear thinning behavior is a beneficial property of oleogels because it allows easy spreadability and flow. Shear-thinning behavior has been reported in many oleogel systems [22–24]. The cooling rate did not show any significant influence on the shear-thinning behavior of the oleogels.

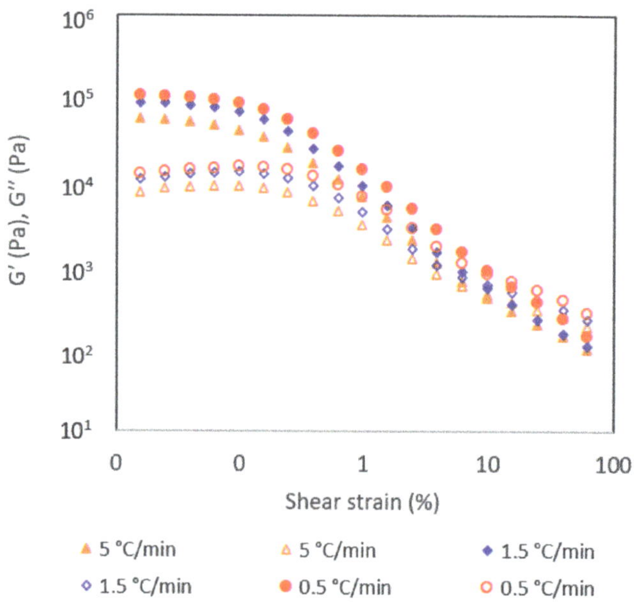

(A)

Figure 2. *Cont.*

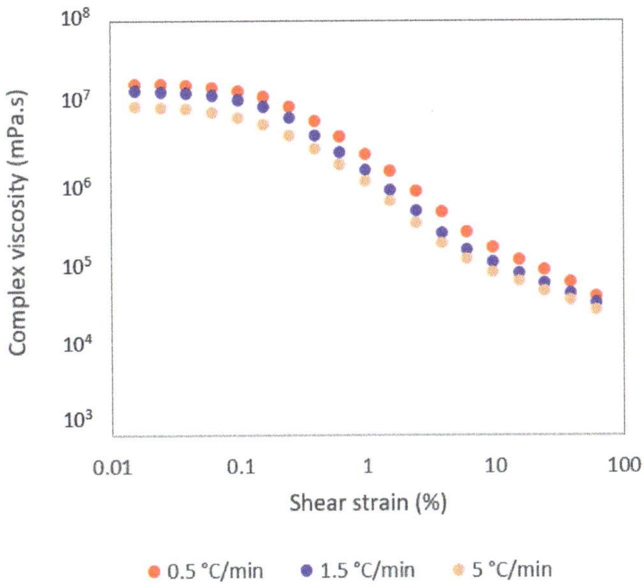

(B)

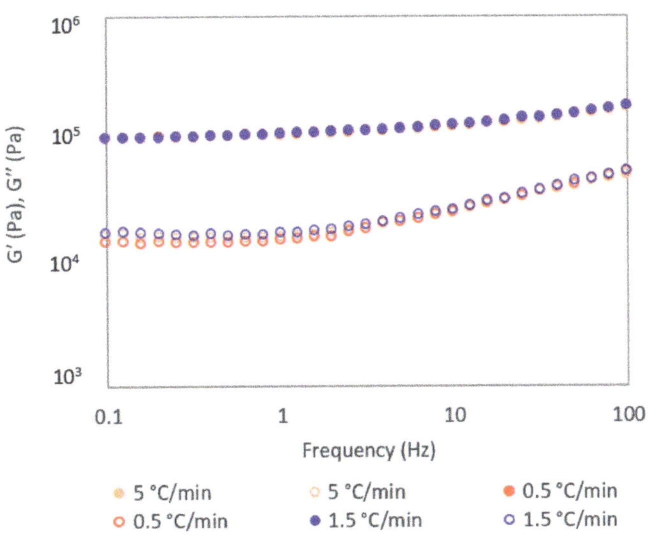

(C)

Figure 2. *Cont.*

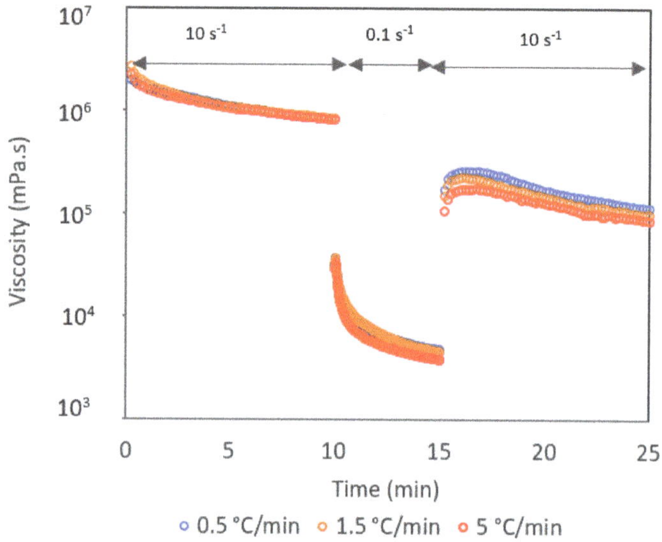

(D)

Figure 2. (**A**) Amplitude sweeps, (**B**) complex viscosity, (**C**) frequency sweeps, and (**D**) thixotropic behavior of oleogels produced at different cooling rates. In (**A**,**C**), solid markers represent G′ (storage modulus) and open markers represent G″ (loss modulus).

Frequency sweeps of the samples are shown in Figure 2C. It is evident that all samples exhibited higher G′ values than G″, confirming their solid-like behavior across all frequencies. The samples showing constant G′ with increasing frequency (frequency-independence) are categorized as strong gels, whereas samples showing increasing G′ with increasing frequency are weak gels (frequency-dependent) [17]. All samples exhibited similar behavior with increasing frequency, indicating that the cooling rate did not influence the behavior of oleogel with changing frequencies. The G′ values of all samples remained almost constant from 0.1 to 10 Hz, indicating a frequency-independent behavior, and a slight increase in the G′ afterward. Similar results were reported for different oleogel systems [19,24–26].

The thixotropic behavior of oleogels refers to the decrease in viscosity over time when subjected to constant stress, followed by a gradual recovery of viscosity when the stress is removed [27]. As shown in Table 1, the structural recovery percentage of the oleogel cooled at the lowest rate was significantly higher than the other two, indicating the significant influence of the cooling rate on the thixotropic properties of oleogel. Despite the same oleogel formula, the differences in the thixotropic properties of the oleogels cooled at different rates could be due to the differences in the gel structure. As indicated in the microscopic structure (below), there were prominent differences in the crystal sizes and spatial arrangements of the oleogels. The differences in the crystal arrangement could be the reason for the structural recovery ability of the oleogels. The higher oil-binding capacity, LVR, G′ at LVR, and structural recovery ability of the oleogel cooled at a slower rate indicate that slow cooling results in more desirable properties on beeswax-stearic acid oleogel compared to fast cooling rates. The studies on the effect of cooling rate on oleogel properties have not studied the influence of cooling rate on the thixotropic properties of oleogels.

2.4. Microstructure

The microstructure of the oleogels is shown in Figure 3. Table 2 shows the fractal dimension, the length of the crystals, and the distance between the crystals in the oleogels.

All samples exhibited needle-like crystals, the same as observed in our previous study with the same formula [11]. The needle-like crystal structure is a characteristic of wax-based oleogels beeswax [28]. The fractal dimension provides more detailed numerical information on the spatial arrangement of the crystals. The higher the fractal dimension, the more uniform distribution of mass with fewer cavities [29]. According to the fractal dimensions, a more uniform distribution of crystals was observed at low cooling rates than at high cooling rates. Studies have documented variations in crystal size and distribution among oleogels cooled at different rates. Notably, an increase in cooling rate resulted in the formation of smaller crystals. These findings align with previous research by Blake and Marangoni [5], Valoppi et al. [10], and Trujillo-Ramírez et al. [4]. Comparing the crystal lengths of oleogels cooled at rates of 1.5 and 5 °C/min, it is evident that both exhibited nearly identical crystal sizes, which were smaller than those observed in the oleogels cooled at 0.5 °C/min. However, the spatial distribution of crystals was more uniform in the oleogels prepared at the cooling rates of 0.5 and 1.5 °C/min compared to oleogel cooled at a faster rate. It can be understood by the distance between the crystals. The findings were consistent with those reported by Scharfe et al. [8] who also found that highly ordered wax crystal structures were formed at low cooling rates.

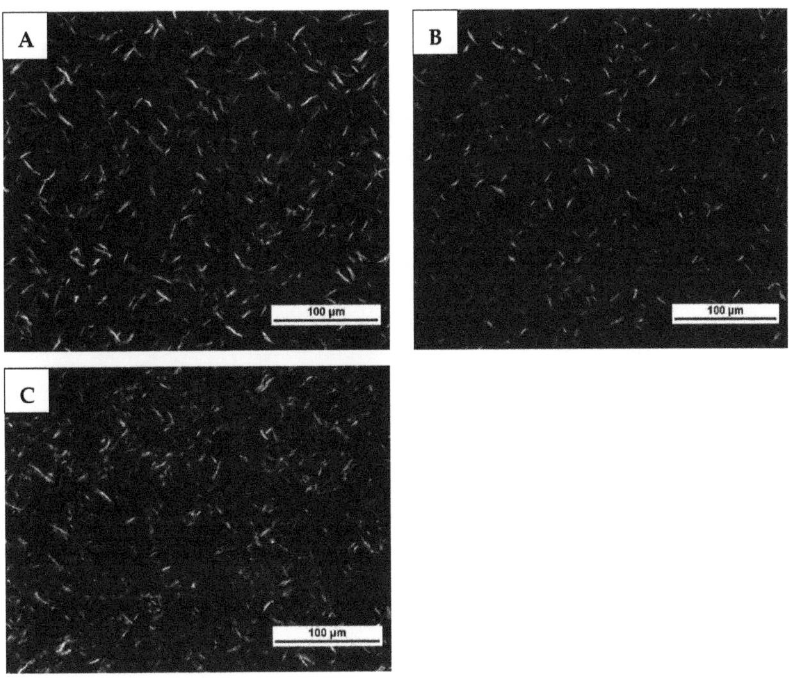

Figure 3. Brightfield polarized light microscopy images of oleogels cooled at 0.5 °C/min (**A**), 1.5 °C/min (**B**), and 5 °C/min (**C**). Images were acquired at 200× magnification. Scale bar 100 μm.

The reason for the poorly ordered crystals at a faster cooling rate could be that there was not enough time for the organization of crystals. In the development of sugar-based organogels, Cui et al. [30] observed that the rapid cooling process did not afford sufficient time for the nuclei of low molecular weight gelators to transition from a metastable to a stable state. The poorly ordered crystals cannot hold oil efficiently, as they may produce softer gels than highly ordered crystals. As previously discussed, the poorly ordered crystals may account for the lower strength, oil-binding capacity, and structural recovery ability of the oleogels cooled at faster rates. A higher fractal dimension increases the oil-binding capacity [20,31]. Therefore, in the present study, the higher oil-binding capacity of the oleogels cooled at a slower rate compared to the faster rate could be attributed to the higher fractal dimensions of the oleogels cooled at a slower rate.

Table 2. Fractal dimension, length of the crystals, and distance between crystals in the oleogels produced at different cooling rates.

Cooling Rate	Fractal Dimension	Length of Crystals (μm)	Distance Between Crystals (μm)
0.5 °C/min	1.976 ± 0.010 ^{a}	18.29 ± 4.58 ^{a}	8.77 ± 4.08 ^{a}
1.5 °C/min	1.567 ± 0.007 ^{b}	13.42 ± 3.46 ^{b}	9.02 ± 2.60 ^{a}
5 °C/min	1.165 ± 0.000 ^{c}	13.49 ± 2.22 ^{b}	12.45 ± 8.21 ^{a}

Distinct superscript letters (a–c) in the same column show a significant difference ($p < 0.05$).

2.5. FTIR Analysis

FTIR spectra of the oleogels cooled at different rates are shown in Figure 4. Within the functional group region of the spectra, three distinct peaks were noted at 1744 cm^{-1} indicative of the carbonyl group, 2853 cm^{-1} representing asymmetric CH_2 stretching, and 2923 cm^{-1} indicative of symmetric CH_2 stretching. Notably, no noticeable spectral alterations were observed across the oleogels, suggesting that the cooling rate did not exert any notable influence on the molecular interactions among the components of the oleogel.

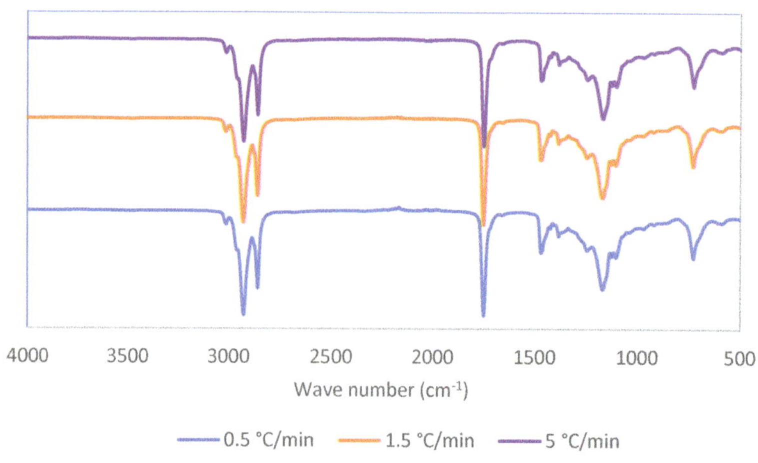

Figure 4. FTIR spectra of oleogels produced at different cooling rates.

2.6. Thermal Properties

Thermal properties of oleogels, such as the onset of melting, onset of crystallization, peak melting, and peak crystallization, were evaluated using DSC. The thermal parameters are presented in Table 3. The melting and crystallization properties of oleogels cooled at different rates did not show any significant differences. This indicates that the cooling rate did not have a significant influence on the melting and crystallization properties of oleogels. Usually, melting and crystallization behaviors are determined by the oleogelators [32]. Since all oleogel samples in this study are prepared using the same formula, the thermal properties are also reported to be the same because during the DSC analysis, initially the samples were heated to 85 °C for 10 min to erase all prior crystalline history (this would have deleted all the crystal structure formed by the different cooling rates) and then the samples underwent the same heating and cooling cycles. Therefore, since the formulation was consistent, similar thermal properties were recorded for all three samples.

Table 3. Thermal properties of oleogels produced at different cooling rates.

Average Cooling Rate	Onset Melting (°C)	Peak Melting (°C)	Onset Crystallization (°C)	Peak Crystallization (°C)
0.5 °C/min	26.05 ± 0.21 [a]	49.75 ± 0.49 [a]	48.00 ± 0.14 [a]	47.55 ± 0.07 [a]
1.5 °C/min	25.40 ± 0.57 [a]	50.80 ± 0.14 [a]	48.65 ± 0.49 [a]	47.90 ± 0.28 [a]
5.0 °C/min	26.35 ± 0.49 [a]	50.25 ± 0.21 [a]	48.65 ± 0.35 [a]	48.15 ± 0.35 [a]

Distinct superscript letter a in the same column shows a non-significant difference ($p < 0.05$).

3. Conclusions

The cooling rate was found to exert a significant impact on the rheological properties, microstructure, and oil-binding capacity of the beeswax-stearic acid oleogels formulated with a blend of sesame oil and rice bran oil. The cooling rate of 0.5 °C/min yielded more desirable oleogel properties compared to fast cooling. Achieving slow cooling at room temperature presents a significant economic advantage due to its energy efficiency and ease of processing, making it a practical and cost-effective method. This study underscores that regulating the cooling rate offers a means to tailor oleogel properties. The insights gained from this research hold potential for the food industry in the manipulation and enhancement of oleogel properties, providing a viable alternative to solid fats rich in trans and saturated fats.

4. Materials and Methods

4.1. Materials

Sesame oil (Changs, Thailand) and rice bran oil (Alfa one, USA) were sourced locally in Brisbane, Australia. As analyzed using gas chromatography (GC) (Shimadzu, Japan), the fatty acid profile of the sesame oil was composed mainly of C14:0, C16:0, C18:1, and C18:2 (0.24 ± 0.03%, 9.8 ± 0.03%, 37.29 ± 0.83%, and 52.66 ± 1.21%, respectively), while the rice bran oil was composed mainly of C16:0, C18:1, and C18:2 (22.54 ± 0.40%, 43.41 ± 0.19% and 33.46 ± 0.85%, respectively). Refined beeswax, stearic acid, and chemicals and reagents were acquired from Sigma Aldrich, Australia. The melting points of beeswax and stearic acid were determined to be 62.2 ± 0.56 °C and 56.30 ± 0.14 °C, respectively, by differential scanning calorimetry.

4.2. Development of Oleogels

Oleogels were prepared through the direct method of oleogelation. The oleogel formulation was prepared using a blend of sesame oil and rice bran oil in a 4:5 ratio (*w*/*w*) and a mixture of beeswax and stearic acid (3:1 (*w*/*w*) ratio at a total concentration of 12% (*w*/*w*)) [11]. The mixture was heated to 80 °C while stirring continuously for 10 min using a magnetic stirrer. Afterward, the liquid mixture was cooled down to 20 °C at three different cooling rates (0.5, 1.5, and 5 °C/min) and stored for 48 h at 20 °C prior to analysis. Cooling rates such as 1.5 and 5 °C/min were achieved by keeping the sample tubes with the hot oleogel mixture at 4 °C (in a water bath) and -18 °C (in a freezer room), respectively, and the cooling rate of 0.5 °C/min was achieved by keeping the sample tubes at the ambient conditions (uncontrolled cooling). The temperature at the thermal center of the samples in identical containers (screwcap polypropylene containers with 25 mL capacity with a dimension of 100 × 65 mm (h × w)), with the same amount of sample (160 g) was measured using a thermocouple and the temperature at the center was recorded every 2 min until the temperature was reached at 20 °C. Samples were prepared in triplicates for each experiment.

4.3. Analytical Methods

4.3.1. Oil-Binding Capacity

The oil-binding capacity of the oleogels was assessed using the method outlined by Sivakanthan et al. [11]. Briefly, 1.5 mL of the heated sample was placed into a 2 mL centrifuge tube (weight of centrifuge tube—M1) and cooled under the same oleogelation conditions at different cooling rates and incubated for 48 h at 20 °C. The weight of the tube with the gel was measured (M2) and centrifuged in a centrifuge (Eppendorf MiniSpin, Eppendorf, Leipzig, Germany) for 15 min at 13,000 rpm at 20 °C. After centrifugation, the tubes were kept inverted for 1 h to drain the liquid oil and weighed again (M3). The following formula was used to calculate the oil-binding capacity.

$$\text{Oil binding capacity } (\%) = \frac{M2 - M3}{M2 - M1} \times 100$$

4.3.2. Rheology

The oleogels were subjected to rheological analysis using a rheometer (Anton Paar MCR302 Rheometer, Graz, Austria) fitted to a Peltier temperature control system and TruStrain™ option, employing a parallel plate (sand-blasted, diameter 50 mm) maintaining an initial fixed gap of 0.5 mm. To secure a proper grip of the plate on the sample, a normal force of 0.1 N was applied. All samples were provided with a recovery time of 5 min after trimming. Triplicate analyses were performed at 20 °C. Data were analyzed using RheoCompass™ version 1.30.999.

Following the procedure outlined by Aguilar-Zárate et al. [33] with minor adjustments, amplitude and frequency sweep experiments were conducted. In the amplitude sweep, a range of strain values (0.01 to 100%) was applied at a consistent frequency of 1 Hz. This allowed for the determination of the Linear Viscoelastic Range (LVR), G' at LVR, and complex viscosity (η^*). The LVR was determined as a plateau observed in both G' and G'' from the amplitude sweeps. Subsequently, frequency sweeps were performed within the LVR, spanning from 0.01 to 100 Hz, at a fixed stress value.

The structural recovery capability of the oleogels was measured by a 3-interval thixotropy test (3ITT), Rot-Rot-Rot method. The oleogels were exposed to successive shear rates: 0.1 s^{-1} for the initial 10 min, then, 10 s^{-1} for 5 min, and then back to 0.1 s^{-1} for the final 10 min as per the methodology outlined by Patel and Dewettinck [34]. The percentage of structure recovery was calculated as described by Tavernier et al. [17].

4.3.3. Thermal Analysis

Melting and crystallization characteristics of oleogels were examined using Differential Scanning Calorimetry (DSC 204 F1 Phoenix, Netzsch, Germany) as described by Sivakanthan et al. [11]. About 10 $\pm$ 1 mg of the sample was placed in aluminum crucibles, sealed and loaded into the DSC, and subjected to heating and cooling cycles under a nitrogen atmosphere. An empty crucible was used as the reference. The samples were heated to 85 °C for 10 min to eliminate any prior crystalline structure, followed by cooling to 0 °C at the rate of 2 °C/min. Then, an isothermal condition was maintained at 0 °C for 20 min, followed by reheating to 85 °C at the rate of 5 °C/min. The data were analyzed using Netzsch Proteus® software (Proteus-80).

4.3.4. Fourier Transform Infrared Spectroscopy (FTIR)

The FTIR spectra of the oleogels were collected using a Fourier transform infrared spectrometer (Nicolet iS50 FT-IR, Thermo Scientific, Waltham, MA, USA). Spectra were obtained by 64 scans (resolution—4 cm^{-1}, range—4000 and 400 cm^{-1}) at room temperature (20 °C). The OMNIC (v1.9) software was used for the analysis of peaks.

4.3.5. Polarized Light Microscopy

The oleogels were cooled at three different rates (0.5, 1.5, and 5 °C/min), and their microstructures were analyzed using a polarized light microscope (Nikon Eclipse LV100ND, Nikon Instruments Inc., USA) equipped with a digital camera (Nikon DS-Fi2). A drop of hot oleogel mixture was placed onto a glass slide (preheated), covered with a coverslip, and cooled at the designated rates. The samples were then stored for 48 h at 20 °C. Then, bright-field imaging was performed at a magnification of 200× at 20 °C. Image processing was performed using ImageJ software version 1.54f (National Institutes of Health, Bethesda, MD, USA). Fractal dimension analysis was conducted using the "fractal box-counting" tool after transforming the images to 8-bit binary format with the box sizes of 2, 3, 4, 6, 8, 12, 16, 32, and 64, and a black background. To measure the distance between crystals and their lengths, the images were converted to 8-bit grayscale. Given the needle-like shape of the crystals, the distance was measured by drawing lines between the centers of the two crystals.

4.4. Statistical Analysis

The data were statistically analyzed using Minitab (Version 21.1) (Minitab, LLC, USA). The results are expressed as mean values with their corresponding standard deviations. To assess the significance of differences between treatments, a one-way ANOVA and Tukey's post hoc test were applied, with a 95% confidence level ($p < 0.05$).

Author Contributions: S.S.: Conceptualization-Lead, Investigation-Lead, Writing—original draft-Lead, Writing—review and editing-Equal. S.F.: Conceptualization-Supporting, Project administration-Supporting, Supervision-Supporting, Writing—review and editing-Supporting. S.M.: Conceptualization-Supporting, Project administration-Supporting, Supervision-Supporting, Writing—review and editing-Supporting. T.M.: Conceptualization-Supporting, Project administration-Supporting, Supervision-Supporting, Writing—review and editing-Supporting. A.K.: Conceptualization-Supporting, Project administration-Lead, Supervision-Lead, Writing—review and editing-Supporting. All authors have read and agreed to the published version of the manuscript.

Funding: This research was funded by the Queensland University of Technology, Brisbane Australia, and Accelerating Higher Education Expansion and Development (AHEAD): a World Bank-funded Sri Lankan government operation (Grant number: AHEAD/PhD/R3/Agri/550).

Institutional Review Board Statement: Not Applicable.

Informed Consent Statement: Not Applicable.

Data Availability Statement: The data presented in this study are available on request from the corresponding author.

Acknowledgments: The authors would like to acknowledge the laboratory facilities from the Central Analytical Research Facility (CARF) and the Centre for Agriculture and the Bioeconomy (CAB) of the Queensland University of Technology, Australia.

Conflicts of Interest: The authors declare no conflicts of interest.

References

1. Ginter, E.; Simko, V. New data on harmful effects of trans-fatty acids. *Bratisl. Lek. Listy.* **2016**, *117*, 251–253. [CrossRef] [PubMed]
2. Gengatharan, A.; Mohamad, N.V.; Zahari, C.N.M.C.; Vijayakumar, R. Oleogels: Innovative formulations as fat substitutes and bioactive delivery systems in food and beyond. *Food Struct.* **2023**, *38*, 100356. [CrossRef]
3. Barroso, N.G.; Santos, M.A.S.; Okuro, P.K.; Cunha, R.L. Composition and process approaches that underpin the mechanical properties of oleogels. *J. Am. Oil Chem. Soc.* **2022**, *99*, 971–984. [CrossRef]
4. Trujillo-Ramírez, D.; Lobato-Calleros, C.; Jaime Vernon-Carter, E.; Alvarez-Ramirez, J. Cooling rate, sorbitan and glyceryl monostearate gelators elicit different microstructural, viscoelastic and textural properties in chia seed oleogels. *Food Res. Int.* **2019**, *119*, 829–838. [CrossRef]
5. Blake, A.I.; Marangoni, A.G. The use of cooling rate to engineer the microstructure and oil binding capacity of wax crystal networks. *Food Biophys.* **2015**, *10*, 456–465. [CrossRef]
6. Yao, Y.; Zhou, H.; Liu, W.; Li, C.; Wang, S. The effect of cooling rate on the microstructure and macroscopic properties of rice bran wax oleogels. *J. Oleo Sci.* **2021**, *70*, 135–143. [CrossRef] [PubMed]
7. Giacomozzi, A.S.; Palla, C.A.; Carrín, M.E.; Martini, S. Physical properties of monoglycerides oleogels modified by concentration, cooling rate, and high-intensity ultrasound. *J. Food Sci.* **2019**, *84*, 2549–2561. [CrossRef]
8. Scharfe, M.; Niksch, J.; Flöter, E. Influence of minor oil components on sunflower, rice bran, candelilla, and beeswax oleogels. *Eur. J. Lipid Sci. Technol.* **2022**, *124*, 2100068. [CrossRef]
9. Valoppi, F.; Calligaris, S.; Marangoni, A.G. Structure and physical properties of oleogels containing peanut oil and saturated fatty alcohols. *Eur. J. Lipid Sci. Technol.* **2017**, *119*, 1600252. [CrossRef]

10. Valoppi, F.; Salmi, A.; Ratilainen, M.; Barba, L.; Puranen, T.; Tommiska, O.; Helander, P.; Heikkilä, J.; Haeggström, E. Controlling oleogel crystallization using ultrasonic standing waves. *Sci. Rep.* **2020**, *10*, 14448. [CrossRef]
11. Sivakanthan, S.; Fawzia, S.; Mundree, S.; Madhujith, T.; Karim, A. Optimization and characterization of new oleogels developed based on sesame oil and rice bran oil. *Food Hydrocoll.* **2023**, *142*, 108839. [CrossRef]
12. Flöter, E.; Wettlaufer, T.; Conty, V.; Scharfe, M. Oleogels: Their applicability and methods of characterization. *Molecules* **2021**, *26*, 1673. [CrossRef]
13. Shi, Y.; Liu, C.; Zheng, Z.; Chai, X.; Han, W.; Liu, Y. Gelation behavior and crystal network of natural waxes and corresponding binary blends in high-oleic sunflower oil. *J. Food Sci.* **2021**, *86*, 3987–4000. [CrossRef]
14. Ghazani, S.M.; Dobson, S.; Marangoni, A.G. Hardness, plasticity, and oil binding capacity of binary mixtures of natural waxes in olive oil. *Curr. Res. Food Sci.* **2022**, *5*, 998–1008. [CrossRef]
15. Pinto, T.C.; Martins, A.J.; Pastrana, L.; Pereira, M.C.; Cerqueira, M.A. Oleogel-based systems for the delivery of bioactive compounds in foods. *Gels* **2021**, *7*, 86. [CrossRef]
16. Okuro, P.K.; Santos, T.P.; Cunha, R.L. Compositional and structural aspects of hydro- and oleogels: Similarities and specificities from the perspective of digestibility. *Trends Food Sci. Technol.* **2021**, *111*, 55–67. [CrossRef]
17. Tavernier, I.; Doan, C.D.; Van der Meeren, P.; Heyman, B.; Dewettinck, K. The potential of waxes to alter the microstructural properties of emulsion-templated oleogels. *Eur. J. Lipid Sci. Technol.* **2018**, *120*, 1700393. [CrossRef]
18. Naeli, M.H.; Milani, J.M.; Farmani, J.; Zargaraan, A. Developing and optimizing low-saturated oleogel shortening based on ethyl cellulose and hydroxypropyl methyl cellulose biopolymers. *Food Chem.* **2022**, *369*, 130963. [CrossRef]
19. Okuro, P.K.; Tavernier, I.; Bin Sintang, M.D.; Skirtach, A.G.; Vicente, A.A.; Dewettinck, K.; Cunha, R.L. Synergistic interactions between lecithin and fruit wax in oleogel formation. *Food Funct.* **2018**, *9*, 1755–1767. [CrossRef]
20. Palla, C.; de Vicente, J.; Carrín, M.E.; Gálvez Ruiz, M.J. Effects of cooling temperature profiles on the monoglycerides oleogel properties: A rheo-microscopy study. *Food Res. Int.* **2019**, *125*, 108613. [CrossRef]
21. Patel, A.R.; Nicholson, R.A.; Marangoni, A.G. Applications of fat mimetics for the replacement of saturated and hydrogenated fat in food products. *Curr. Opin. Food Sci.* **2020**, *33*, 61–68. [CrossRef]
22. Zhang, Y.; Xu, J.; Tang, C.; Li, Y. Crystallization behavior and physical properties of monoglycerides-based oleogels as function of oleogelator concentration. *Foods* **2023**, *12*, 345. [CrossRef] [PubMed]
23. Gao, Y.; Lei, Y.; Wu, Y.; Liang, H.; Li, J.; Pei, Y.; Li, Y.; Li, B.; Luo, X.; Liu, S. Beeswax: A potential self-emulsifying agent for the construction of thermal-sensitive food W/O emulsion. *Food Chem.* **2021**, *349*, 129203. [CrossRef] [PubMed]
24. Martins, A.J.; Cerqueira, M.A.; Fasolin, L.H.; Cunha, R.L.; Vicente, A.A. Beeswax organogels: Influence of gelator concentration and oil type in the gelation process. *Food Res. Int.* **2016**, *84*, 170–179. [CrossRef]
25. Thomas, P.E.; Saravanan, M.; Prabhasankar, P. Virgin coconut oil oleogel: Gelation mechanism, rheological, structural and thermal properties. *Int. J. Food Sci. Technol.* **2023**, *58*, 1434–1443. [CrossRef]
26. Abdolmaleki, K.; Alizadeh, L.; Nayebzadeh, K.; Baranowska, H.M.; Kowalczewski, P.Ł.; Mousavi Khaneghah, A. Potential application of hydrocolloid-based oleogel and beeswax oleogel as partial substitutes of solid fat in margarine. *Appl. Sci.* **2022**, *12*, 12136. [CrossRef]
27. Si, H.; Cheong, L.-Z.; Huang, J.; Wang, X.; Zhang, H. Physical properties of soybean oleogels and oil migration evaluation in model praline system. *J. Am. Oil Chem. Soc.* **2016**, *93*, 1075–1084. [CrossRef]
28. Sivakanthan, S.; Fawzia, S.; Madhujith, T.; Karim, A. Synergistic effects of oleogelators in tailoring the properties of oleogels: A review. *Compr. Rev. Food Sci. Food Saf.* **2022**, *21*, 3507–3539. [CrossRef]
29. Frolova, Y.; Sarkisyan, V.; Sobolev, R.; Makarenko, M.; Semin, M.; Kochetkova, A. The influence of edible oils' composition on the properties of beeswax-based oleogels. *Gels* **2022**, *8*, 48. [CrossRef]
30. Cui, J.; Liu, A.; Guan, Y.; Zheng, J.; Shen, Z.; Wan, X. Tuning the helicity of self-assembled structure of a sugar-based organogelator by the proper choice of cooling rate. *Langmuir* **2010**, *26*, 3615–3622. [CrossRef]
31. Mert, B.; Demirkesen, I. Evaluation of highly unsaturated oleogels as shortening replacer in a short dough product. *LWT* **2016**, *68*, 477–484. [CrossRef]
32. da Silva, T.L.T.; Arellano, D.B.; Martini, S. Use of high-intensity ultrasound to change the physical properties of oleogels and emulsion gels. *J. Am. Oil Chem. Soc.* **2019**, *96*, 681–691. [CrossRef]
33. Aguilar-Zárate, M.; Macias-Rodriguez, B.A.; Toro-Vazquez, J.F.; Marangoni, A.G. Engineering rheological properties of edible oleogels with ethylcellulose and lecithin. *Carbohydr. Polym.* **2019**, *205*, 98–105. [CrossRef] [PubMed]
34. Patel, A.R.; Dewettinck, K. Comparative evaluation of structured oil systems: Shellac oleogel, HPMC oleogel, and HIPE gel. *Eur. J. Lipid Sci. Technol.* **2015**, *117*, 1772–1781. [CrossRef]

gels

MDPI

Article

Textural, Color, and Sensory Analysis of Cookies Prepared with Hemp Oil-Based Oleogels

Ana Leahu, Cristina Ghinea *, Sorina Ropciuc and Cristina Damian

Faculty of Food Engineering, Stefan cel Mare University of Suceava, 720229 Suceava, Romania; analeahu@fia.usv.ro (A.L.); sorina.ropciuc@fia.usv.ro (S.R.); cristinadamian@fia.usv.ro (C.D.)
* Correspondence: cristina.ghinea@fia.usv.ro

Abstract: The amount of saturated fat in cookies can be reduced by replacing margarine with oleogel, resulting in healthier products. In this study, the rheological and textural profile of cookies formulated with oleogel as the main margarine substitute was evaluated. Hemp seed vegetable oil was oleogelized with four types of waxes: beeswax (BW), carnauba wax (CW), candelilla wax (DW), rice bran wax (RW), and three oleogeling agents, sitosterol (S), pea protein (PP), and xanthan gum (XG), respectively. The textural and rheological properties of the oleogel dough samples were analyzed using the PertenTVT-6700 texturometer (Perten Instruments, Sweden) and the Haake rheometer. The results showed an increase in the hardness of cookie doughs with oleogels. The values of the elastic component (G') and the viscous component (G'') increased, which means that the oleogels used affected the rheological behavior at 25 °C, causing an increase in the dough consistency. Sensory attributes, texture, and color parameters of cookies with oleogels were determined. The cookies' hardness increased significantly from 4409.83 ± 0.13 g (control sample) to 7085.33 ± 0.15 g in the cookie sample prepared with hemp oil sitosterol oleogel, whereas the sample with candelilla wax had the lowest hardness value of 4048.09 ± 0.14 g. The color of the oleogel cookies was darker than that of the control cookies. The cookie sample with hemp oil and beeswax oleogel was the most appreciated by the evaluators among the oleogel cookie samples. The findings suggest that hemp seed oil oleogel is an effective fat substitute in cookies, promoting the application of this vegetable oil in food products.

Keywords: dough; fat substitution; oleogel; hemp seed oil; rheology; texture analysis

1. Introduction

The globalization of markets has transformed food systems and led to major changes in food consumption behavior (such as replacing family meals with popular fast foods and reducing cooking time), which can be influenced by concern for the health and nutritional status of the population. High-fat snacks, like cookies, are very popular among consumers. According to Drewnowski and Almiron-Roig [1], fats contribute to the texture, flavor, and aroma of food products. The consumption of high-fat snacks (containing trans and saturated fatty acids and sugar) has been linked to an increased risk of obesity [2], diabetes [3], colorectal cancer [4], and cardiovascular disease [5]. Childhood obesity is the most common chronic disease in children and adolescents, and changes in eating and physical activity patterns have led to a 30% higher childhood obesity rate in developing countries than in developed countries [6]. Simple biscuits (without cream and ingredients such as peanuts, fruits, etc.) are ready to eat, affordable, and have a long shelf life and a wide range of tastes. They represent a self-sufficient part of the bakery industry [7] and are usually used for "snacks" by both children and adults [8]. Cookies (from soft dough), mainly containing flour, fat, and sugar, are popular bakery snacks due to their low cost, good taste, texture, and storage capacity [9]. The United States, India, and China are the biggest markets for cookies, while in Western Europe, Italy and Spain are the main demand markets [10]. Cookies are highly appreciated in the northeast of Romania (the Bucovina region), where there are small processing units for pastry products. Although cookies are commonly consumed in the rural areas of northern Romania, today, these products, whose sales are growing significantly, are becoming fashionable in urban cities. In bakery products, but especially in cookies, shortening and margarine are some of the most used fat sources that have a flexible base fat system that provides the desired sensory and technological qualities of a certain product. Thus, alternatives have been developed, considering polymers, polysaccharides, proteins, or lipids to replace saturated fats [11]. In the food industry, oleogels offer properties similar to solid fats with low levels of trans and saturated fat and can be carriers of bioactive compounds, maintaining desirable functional properties of fats, such as mouthfeel, creaminess, and stability [12]. Oleogels, characterized by their semisolid matrix, offer a promising alternative to trans and saturated fats in food products, which are known to be harmful to health, mimicking their texture and functionality [13]. Oleogels can protect bioactive compounds from degradation and improve the nutritional profile and texture of spreads, bakery, confectionery, dairy, and meat products by increasing the bioavailability of nutrients and reducing saturated fat content.

Cannabis sativa L. is an annual plant, which is widely spread around different geographical zones that produces small seeds. Cold pressing of these seeds (35% oil) allows obtaining hemp seed oil, which is mainly used as food [14,15]. Hemp seed oil (*Cannabis sativa* L.) is considered a healthy oil and contains a unique ω-6/ω-3 fatty acid ratio of 3:1, which can be optimal for human nutrition and, therefore, suitable for both human consumption and utilization in the pharmaceutical and cosmetic industries [16–18]. People from Mediterranean areas have low rates of coronary heart disease because the n-6/n-3 ratio of hemp seed oil is similar to the n-6/n-3 ratio found in their diets [19]. Cold-pressed hemp oil, a natural product, is an excellent source of protein, fatty acids, enzymes, biologically active compounds, and vitamins associated with hemp seed components [20]. The bioactive compounds, but also the unsaponifiable fraction, offer health benefits, such as anti-cancer and anti-cardiovascular diseases, antimicrobial, anti-inflammatory, anti-thrombotic, anti-arterogenic, and hypocholesterolemic [18,21].

In this work, cold-pressed hempseed oil oleogels were prepared with 9% natural waxes, including beeswax (BW), carnauba wax (CW), candelilla wax (DW), rice bran wax (RW), and three oleogelation agents, respectively, sitosterol (S), pea protein (PP), and xanthan gum (XG), to evaluate their potential as solid fat replacements in cookies. In previous studies, oleogels elaborated with several vegetable oils and waxes of different types were analyzed [13,22,23]. Considering the technological feasibility, oil binding capacities, and fatty acid composition, the use of hemp seed oil was chosen for the development of oleogels and the reformulation of the cookie recipe. The objective of the research was to analyze the effect of replacing saturated fats with new oleogels prepared with hemp seed oil on the textural and sensory properties of cookies.

2. Results and Discussion

2.1. Texture Profile Analysis of Dough and Cookies

The results of the cookie dough texture analysis are shown in Table 1. The hardness increased with the replacement of margarine by oleogels, with the highest increase

(4052.09 ± 0.12) in sample S5_S_HO with sitosterol as an oleogelation agent. In the case of samples of dough with oleogels obtained from hemp oil and natural wax, an increase in hardness values was observed for all samples, and the highest values were recorded for the sample with rice bran wax (S4_RW_HO), followed by the sample with beeswax (S1_BW_HO). The increase in the hardness of cookie doughs with oleogels was also observed by Tanislav et al. [24]. The hardness of cookie doughs with oleogels is influenced by the hardness of the oleogels, as harder oleogels can adversely affect the processing of the dough. The addition of 9% wax resulted in oleogels with higher hardness and lower stickiness [23]. The increase in hardness and decrease in adhesiveness may be due to the higher level of saturated fatty acids in these types of oleogels. Fully saturated with hydrogen and showing a more rigid and fixed structure, these are more solid and compact in the dough system [23]. At the same time, the hardness of oleogels indicates that they may be an alternative to saturated fats [24]. Oleogels form strong interactions with interacting surfaces, thus promoting adhesion. Adhesivity is the force required to remove food adhering to a particular surface, and the negative values shown in Table 1 signified a downward-directed stress [25]. Oleogels form strong interactions with interacting surfaces, thus promoting adhesion according to Muskat et al. [26]. The dough of the cookie sample with oleogel from hemp oil and rice bran wax (S4_RW_HO) had the highest adhesion, followed by sample S5_S_HO (with sitosterol as an oleogelation agent). Adhesivity is influenced both by the type of the oleogelation agent but also by the type of oil used to obtain the oleogels added to cookies; for example, Manohar and Rao [27] showed that dough obtained with rapeseed oil oleogels and structuring agents (ethylcellulose, candelilla wax, rice bran wax, white beeswax, yellow beeswax, and monoacylglycerol) had greater adhesiveness compared to dough with palm oil.

Table 1. Texture properties of the cookie dough with oleogels.

Sample	Hardness (g)	Adhesiveness (J)	Springiness (mm)	Cohesiveness	Gumminess	Stickiness
C	1589.99 ± 0.19 ^{h}	−50.43 ± 0.13 ^{b}	1.00 ± 0.11 ^{a}	0.27 ± 0.13 ^{a}	423.44 ± 0.14 ^{f}	−9.47 ± 0.16 ^{a}
S1_BW_HO	2146.78 ± 0.18 ^{d}	−450.38 ± 0.18 ^{f}	1.00 ± 0.19 ^{a}	0.18 ± 0.12 ^{a}	384.73 ± 0.13 ^{g}	−67.72 ± 0.11 ^{f}
S2_DW_HO	1598.77 ± 0.17 ^{g}	9.00 ± 0.12 ^{a}	0.99 ± 0.19 ^{a}	0.32 ± 0.11 ^{a}	503.65 ±0.14 ^{e}	−83.75 ± 0.14 ^{g}
S3_CW_HO	1981.84 ± 0.14 ^{e}	−259.66 ± 0.16 ^{e}	1.00 ± 0.18 ^{a}	0.19 ± 0.12 ^{a}	384.91 ±0.03 ^{g}	−51.45 ± 0.15 ^{b}
S4_RW_HO	2343.85 ± 0.15 ^{c}	−856.27 ± 0.17 ^{h}	1.00 ± 0.12 ^{a}	0.22 ± 0.13 ^{a}	507.00 ± 0.19 ^{d}	−57.51 ± 0.11 ^{c}
S5_S_HO	4052.09 ± 0.12 ^{a}	−466.66 ±0.16 ^{g}	0.99 ± 0.14 ^{a}	0.18 ± 0.13 ^{a}	734.47 ± 0.17 ^{b}	−63.29 ± 0.19 ^{d}
S6_PP_HO	2998.8 ± 0.18 ^{b}	−115.71 ±0.11 ^{c}	1.00 ± 0.19 ^{a}	0.26 ± 0.11 ^{a}	792.04 ± 0.04 ^{a}	−66.25 ± 0.14 ^{e}
S7_XG_HO	1711.82 ± 0.12 ^{f}	−185.41 ± 0.14 ^{d}	1.00 ± 0.07 ^{a}	0.35 ± 0.13 ^{a}	601.30 ± 0.13 ^{c}	−94.73 ± 0.13 ^{h}

Values are mean ± standard deviation (n = 3). Different superscript letters (a–h) in the same column indicate significant differences between values at the $p < 0.05$ level.

The springiness index correlated with stickiness tells us the recovery properties of the dough (a value of 1 indicates a fully elastic material, and 0 indicates a fully viscous material) [28]. In the cookie manufacturing process, the samples containing oleogels with hemp oil (HO), rice bran wax (RW), and xanthan gum (XG) (S4_RW_HO, S7_XG_HO) behaved best during rolling and shaping operations, although the springiness index values are very close. The results obtained for the springiness index showed that the values are not statistically significant. Cohesiveness showed opposite trends, indicating that a more cohesive dough has a lower recovery capacity after being subjected to stress, thus increasing its value. Cohesiveness decreased for the dough samples with oleogels from natural wax, except for the sample with candelilla wax (S2_DW_HO), but the values are not statistically significantly different. Sample S7_XG_HO (with xanthan gum as an oleogelation agent) had the highest value (0.35 ± 0.13) for cohesiveness, but was also not statistically different from the values of the other samples. The decrease in cohesiveness values indicates the lowest internal resistance of the dough samples with oleogels [29]. The obtained values of gumminess varied significantly, which means that the overall effect of the energy required to break down the internal forces of the dough varies considerably [30]. Adding oleogels to foods can lead to a gummy texture [31]. The increased gumminess of cookie samples with oleogel (especially from pea protein and hemp seed oil) may be due to the ability of pea protein to bind water and fat [32]. According to Tanislav et al. [29], fat reduces the stickiness of the dough surface and also oleogels based on the results of the present study. The results obtained using Pearson correlation for dough textural parameters indicated that hardness showed negative correlations with adhesiveness (r = −0.379), springiness (r = −0.199), cohesiveness (r = −0.529), and stickiness and positive correlations with gumminess (r = 0.716), while adhesiveness showed a negative correlation with springiness (r = −0.206) and positive correlations with cohesiveness (r = 0.609), gumminess, and stickiness. Springiness had a negative correlation with gumminess (r = −0.076) and positive correlations with cohesiveness and stickiness. Cohesiveness had a positive correlation with gumminess and a negative correlation with stickiness, while gumminess had a negative correlation with stickiness.

Texture analysis of cookies was performed by measuring the hardness, strength, and fracturability of cookies (Table 2). The hardness increased significantly from 4409.83 ± 0.13 g (control sample) to 7085.33 ± 0.15 g in sample S5_S_HO, a cookie prepared with hemp oil sitosterol oleogel, whereas sample S2_DW_HO had the lowest hardness value of 4048.09± 0.14 g. These findings are similar to other research [33], in which the textural properties of cookies made with oleogels produced by combining soy wax and rice bran oil were evaluated. Pradhan et al. [33] reported hardness values between 4400 and 4800 g for oleogel and butter cookie samples. In another study, the hardness of cookies prepared with commercial corn oil oleogels and four kinds of emulsifiers was significantly higher than that of the shortening cookies [34]. Excessive hardness will reduce the texture of the cookies [35]; a higher hardness was obtained for cookie samples with xanthan gum and sitosterol oleogels (samples S7_XG_HO and S5_S_HO). A significantly higher hardness of sitosterol oleogels attributed to superior structuring ability was observed by Tanislav al. [24]. The type of oleogelling agents and the proportions of oleogel influence the hardness of the baked product [36]. Fracturability refers to how easily the cookies will break [24]. In the present study, the fracturability values varied from 4.27 ± 0.13 mm (sample S6_PP_HO) to 14.99 ± 0.11 mm (sample S7_XG_HO). The cookie sample S7_XG_HO, with hemp oil oleogel and xanthan gum, had a higher fracturability value than the control sample, while the other samples had lower values compared to the control sample.

Table 2. Texture parameters of the cookies with oleogels.

Sample	Hardness (g)	Strength (g/mm)	Fracturability (mm)
C	4409.83 ± 0.13 ^{e}	220.49 ± 0.13 ^{e}	11.42 ± 0.10 ^{b}
S1_BW_HO	4387.91 ± 0.17 ^{f}	219.39 ± 0.15 ^{f}	5.28 ± 0.13 ^{e}
S2_DW_HO	4048.09 ± 0.14 ^{h}	202.40 ± 0.14 ^{h}	7.41 ± 0.09 ^{c}
S3_CW_HO	5284.50 ± 0.10 ^{d}	264.22 ± 0.17 ^{c}	6.03 ± 0.14 ^{d}
S4_RW_HO	5287.50 ± 0.15 ^{c}	226.68 ± 0.11 ^{d}	5.59 ± 0.12 ^{e}
S5_S_HO	7085.33 ± 0.15 ^{a}	354.26 ± 0.13 ^{a}	4.77 ± 0.15 ^{f}
S6_PP_HO	4074.84 ± 0.17 ^{g}	203.74 ± 0.18 ^{g}	4.27 ± 0.13 ^{g}
S7_XG_HO	6383.59 ± 0.12 ^{b}	319.17 ± 0.16 ^{b}	14.99 ± 0.11 ^{a}

Values are mean ± standard deviation (n = 3). Different superscript letters (a–h) in the same column indicate significant differences between values at the $p < 0.05$ level.

2.2. Rheological Characterisation of Dough with Oleogels

The viscoelastic properties of the soft dough measured as a function of frequency are shown in Figure 1a,b.

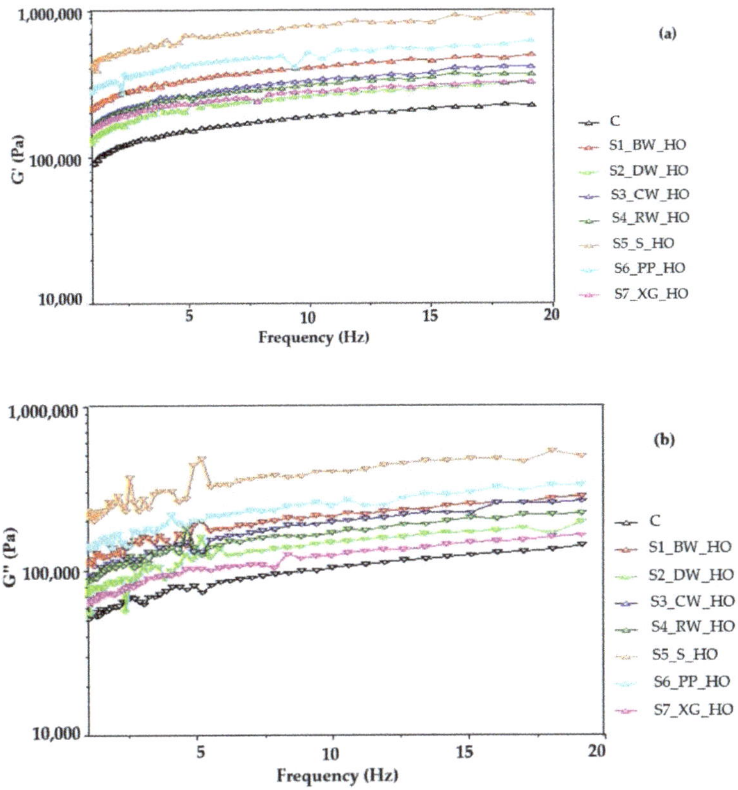

Figure 1. Variation of the modulus of (**a**) elasticity (G′) and modulus of (**b**) viscosity (G″) with the angular frequency for hemp oil (HO) oleogel dough samples.

Rheological measurements indicated that all the dough samples had an elastic solid type of behavior, and they presented higher values of G′ (storage mode) than G″ (loss mode) in the frequency range tested. The addition of oleogel increased the storage (G′) and loss moduli (G″) of the cookie dough compared to the control sample, indicating a more elastic and viscous character and thus stiffer composite dough networks. The higher lipid content of the oleogels contributes to the higher temporal stability of the dough network [37]. The cookies prepared in this study were classified as a soft dough product; under dough mixing conditions, the rheological properties of oleogels change, and they can quickly lose their solid structure as a result of shear forces and become much softer than shortening [38]. Similarly, variations in the modulus of elasticity (G′) and modulus of viscosity (G″) with temperature are shown in Figure 2a,b. As the temperature increased, the values of G′ and G″ decreased slightly, possibly due to hydrolysis of damaged starch by amylase according to Tang and Liu [39]. By analyzing the changes in the G′ and G″ modulus values as a function of temperature, it was found that there was a common trend of progressive decrease until a certain temperature was reached, which indicated a decrease in interactions in the system [40]. The doughs with oleogel obtained with xanthan gum and the doughs with sitosterol show low viscoelastic behavior. Xanthan gum did not produce high elasticity for short sugar cookie doughs, and the same behavior was observed for the doughs with sitosterol. The cookies obtained from these doughs showed the highest hardness values, and the maximum force point was the highest for the cookies with gum and those with sitosterol.

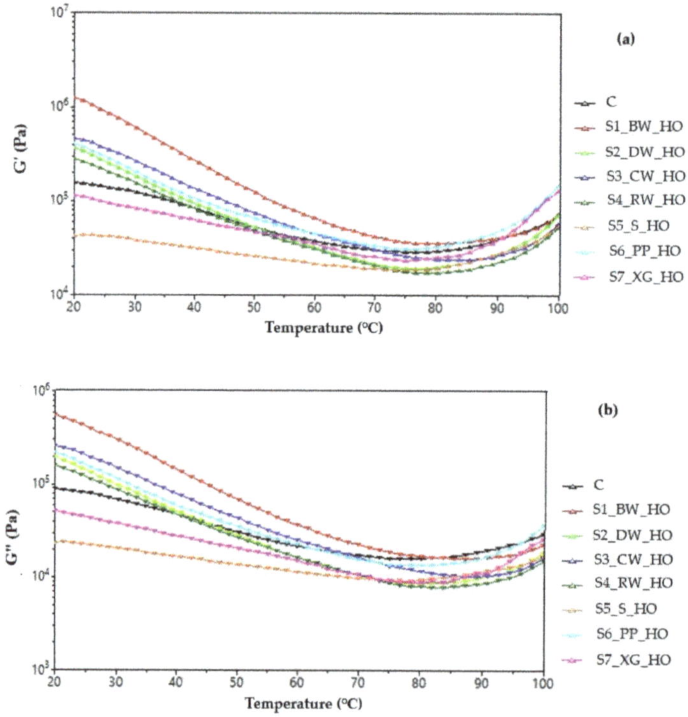

Figure 2. Variation of the modulus of (**a**) elasticity (G′) and modulus of (**b**) viscosity (G″) with the temperature for hemp oil (HO) oleogel dough samples.

2.3. Cookies' Color

Maillard and caramelization reactions are responsible for the coloring and flavor of cookies [41]. There are significant differences between the color of the control sample and the color of cookies with oleogels (Table 3). The lightness (L*) values of the cookie samples with oleogels decreased compared to the control sample. According to Sobolev et al. [42], in order not to interfere with food color perception, a high value of L* is preferable. In the case of cookies with oleogel from hemp oil and natural wax, the highest L* value was for the cookie sample with beeswax oleogel (59.94 ± 0.13), while for the cookie samples with oleogelation agents, the cookie sample with oleogel from hemp oil and xanthan gum had the highest value (63.36 ± 0.38). It was observed that the cookie samples with oleogels from hemp oil and natural waxes are darker than cookie samples with oleogels from hemp oil and oleogelation agents, like xanthan gum and pea protein. Pang et al. [35] reported a decrease in the L* values of the oleogel cakes from rice bran oil (RBO) with the following oleogelling agents: candelilla wax (CDW), beeswax (BW), rice bran wax (RBW), and carnauba wax (CRW) [30]. Also, Zulfiqar et al. [28] indicated that the L* value decreased with an increasing percentage of using a blend of corn oil and sunflower oil oleogel with beeswax as an oleogelator in cookies. The a* values of the oleogel cookie samples decreased compared to the control sample, with the exception of the hemp oil and pea protein oleogel cookie sample (S6_PP_HO). All values were positive, indicating that the red tone dominates the green tone in all samples of oleogel-based cookies. Also, positive values were obtained for the b* color parameter for all cookie samples, indicating that the yellow tone dominates the blue tone. Cookies formulated with oleogel can be characterized by a shade of red and yellow according to Tanislav et al. [29]. The results obtained for total color difference (ΔE) indicate major color differences between the oleogel-based cookies and the control sample. The higher color differences were observed for oleogel-based cookies with hemp oil and candelilla wax (S2_DW_HO) and with rice wax (S4_RW_HO) and for cookie samples with sitosterol used as an oleogelation agent (S5_S_HO). The differences between the L* color parameter values of margarine and margarine cookie samples were about 6%, with the L* values decreasing in the margarine cookie sample (C), which meant that the control sample was darker in color than the margarine. The L* values of S6_PP_HO, S1_BW_HO, and S5_S_HO cookie samples increased by 73.25%, 50.63%, and 27.62% compared to the L* values of PP, BW, and P with HO oleogel samples. The PP_HO, BW_HO, and S_HO oleogel samples are the darkest in color among all the investigated oleogel samples, and the addition of oleogels to the cookie samples leads to the opening of their color. For the S2_DW_HO sample, no color changes were observed regarding the L* parameter, while for the S3_CW_HO sample, a decrease in the L* values of approximately 2% was determined compared to the oleogel sample. If in the case of the S4_RW_HO cookie sample the L* values increased by about 15% compared to the L* value for the oleogels obtained from RW and HO, in the case of the S7_XG_HO cookie sample, the color parameter L* values decreased by 16%. The oleogel sample with XG and HO is the lightest in color among the oleogel samples, but the addition of the oleogel in the cookies leads to the darkening of their color. All the oleogel samples, as well as the margarine sample, had negative values for the color parameter a*, indicating green-colored samples, while all the cookie samples had positive values for this parameter, indicating reddish-colored samples. For all samples (margarine, oleogels, oleogel cookies), positive values for the color parameter b* were determined, indicating a yellow color of the samples. The addition of oleogels in cookies from BW, DW, RW, and XG with HO led to an increase in the values of the color parameter b*, while the addition of oleogels from CW, S, and PP with HO led to a decrease in the values of this parameter compared to the values obtained for the respective oleogels.

Table 3. Color parameters of the cookie samples.

Samples	L*	a*	b*	ΔE
C	68.31 ± 1.52 ^{a}	3.35 ± 0.21 ^{b}	36.25 ± 1.16 ^{a}	-
S1_BW_HO	59.94 ± 0.13 ^{c}	2.23 ± 0.05 ^{c}	31.96 ± 0.21 ^{bc}	9.55 ± 1.05 ^{bc}
S2_DW_HO	53.63 ± 0.09 ^{d}	1.21 ± 0.03 ^{d}	31.57 ± 0.67 ^{c}	15.55 ± 1.40 ^{a}
S3_CW_HO	55.66 ± 0.69 ^{d}	1.31 ± 0.03 ^{d}	30.42 ± 0.24 ^{cd}	14.14 ± 1.62 ^{ab}
S4_RW_HO	53.88 ± 1.64 ^{d}	1.98 ± 0.09 ^{c}	31.71 ± 0.39 ^{c}	15.23 ± 2.80 ^{a}
S5_S_HO	55.12 ± 0.82 ^{d}	0.89 ± 0.04 ^{e}	28.86 ± 0.65 ^{d}	15.36 ± 1.78 ^{a}
S6_PP_HO	62.44 ± 0.42 ^{bc}	3.87± 0.05 ^{a}	36.02 ± 0.07 ^{a}	5.96 ± 1.92 ^{c}
S7_XG_HO	63.36 ± 0.38 ^{b}	2.03 ± 0.04 ^{c}	33.41 ± 0.30 ^{b}	6.02 ± 1.15 ^{c}

Values are mean ± standard deviation (n = 3). Different superscript letters (a–e) in the same column indicate significant differences between values at the $p < 0.05$ level.

2.4. Sensory Evaluation

Sensory attributes for cookies with oleogels from hemp seed oil and natural waxes (beeswax, carnauba wax, candelilla wax, rice bran wax) and other oleogelation agents (sitosterol, pea protein, and xanthan gum) are illustrated in Figure 3. Consumers scored the appearance, color, texture, flavor, and overall acceptability of the eight cookie samples presented with number codes. The group determined the sensory attribute "flavor", which defines the overall perception of the aroma and taste associated with cookies after they are chewed and swallowed, as indicated by Yılmaz and Öğütcü [43]. The control cookie sample (with margarine) scored highest for all sensory attributes. The cookie sample with oleogel from hemp seed oil and beeswax (S1_BW_HO) was more appreciated for appearance and texture, while the cookie sample with oleogel from hemp seed oil and candelilla wax (S2_DW_HO) and the cookie sample with oleogel from hemp seed oil and xanthan gum (S7_XG_HO) were more appreciated for flavor and color, respectively. The least appreciated samples were S4_RW_HO for appearance and color, S6_9PP_HO for texture, and S5_9S_HO for flavor. Regarding the overall acceptability, the highest scores for cookies with oleogels were obtained by cookie samples S1_BW_HO and S2_DW_HO, while the lowest score was obtained by cookie sample S6_9PP_HO. The color of cookies with hemp seed oleogels was darker than that of control cookies. Similar results were reported by Flores-García et al. [44] using organic candelilla wax and extra-virgin linseed oil. The familiar golden-brown color that the cookie samples exhibited may be one of the reasons why evaluators liked these samples; the dark-brown colors resulting from the intense Maillard reaction may appeal much less to consumers. Evaluators downgraded the aroma and taste of the products due to the hemp oil having a slightly bitter taste and specific smell. With a somewhat bitter aftertaste, hemp oil can be used as a substitute for olive oil because it has a color similar to that of olive oil [45]. The oxidation of fats from cookies may influence consumer acceptability. It was observed that the oleogel's oxidative stability increased with the addition of different types of wax [13]. The peroxide value for hemp seed oil can be between 2.18 and 7.73 meqO_2/kg, and the addition of candelilla wax (9%) can decrease the peroxide value up to 1.25 meqO_2/kg [46]. β-sitosterol exhibits antioxidant properties and acts as a free radical scavenger [47]. β-sitosterol donates hydrogen atoms faster to free radicals and reduces the oxidation of polyunsaturated fatty acids. β-sitosterol in higher concentrations could not only reduce lipid oxidation but also form a stronger gel [48]. Panagiotopoulou et al. reported similar results for lipid oxidation in frankfurters prepared with phytosterol-based organogels [49].

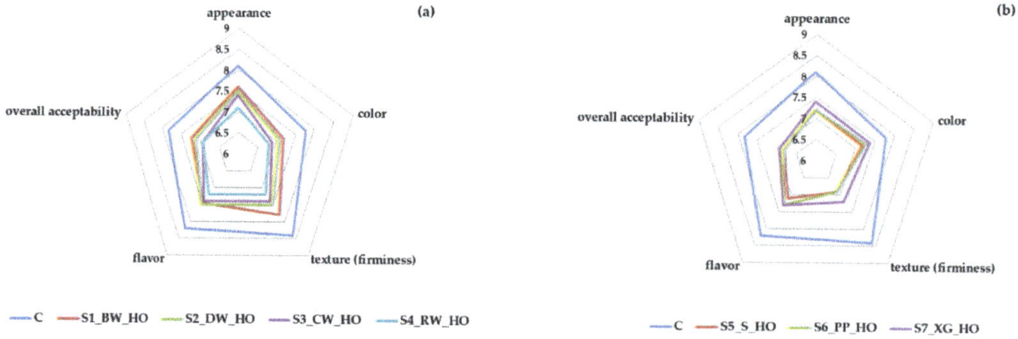

Figure 3. Sensory evaluation of cookies with hemp seed oil oleogels: (**a**) natural waxes and (**b**) other oleogelation agents.

2.5. Principal Component Analysis

The relationship between textural and color parameters of cookie samples with oleogels is shown in Figure 4. Principal component 1 (PC1) accounted for 58.9% of the total variation, while PC2 accounted for 20.2% of the total variation. All investigated parameters had positive loadings on PC1, except for hardness (−0.224) and strength (−0.194), which had negative loadings on PC1. Also, all investigated parameters had positive loadings on PC2, except for color parameters a* (−0.211) and b* (−0.122). Small angles between flavor, texture, appearance, overall acceptability, L*, and color reflected a great degree of correlation between these variables. These attributes were dominant in the control cookie sample, while strength and hardness predominated in cookies with oleogels from hemp seeds oil with xanthan gum and sitosterol. Additionally, color parameters, such as a* and b*, were key parameters in cookie samples with oleogels from hemp seed oil and oleogelling agents, like beeswax and pea protein. The control sample (C) had positive values for both PC1 and PC2, while samples S1_BW_HO and S6_PP_HO (situated to the right in the score biplot according to Figure 4) had positive values for PC1 and negative values for PC2. The other cookie samples are situated to the left in the score biplot. Three samples (S2_DW_HO, S3_CW_HO, and S4_RW_HO) had negative values for PC1 and PC2, and the other two samples (S5_S_HO, S7_XG_HO) had negative values for PC1 and positive values for PC2.

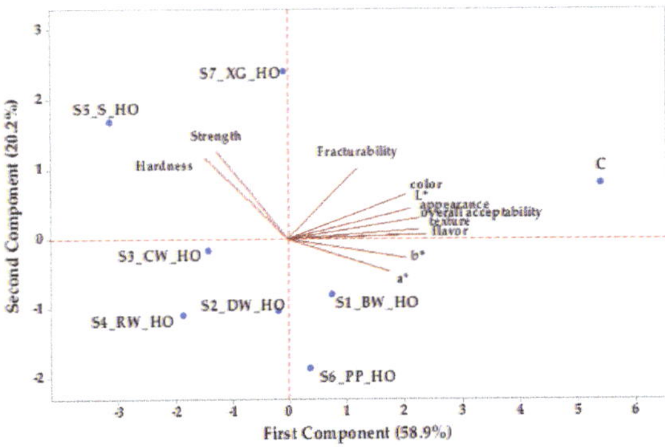

Figure 4. Principal component analysis biplot depicting the relationship between parameters of cookie samples (HO—hemp seed oil, oleogelling agents: BW—beeswax; DW—candelilla wax; CW—carnauba wax; RW—rice bran wax; S—sitosterol; PP—pea protein; XG—xanthan gum; L*, a*, and b*—color parameters).

3. Conclusions

The possibility of using oleogels, obtained from hemp seed oil and different oleogeling agents in cookie production, was investigated. This study showed that the type of oleogel used, as a margarine substitute, affects the textural parameters (adhesiveness, springiness, cohesiveness, and stickiness) of the obtained cookie dough. Replacing fat with oleogels led to an increase in the hardness of cookie doughs, and also, the addition of oleogel increased the viscoelastic moduli of the cookie doughs. Cookies with oleogels are darker in color, with a red and yellow hue. The sensory results indicate that consumers may choose oleogel cookies (made from hemp oil and wax (especially beeswax and candelilla), which are more popular than the others). Hemp seed oleogels with different gelators have practical solid fat replacement applications and present possible opportunities to formulate healthier cookies.

4. Materials and Methods

4.1. Materials

Hemp seed oil (HO) was purchased from a grocery store (Suceava, Romania), while candelilla wax (DW) and carnauba wax (CW) were supplied by Sigma-Aldrich (Hamburg, Germany). Beeswax (BW) and rice bran wax (RW) were obtained from Merck,

Bucharest, Romania, while sitosterol (S), pea protein (PP), and xanthan gum (XG) were purchased from a health food store (Suceava, Romania). Wheat flour (type 480), sugar, milk powder, eggs, salt, and sodium bicarbonate were purchased from a local supermarket (Suceava, Romania).

4.2. Oleogel Preparation

Oleogels were obtained, according to the method described by Ropciuc et al. [13], using hemp seed oil and oleogelation agents (beeswax, candelilla wax, carnauba wax, sitosterol, pea protein, xanthan gum) in the percentage of 9%. A hot plate with an orbital stirrer was used to heat the oil and wax to 80 °C until the wax was completely dissolved. A mixture was obtained and poured into tubes for solidification and was then stored under refrigerated conditions at 4 °C for five days until analysis [23]. In the case of pea protein and xanthan gum, stable foams were first prepared according to Mohanan et al. [50] and Mohanan et al. [51]. Afterwards, oleogels were prepared by adding a hot mixture (80 °C) of hemp seed oil to freeze-dried foams.

4.3. Sample Preparation

Different doughs were prepared starting from wheat flour type 480 to obtain a final quantity of about 200 g. The ingredients were cookie mixture (105 g wheat flour, 30 g sugar, 10.5 g milk powder, 33.6 g mixing eggs, 0.6 g salt, 0.3 g sodium bicarbonate), oleogel, and margarine (for the control sample), and they were added as indicated in the experimental design model (Table 4). The dough samples have been formulated so that the fat content is the same in all samples. After baking and cooling, the fat content is not changed; only the water evaporates from the product.

Table 4. Cookie dough recipe formulations.

Samples	Cookie Mixture [1]	Margarine, g	Oleogel, g
C	180	20	-
S1_BW_HO	180	-	20
S2_DW_HO	180	-	20
S3_CW_HO	180	-	20
S4_RW_HO	180	-	20
S5_S_HO	180	-	20
S6_PP_HO	180	-	20
S7_XG_HO	180	-	20

[1] Cookie mixture: 105 g wheat flour, 30 g sugar, 10.5 g milk powder, 33.6 g mixing eggs, 0.6 g salt, 0.3 g sodium bicarbonate. HO—hemp seed oil; oleogelling agents: BW—beeswax; DW—candelilla wax; CW—carnauba wax; RW—rice bran wax; S—sitosterol; PP—pea protein; XG—xanthan gum.

The used sample codes were as follows: the control sample (C) with vegetable shortening and cookie samples with 10% oleogel: S1_BW_HO (oleogel from hemp seed oil and 9% beeswax), S2_DW_HO (oleogel from hemp seed oil and 9% candelilla wax), S3_CW_HO (oleogel from hemp seed oil and 9% carnauba wax), S4_RW_HO (oleogel from hemp seed oil and 9% rice bran wax), S5_S_HO (oleogel from hemp seed oil and 9% sitosterol), S6_PP_HO (oleogel from hemp seed oil and 9% pea protein), and S7_XG_HO (oleogel from hemp seed oil and 9% xanthan gum). All ingredients were homogenized in a bowl with a kitchen mixer (KitchenAid Professional Mixer, Greenville, OH, USA) for 13 min at speed 4. The cookie dough was spread to a thickness of 8 mm, and 5 cm diameter discs were cut using a circular mold. The cookies were placed on a tray and baked in an electric oven at 200 °C for 25 min (Figure 5). After cooling to room temperature, the cookie samples were packed in polypropylene bags and stored at room temperature until analysis.

Figure 5. Visual appearance of dough samples.

4.4. Texture Profile Analysis of Dough and Cookies

After kneading, the dough was allowed to stand for 12 h under refrigerated conditions. The textural properties of the dough were determined at 20–22 °C. The textural properties of the dough and cookies were measured with the Perten TVT 6700 texturometer (Perten Instruments, Hägersten, Sweden). The textural properties were determined as described by Leahu et al. [52] with minor modifications. To determine the textural characteristics, the compression test was applied using 45 mm Diameter, Stainless Steel (P-CY45S), the compression percentage per sample was 30%, and the compression force was 10 g. The characteristics of dough hardness, stickiness, cohesiveness, elasticity, adhesiveness, and gumminess were analyzed in triplicate. Cookie samples were placed on the cutting table 6 h after baking and cooling. Cutting samples were performed on 1 cm high biscuits, previously measured with a cookie cutter. A Craft Blade Knife (P-CBK) with a cutting force of 25 g was used. The maximum point of force was noted, which expresses the breaking point of the biscuit. The test carried out for the cookies was one of fracturability (a measure of the tendency to fracture, in response to force). The sample was placed on a support with 2 parallel plates separated by 20 mm. The knife descended on the sample until it fractured. The initial descent speed of the cylindrical part was 1 mm/s, and the speed during the test was 3 mm/s. The force applied to the samples was 50 g. The blade cuts through the cookies, and the force required to do so is recorded as cookie strength [gmm]. Three repeated measurements were performed on each sample.

4.5. Rheological Analysis of Dough

The dynamic rheometer, model Haake Mars 40 (Thermo Fisher Scientific, Karlsruhe, Germany), was used to determine dough viscosity. Viscosity as a function of time (10 min) at a constant shear rate of 100 s^{-1} and a frequency of 1–100 Hz and graphical representations of viscosity curves were determined using Haake Rheo Win Data Manager 4 Software for Haake Mars 40 Rheometer (Thermo Fisher Scientific, Karlsruhe, Germany) according to the method previously described by Ropciuc et al. [23]. Dough samples were kept in plastic bags to prevent drying. The viscoelastic properties of the dough were determined at varying frequency and temperature. Plate/plate geometry (40/80 mm) with a 3 mm gap was used. The viscoelastic modulus was determined at a frequency of 1–20 Hz and a temperature of 20 °C. In order to know the baking behavior of dough with different oleogelating agents, the behavior of the viscoelastic modulus at variable temperatures was determined. The viscoelastic and elastic properties of biscuit dough at variable temperatures between 20 and 90 °C were determined.

4.6. Determination of Color Parameters

Color parameters were analyzed using a Minolta Chroma Meter (Model CR 310, Minolta Camera Co. Ltd., Tokyo, Japan). The results were expressed in terms of L* (lightness/darkness), a* (redness/greenness), and b* (yellowness/blueness) using the average of 3 readings per sample on different points of the cookie's surface [21]. Total color difference (ΔE) was calculated by the following equation [41]:

$$\Delta E = \sqrt{\left(L^* - L_0^*\right)^2 + \left(a^* - a_0^*\right)^2 + \left(b^* - b_0^*\right)^2} \quad (1)$$

where L^*, a^*, and b^* are color parameters of cookies with oleogels, while L_0^*, a_0^*, and b_0^* are color parameters of control cookies. According to Xiong et al. [41], if $\Delta E < 1$, there are no differences between the samples; $1 < \Delta E < 3$ indicates minor color differences, and $\Delta E > 3$ indicates major color differences.

4.7. Determination of the Cookies' Sensory Attributes

The cookie samples were proposed for sensory analysis to a group of 45 members selected from the students of the Faculty of Food Engineering using a defined sensory language (taste, flavor, texture, color, and overall acceptability) on a 9-point hedonic scale, where 9 was extremely like and 1 was extremely dislike, to understand consumer acceptance [53]. In hedonic tests, untrained consumers are used as testers. The coded samples were offered to the tasters, and the sensory evaluation was performed by using the preference test. The preference or ranking test was carried out according to the ISO 8587:2006 standard [54].

4.8. Statistical Analysis

Minitab 17 software (Minitab, Inc., State College, PA, USA) was used for statistical analysis. One-way analysis of variance (ANOVA) with a 95% confidence interval ($p < 0.05$) and Tukey's test were considered to compare the results obtained; also, principal component analysis and Pearson correlation were performed. ANOVA shows whether there is a statistically significant difference between the means of three or more groups of data (when the null hypothesis is rejected, $p \leq 0.05$), while Tukey's test (one of the post hoc tests) indicates which of these are different. Principal component analysis (PCA) was conducted to assess the relationship between textural, color, and sensory attributes and to evaluate similarities and differences between cookie samples.

Author Contributions: Conceptualization, S.R., A.L., C.G. and C.D.; methodology, S.R., A.L. and C.G.; software, S.R. and C.G.; validation, S.R., A.L. and C.D.; formal analysis, S.R., A.L., C.G. and C.D.; investigation, S.R., A.L. and C.G.; resources, A.L. and S.R.; data curation, S.R.; writing—original draft preparation, A.L.; writing—review and editing, C.G. and A.L.; visualization, A.L. and C.G.; supervision, S.R. and A.L.; project administration, S.R. and A.L.; funding acquisition, S.R., A.L. and C.G. All authors have read and agreed to the published version of the manuscript.

Funding: This research received no external funding.

Institutional Review Board Statement: This study involves only a one-time, short-term evaluation of the sensory attributes of yogurt and does not constitute a long-term functional product test. The study has not involved ethical reviews for human trials.

Informed Consent Statement: Informed consent was obtained from all subjects involved in the study.

Data Availability Statement: The original contributions presented in the study are included in the article; further inquiries can be directed to the corresponding author.

Conflicts of Interest: The authors declare no conflicts of interest.

References

1. Drewnowski, A.; Almiron-Roig, E. Human Perceptions and Preferences for Fat-Rich Foods. In *Fat Detection: Taste, Texture, and Post Ingestive Effects*; Montmayeur, J.P., Le Coutre, J., Eds.; CRC Press/Taylor & Francis: Boca Raton, FL, USA, 2010; Chapter 11, pp. 265–291.
2. Dias, F.D.S.L.; Passos, M.E.A.; do Carmo, M.D.G.T.; Lopes, M.L.M.; Mesquita, V.L.V. Fatty acid profile of biscuits and salty snacks consumed by Brazilian college students. *Food Chem.* **2015**, *171*, 351–355. [CrossRef] [PubMed]
3. Rosales Chavez, J.B.; Bruening, M.; Ohri-Vachaspati, P.; Lee, R.E.; Jehn, M. Street food stand availability, density, and distribution across income levels in Mexico City. *Int. J. Environ. Res. Public Health* **2021**, *18*, 3953. [CrossRef] [PubMed]
4. Pietrzyk, Ł. Food properties and dietary habits in colorectal cancer prevention and development. *Int. J. Food Prop.* **2017**, *20*, 2323–2343. [CrossRef]
5. Silva, T.J.; Barrera-Arellano, D.; Ribeiro, A.P.B. Oleogel-based emulsions: Concepts, structuring agents, and applications in food. *J. Food Sci.* **2021**, *86*, 2785–2801. [CrossRef]
6. Pantea Stoian, A.; Andronache, L.F.; Hainarosie, R.; Paduraru, D.N.; Badiu, C.D.; Arsene, A.; Mehedintu, C.; Ditu, G.; Pituru, S.; Orlov, C.; et al. Dietary habits and lifestyle in school-aged children from Bucharest. *J. Mind Med. Sci.* **2018**, *5*, 85–92. [CrossRef]
7. Chiş, M.S.; Pop, A.; Păucean, A.; Socaci, S.A.; Alexa, E.; Man, S.M.; Bota, M.; Muste, S. Fatty acids, volatile and sensory profile of multigrain biscuits enriched with spent malt rootles. *Molecules* **2020**, *25*, 442. [CrossRef]
8. Grasso, S.; Asioli, D. Consumer preferences for upcycled ingredients: A case study with biscuits. *Food Qual. Pref.* **2020**, *84*, 103951. [CrossRef]
9. Martínez, E.; García-Martínez, R.; Álvarez-Ortí, M.; Rabadán, A.; Pardo-Giménez, A.; Pardo, J.E. Elaboration of Gluten-Free Cookies with Defatted Seed Flours: Effects on Technological, Nutritional, and Consumer Aspects. *Foods* **2021**, *10*, 1213. [CrossRef]
10. Market Intelligence Report for Biscuits. Agricultural and Processed Food Products Export Development Authority, APEDA, New Delhi, India. 2024. Available online: https://agriexchange.apeda.gov.in/Weekly_eReport/Biscuit_%20Report.pdf (accessed on 22 November 2024).
11. Nourmohammadi, N.; Austin, L.; Chen, D. Protein-based fat replacers: A focus on fabrication methods and fat-mimic mechanisms. *Foods* **2023**, *12*, 957. [CrossRef] [PubMed]
12. Perța-Crișan, S.; Ursachi, C.Ș.; Chereji, B.D.; Tolan, I.; Munteanu, F.D. Food-grade oleogels: Trends in analysis, characterization, and applicability. *Gels* **2023**, *9*, 386. [CrossRef] [PubMed]
13. Ropciuc, S.; Dranca, F.; Oroian, M.A.; Leahu, A.; Codină, G.G.; Prisacaru, A.E. Structuring of cold pressed oils: Evaluation of the physicochemical characteristics and microstructure of white beeswax oleogels. *Gels* **2023**, *9*, 216. [CrossRef] [PubMed]
14. Izzo, L.; Pacifico, S.; Piccolella, S.; Castaldo, L.; Narváez, A.; Grosso, M.; Ritieni, A. Chemical analysis of minor bioactive components and cannabidiolic acid in commercial hemp seed oil. *Molecules* **2020**, *25*, 3710. [CrossRef] [PubMed]
15. Tura, M.; Ansorena, D.; Astiasarán, I.; Mandrioli, M.; Toschi, T.G. Evaluation of hemp seed oils stability under accelerated storage test. *Antioxidants* **2022**, *11*, 490. [CrossRef]
16. Devi, V.; Khanam, S. Comparative study of different extraction processes for hemp (*Cannabis sativa*) seed oil considering physical, chemical and industrial-scale economic aspects. *J. Cleaner Prod.* **2019**, *207*, 645–657. [CrossRef]
17. Apostol, L. Studies on using hemp seed as functional ingredient in the production of functional food products. *J. Ecoagritourism* **2017**, *13*, 12–17.
18. Blasi, F.; Tringaniello, C.; Verducci, G.; Cossignani, L. Bioactive minor components of Italian and Extra-European hemp seed oils. *LWT* **2022**, *158*, 113167. [CrossRef]
19. Teh, S.S.; Birch, J. Physicochemical and quality characteristics of cold-pressed hemp, flax and canola seed oils. *J. Food Compos. Anal.* **2013**, *30*, 26–31. [CrossRef]
20. Tănase Apetroaei, V.; Pricop, E.M.; Istrati, D.I.; Vizireanu, C. Hemp Seeds (*Cannabis sativa* L.) as a Valuable Source of Natural Ingredients for Functional Foods—A Review. *Molecules* **2024**, *29*, 2097. [CrossRef] [PubMed]
21. Crescente, G.; Piccolella, S.; Esposito, A.; Scognamiglio, M.; Fiorentino, A.; Pacifico, S. Chemical composition and nutraceutical properties of hempseed: An ancient food with actual functional value. *Phytochem. Rev.* **2018**, *17*, 733–749. [CrossRef]
22. Ropciuc, S.; Codina, G.G.; Oroian, M.A.; Dranca, F.; Leahu, A.; Prisacaru, A.E. Formulation of oleogels based on candelilla wax: Physicochemical and rheological characterization. *Int. Multidiscip. Sci. GeoConference SGEM* **2023**, *23*, 239–246.
23. Ropciuc, S.; Dranca, F.; Oroian, M.A.; Leahu, A.; Prisacaru, A.E.; Spinei, M.; Codină, G.G. Characterization of Beeswax and Rice Bran Wax Oleogels Based on Different Types of Vegetable Oils and Their Impact on Wheat Flour Dough Technological Behavior during Bun Making. *Gels* **2024**, *10*, 194. [CrossRef]
24. Tanislav, A.E.; Pușcaș, A.; Păucean, A.; Mureșan, A.E.; Semeniuc, C.A.; Mureșan, V.; Mudura, E. Evaluation of structural behavior in the process dynamics of oleogel-based tender dough products. *Gels* **2022**, *8*, 317. [CrossRef]

25. Zheng, Y.; Liu, Z.; Mo, B. Texture profile analysis of sliced cheese in relation to chemical composition and storage temperature. *J. Chem.* **2016**, *2016*, 8690380. [CrossRef]
26. Muskat, L.C.; Jiang, L.; Brikmann, J.; Rostás, M.; Patel, A.V. Development of a Self-Adhesive Oleogel Formulation Designed for the Slow Release of Semiochemicals. *Macromol. Mater. Eng.* **2022**, *307*, 2200276. [CrossRef]
27. Manohar, R.S.; Rao, P.H. Interrelationship between rheological characteristics of dough and quality of biscuits; use of elastic recovery of dough to predict biscuit quality. *Food Res Int.* **2002**, *35*, 807–813. [CrossRef]
28. Zulfiqar, A.; Shabbir, M.A.; Tahir, F.; Khan, M.R.; Ahmed, W.; Yıkmış, S.; Manzoor, M.F.; Abdi, G.; Aadil, R.M. Development of oleogel by structuring the blend of corn oil and sunflower oil with beeswax to replace margarine in cookies. *Food Chem. X* **2024**, *23*, 101676. [CrossRef]
29. Tanislav, A.E.; Șandru, B.; Man, S.M.; Pușcaș, A.; Mureșan, A.E.; Păucean, A.; Mureșan, V.; Mudura, E. Investigating the complete replacement of conventional fat with oleogel on the structural behavior of five different pastry products. *Eur. Food Res. Technol.* **2024**, *250*, 1933–1947. [CrossRef]
30. Mohamed, A.A.; Alamri, M.S.; Hussain, S.; Ibraheem, M.A.; Qasem, A.A.A.; Shamlan, G.; Ababtain, I.A. Effect of Ziziphus and Cordia Gums on Dough Properties and Baking Performance of Cookies. *Molecules* **2022**, *27*, 3066. [CrossRef] [PubMed]
31. Silva, R.C.d.; Ferdaus, M.J.; Foguel, A.; da Silva, T.L.T. Oleogels as a Fat Substitute in Food: A Current Review. *Gels* **2023**, *9*, 180. [CrossRef] [PubMed]
32. Shanthakumar, P.; Klepacka, J.; Bains, A.; Chawla, P.; Dhull, S.B.; Najda, A. The Current Situation of Pea Protein and Its Application in the Food Industry. *Molecules* **2022**, *27*, 5354. [CrossRef] [PubMed]
33. Pradhan, A.; Anis, A.; Alam, M.A.; Al-Zahrani, S.M.; Jarzebski, M.; Pal, K. Effect of Soy Wax/Rice Bran Oil Oleogel Replacement on the Properties of Whole Wheat Cookie Dough and Cookies. *Foods* **2023**, *12*, 3650. [CrossRef] [PubMed]
34. Li, S.; Zhu, L.; Wu, G.; Jin, Q.; Wang, X.; Zhang, H. Relationship between the microstructure and physical properties of emulsifier based oleogels and cookies quality. *Food Chem.* **2022**, *377*, 131966. [CrossRef] [PubMed]
35. Pang, M.; Kang, S.; Liu, L.; Ma, T.; Zheng, Z.; Cao, L. Physicochemical properties and cookie-making performance as fat replacer of wax-based rice bran oil oleogels. *Gels* **2022**, *9*, 13. [CrossRef] [PubMed]
36. Espert, M.; Wang, Q.; Sanz, T.; Salvador, A. Sunflower oil-based oleogel as fat replacer in croissants: Textural and sensory characterisation. *Food Bioprocess Technol.* **2023**, *16*, 1943–1952. [CrossRef]
37. Troncoso, R.; Torrado, A.; Pérez-Guerra, N.; Tovar, C.A. Viscoelastic Properties of Biscuit Doughs with Different Lipidic Profiles Fortified with a Casein Hydrolysate. *Macromol* **2024**, *4*, 437–447. [CrossRef]
38. Mert, B.; Demirkesen, I. Evaluation of highly unsaturated oleogels as shortening replacer in a short dough product. *LWT-Food Sci. Technol.* **2016**, *68*, 477–484. [CrossRef]
39. Tang, X.; Liu, J. A comparative study of partial replacement of wheat flour with whey and soy protein on rheological properties of dough and cookie quality. *J. Food Qual.* **2017**, *2017*, 2618020. [CrossRef]
40. Salvador, A.; Sanz, T.; Fiszman, S.M. Dynamic rheological characteristics of wheat flour–water doughs. Effect of adding NaCl, sucrose and yeast. *Food Hydrocoll.* **2006**, *20*, 780–786. [CrossRef]
41. Xiong, S.; Li, Y.; Chen, G.; Zou, H.; Yu, Q.; Si, J.; Yang, H. Incorporation of Huangjing flour into cookies improves the physicochemical properties and in vitro starch digestibility. *LWT* **2023**, *184*, 115009. [CrossRef]
42. Sobolev, R.; Frolova, Y.; Sarkisyan, V.; Kochetkova, A. Waxy Oleogels for Partial Substitution of Solid Fat in Margarines. *Gels* **2023**, *9*, 683. [CrossRef]
43. Yılmaz, E.; Öğütcü, M. The texture, sensory properties and stability of cookies prepared with wax oleogels. *Food Funct.* **2015**, *6*, 1194–1204. [CrossRef]
44. Flores-García, C.L.; Medina-Herrera, N.; Rodríguez-Romero, B.A.; Martínez-Ávila, G.C.G.; Rojas, R.; Meza-Carranco, Z. Impact of Fat Replacement by Using Organic-Candelilla-Wax-Based Oleogels on the Physicochemical and Sensorial Properties of a Model Cookie. *Gels* **2023**, *9*, 636. [CrossRef]
45. Cerino, P.; Buonerba, C.; Cannazza, G.; D'Auria, J.; Ottoni, E.; Fulgione, A.; Di Stasio, A.; Pierri, B.; Gallo, A. A review of hemp as food and nutritional supplement. *Cannabis Cannabinoid Res.* **2021**, *6*, 19–27. [CrossRef]
46. Ropciuc, S.; Ghinea, C.; Leahu, A.; Prisacaru, A.E.; Oroian, M.A.; Apostol, L.C.; Dranca, F. Development and Characterization of New Plant-Based Ice Cream Assortments Using Oleogels as Fat Source. *Gels* **2024**, *10*, 397. [CrossRef] [PubMed]
47. Ayaz, M.; Junaid, M.; Ullah, F.; Subhan, F.; Sadiq, A.; Ali, G.; Ovais, M.; Shahid, M.; Ahmad, A.; Wadood, A.; et al. Anti-Alzheimer's studies on β-sitosterol isolated from *Polygonum hydropiper* L. *Front. Pharmacol.* **2017**, *8*, 697. [CrossRef]
48. Islam, J.; Mahmud, N.; Oyom, W.; Aminzare, M.; Tahergorabi, R. Effects of B-Sitosterol Oleogel on Fat Uptake Reduction, Physicochemical Properties, Oxidative Stability, and Sensory Characteristics of Fried Surimi Fish Balls. *J. Agric. Food Res.* **2024**, *17*, 101240. [CrossRef]

49. Panagiotopoulou, E.; Moschakis, T.; Katsanidis, E. Sunflower oil organogels and organogel-in-water emulsions (part II): Implementation in frankfurter sausages. *LWT* **2016**, *73*, 351–356. [CrossRef]
50. Mohanan, A.; Tang, Y.R.; Nickerson, M.T.; Ghosh, S. Oleogelation using pulse protein-stabilized foams and their potential as a baking ingredient. *RSC Adv.* **2020**, *10*, 14892–14905. [CrossRef] [PubMed]
51. Mohanan, A.; Harrison, K.; Cooper, D.M.L.; Nickerson, M.T.; Ghosh, S. Conversion of Pulse Protein Foam-Templated Oleogels into Oleofoams for Improved Baking Application. *Foods* **2022**, *11*, 2887. [CrossRef] [PubMed]
52. Leahu, A.; Ghinea, C.; Ropciuc, S. Rheological, Textural, and Sensorial Characterization of Walnut Butter. *Appl. Sci.* **2022**, *12*, 10976. [CrossRef]
53. Leahu, A.; Ghinea, C.; Ropciuc, S. Apple Pomace: Influence on the Physico-Chemical and Textural Characteristics of Gluten-Free Biscuits. In *International Conference on e-Health and Bioengineering*; Springer Nature: Cham, Switzerland, 2023; pp. 542–550.
54. *ISO 8587:2006*; Sensory analysis–Methodology–Ranking. International Organization for Standardization: Geneva, Switzerland, 2006.

gels

MDPI

Article

Injectable and Conductive Polyurethane Gel with Load-Responsive Antibiosis for Sustained Root Canal Disinfection

Bo Mu [1], Xiaoyu Lei [1], Yinglong Zhang [1], Jingzheng Zhang [1], Qingda Du [1], Yuping Li [1], Dongyu Huang [1], Li Wang [2], Jidong Li [1], Yubao Li [1] and Yi Zuo [1,*]

[1] Research Center for Nano Biomaterials, Analytical & Testing Center, Sichuan University, Chengdu 610064, China; mubo1205@163.com (B.M.); zehn0906@163.com (X.L.); zhangyinglong1996@163.com (Y.Z.); scuzjz@163.com (J.Z.); duqingda0309@163.com (Q.D.); liyuping09115@163.com (Y.L.); h_dongy@163.com (D.H.); nic1979@scu.edu.cn (J.L.); nic7504@scu.edu.cn (Y.L.)

[2] School of Big Health and Intelligent Engineering, Chengdu Medical College, Chengdu 610500, China; liwang@cmc.edu.cn

* Correspondence: zoae@scu.edu.cn

Abstract: To address the limitations of conventional antibacterial therapies, we developed an injectable, conductive polyurethane-based composite gel system for sustained root canal disinfection. This gel incorporates piezoelectric nanoparticles (n-$BaTiO_3$) and conductive segments (aniline trimer, AT) within a polyurethane matrix, which synergistically interact with a static antimicrobial agent (n-ZnO) to achieve dynamic, mechano-responsive antibacterial activity. Under cyclic compression (simulating mastication), the piezoelectric gels exhibited enhanced electroactivity via the mechano-electric coupling effect, generating 2-fold higher voltage and a 1.8–1.9× increase in current compared to non-piezoelectric controls. The dynamic electroactivity of the gels enabled superior long-term performance, achieving 92–97% biofilm eradication, significantly higher than the static n-ZnO-only gel (88%). XPS and UV-vis spectroscopy analyses confirmed mechano-electrochemically amplified reactive oxygen species (ROS) generation, which contributed to improved biofilm disruption. The ISO-compliant gel provides durable, load-responsive disinfection while maintaining good biocompatibility, offering a promising solution to prevent post-treatment reinfection.

Keywords: root canal therapy; piezoelectric antibacterial gel; mechano-electric coupling; biofilm eradication; sustained antibacterial

1. Introduction

As the second largest microbial ecosystem in humans, the oral cavity harbors over 700 bacterial species and approximately 100 fungal species. Recent metagenomic studies have revealed that 296 species-level microbial populations typically coexist within this habitat [1,2]. Generally, the overgrowth of pathogenic bacteria contributes to various oral diseases, including dental caries, gingivitis, periodontitis, candidiasis, and diverse infections [3–5]. Disinfection is a clinical intervention aimed at controlling bacterial proliferation. However, excessive antibiotic use may disrupt the balance of the oral microbiota and foster antibiotic resistance, potentially contributing to severe systemic diseases [6–8]. Root canal therapy becomes essential when tooth integrity is compromised and pulp removal becomes necessary [9]. Root canal filling often represents the last attempt to salvage a tooth affected by irreversible inflammation or necrosis due to caries, structural defects (e.g., cracks or chips), or traumatic dental injuries [10]. The failure of root canal therapy is frequently attributed to bacterial reinfection following the decline in antibacterial activity over time, or more critically, inadequate obturation caused by the shrinkage of filling materials [9]. Furthermore, the oral environment is harsh and intricate, providing an ideal breeding ground for bacteria to form biofilms [11]. Within these dense biofilms, bacteria can evade the effects of antibiotics [12], posing a significant challenge to achieving sustained antibacterial efficacy in the treatment of infected root canals.

Naturally, the tooth serves as a direct tool for performing the chewing function. During chewing, the contraction of masticatory muscles induces occlusion between the upper and lower teeth [13,14]. According to reports, the theoretical maximum bite force between opposing teeth can reach 200–500 kg [15]. Chewing force enhances the efficiency of food digestion and facilitates the absorption and utilization of nutrients. During mastication, occlusal forces generate localized stress concentrations at specific anatomical sites, such as tooth interfaces and periodontal tissues [15]. This stress is transmitted to the jaw through the teeth, aiding in maintaining the alignment and position of the teeth and jawbone. Furthermore, the stress stimulates periodontal tissues, promoting the differentiation of periodontal cells and eliminating inflammatory factors induced by interleukin-1β, thus exerting an anti-inflammatory effect [16–18]. Stress stimulation supports the blood circulation and metabolism of periodontal tissues, thereby maintaining their health and ensuring dental stability [17]. Consequently, the development of a new type of filling cement has been proposed, leveraging the mechanical environment of the oral cavity to achieve sustained antibacterial properties, which could significantly reduce the proliferation of harmful bacteria in root canal therapy.

The efficacy of endodontic usage testing is required to last longer than 28 ± 3 days according to ISO 7405:2018 [19]. However, the antibacterial effects of some root canal filling materials cannot persist beyond one week due to the limited release of antibacterial substances or temporary alterations in pH. Examples include Epiphany SE root canal filling, Pulp Canal Sealer EWT, and EndoSequence BC root canal filling [10]. Incorporating inorganic antibacterial components into root canal filling materials has emerged as an effective strategy to enhance their antibacterial properties, such as silver and copper salts, as well as zinc and titanium oxides [20–23]. Nevertheless, the action mechanisms of these inorganic antibacterial agents—whether through contact inhibition or diffusional bacteriostasis—are highly dependent on environmental conditions. Consequently, their sustained release is often restricted in the dry and dark root canal environment, which is essential for hard curing [24].

Few root canal filling materials have been considered, despite the significant compressive stress generated daily in the oral environment. It has been reported that piezoelectric membranes can continuously generate electrical charges under oral pressure, contributing to bacterial elimination and wound healing [25]. This class of materials exhibits clinical utility across various biomedical domains, facilitating advancements in osseous regeneration, controlled therapeutic release, and engineered tissue reconstruction [26]. Dentistry could benefit from the piezoelectric effect, given an average of 500,000 chewing cycles per year [27]. Compared to lead zirconate titanate, barium titanate nanoparticles (n-$BaTiO_3$) are a simpler chemical compound with excellent piezoelectric properties and good biocompatibility, making them suitable for the complex oral environment [28]. Incorporating piezoelectric $BaTiO_3$ nanoparticles into a Bis-GMA/TEGDMA (1:1) dental composite matrix demonstrates piezoresponsive behavior under mechanical loading. This results in significantly reduced bacterial growth and pathogenic biofilm formation through the production of reactive oxygen species (ROS). The composite was specifically developed for restorative dental applications, such as tooth surface repairs, to prevent secondary caries at the bonded interface [29,30]. However, the dipole rearrangement-generated charges are confined to the surface of inorganic particles, which can be easily masked by insulating polymers during synthesis, thereby weakening the electric signals [31]. Consequently, the piezoelectric potential is limited in its efficacy for electronic transmission on the outer surface of filling materials.

In our prior research, we fabricated a series of conductive polyurethane (CPU) fibrous membranes incorporating aniline trimer (AT), inspired by the periosteum, to promote bone regeneration [32]. The fibrous membrane shows good biocompatibility after subcutaneous implantation. However, due to the formation of a crosslinked network within the matrix, the conductive polymer cannot serve as a self-curing root canal filling material because it results in an organogel with inadequate injectability [33]. In this study, a series of injectable conductive polyurethane gels incorporating piezoelectric n-$BaTiO_3$ and other functional components were developed as a sustained bacteriostasis strategy for root canal therapy (Figure 1). These injectable conductive gels exhibit varying electroactivity under cyclic compressive loading conditions as well as in the absence of mechanical stress. Furthermore, the antibacterial efficacy of these electrically activated gels was specifically evaluated against bacteria and biofilms under different cyclic loading forces. Inspired by their high piezoelectric activity under pressure, in vitro studies revealed that the novel self-powered gels based on conductive polyurethane significantly enhanced antibacterial performance compare to chemical bactericide. The dynamic strategy is expected to achieve a sustained antibacterial effect in the oral chewing environment through the mechano-electric integrated effect.

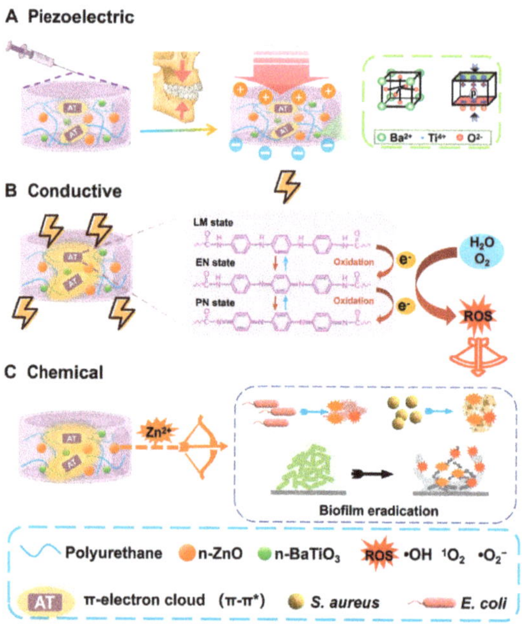

Figure 1. Schematic illustration of the injectable and conductive polyurethane gel for root canal therapy: (**A**) Piezoelectric nanoparticles (n-$BaTiO_3$) synergistically cooperated (**B**) conductive AT segments in polyurethane chains with (**C**) chemical antimicrobial agent (n-ZnO) to develop a series of mechano-powered antibacterial gels with dynamically sustained bacteriostasis as root canal filling.

2. Results and Discussion

2.1. Synthesis of AT and CPU-Based Prepolymers

The reaction diagram of the conductive polyurethane is shown in Figure 2A. As the basic segment of conductive polyurethane (CPU), AT was synthesized and the molecular structure was investigated by ^{1}H NMR and FT-IR spectra. As shown in Figure 2B, the absorption peaks are 6.90 ppm (s, 2H, Ar H), 6.81 ppm (s, 2H, Ar H), and 6.62 ppm (s, 2H, Ar H), which correspond to proton hydrogen on the benzene and quinone rings in the AT structure [34]. Moreover, the absorption peak of 5.43 ppm (s, 4H, NH_2) corresponded to the terminal amino proton peak in the AT structure [35]. As shown in the FT-IR spectra in Figure S3, three characteristic absorption peaks of 3427 cm^{-1}, 3334 cm^{-1}, and 3240 cm^{-1} in the range of 3427–3240 cm^{-1} were attributed to the terminal amino group (-NH_2) in the structure of AT. The characteristic peaks indicated that amino-terminated aniline trimers (AT) had been successfully synthesized and grafted in the CPU prepolymer.

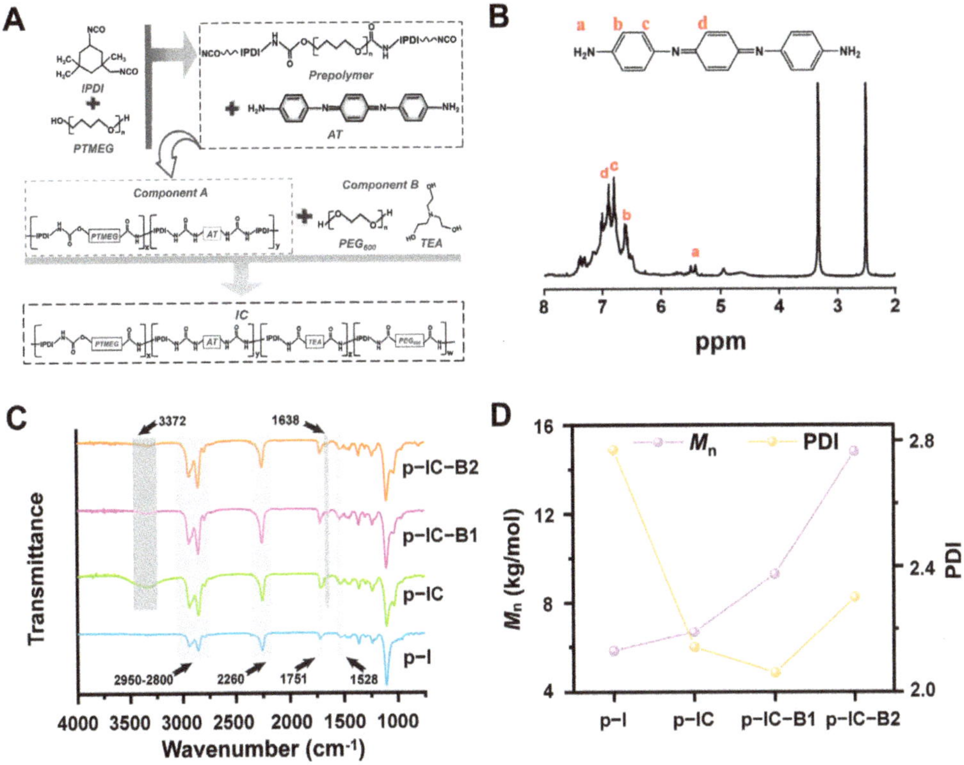

Figure 2. Synthesis and Characterization of CPU-Based Prepolymers: (**A**) Reaction scheme for synthesizing conductive polyurethane (CPU) prepolymers. (**B**) ^{1}H NMR spectrum of the synthesized aniline trimer (AT)—the peaks at 6.90 ppm (Ar-H), 6.81 ppm (Ar-H), and 5.43 ppm (-NH_2) confirm the successful synthesis of AT. (The letters respectively represent the type of H corresponding to the peak positions on the spectral graph) (**C**) FT-IR spectra of different CPU-based pre-polymers—key functional groups, including the peak at 3372 cm^{-1} (-NHCONH-) and 1638 cm^{-1} (C=O), confirm the successful grafting of AT onto the polymer backbone. (**D**) Average molecular weight and polydispersion index (PDI) of p-I and p-IC prepolymers—the p-IC-B2 group exhibits the highest M_n (2.53 times that of p-I) and the lowest PDI (2.06), indicating enhanced homogeneity due to the incorporation of AT and n-$BaTiO_3$.

Prepolymers as component A were copolymerized by PTMEG, IPDI, and/or amine-capped AT, added with functional inorganic particles including n-HA, Bi_2O_3, n-ZnO, and piezoelectric n-$BaTiO_3$. The FT-IR spectra of the CPU-based prepolymers in Figure 2C showed that the characteristic peaks attributed to carbamate (-NHC=O), carbonyl (C=O) on the ester group, and methylene (-CH-) appeared at 1528 cm^{-1}, 1751 cm^{-1}, and 2800–2950 cm^{-1}, respectively. For all the CPU-based prepolymers, the characteristic peaks attributed to the benzene rings of AT were observed at 1502 and 1164 cm^{-1} in Figure S3, indicating the successful linkage of AT to the backbone of conductive prepolymers (IC and IC-B). After copolymerization, the AT fragment was successfully embedded into the conductive polyurethane gels (IC, IC-B1, and IC-B2). However, the characteristic peak of the isocyanate group (-NCO) appeared at 2260 cm^{-1} in the spectra of all the prepolymers, indicating that these prepolymers contain a significant amount of active isocyanate groups. In Figure 2D, the number–average molecular weight (M_n) of the p-IC group was affected by grafted AT higher than that of the p-I group, while the M_n of the p-IC-B1 and p-IC-B2 groups increased largely with the addition of n-$BaTiO_3$ content. Accordingly, the M_n of the p-IC-B2 group was about 2.53 times over that of the p-I group. However, the polydispersion index (PDI) was decreased with AT and n-$BaTiO_3$ added from 2.77 (p-I) down to 2.30 in the p-IC-B2 group, while the lowest value was 2.06 for the p-IC-B1 group at a narrow molecular distribution (Figure 2D). The high molecular weight and low polydispersity mean the addition of AT segment and n-$BaTiO_3$ particles could help a homogenous polymeric structure forming in the CPU composites, but more nanoparticles in the p-IC-B2 group have inhibited the polymerization.

2.2. *Physicochemical Properties of Different Composite Gels for Root Canal Therapy*

After being mixed, component A (prepolymers) and component B underwent a chain extension reaction to prepare injectable CPU gels. In Figure 3A, characteristic peaks corresponding to n-HA, Bi_2O_3, and n-ZnO appeared in the XRD patterns for all the groups of samples, which proved that the three functional nanoparticles were successfully loaded in the injectable conductive polyurethane matrix ((I, IC, IC-B1, and IC-B2)). Further, the IC-B1 and IC-B2 groups showed characteristic peaks corresponding to n-$BaTiO_3$, indicating that piezoelectrical n-$BaTiO_3$ was successfully loaded in the IC-B1 and IC-B2 groups.

Evidently from IR spectra in Figure 3B, the characteristic peak of isocyanate was not observed at 2260 cm^{-1} for all groups after mixing, indicating that the isocyanate groups had been completely consumed after gels curing. The NCO conversion degree has been calculated by the characteristic peak areas of FTIR spectra (at 2200–2300 cm^{-1}) of gels (Figure S4A), wherein the NCO group of the IC-B2 gel decayed fast at the beginning of curing and then stabled slowly after 2 h at the conversion rate of 92.50% until 12 h at the conversion rate of 99.99% (Figures 3C and S4B). Other groups presented a similar trend except the conversion rate of the I gel with only 87.69% at 12 h. Isocyanate monomers of the CPU gels (IC, IC-B1, and IC-B2 groups) were almost consumed after 12 h which is good for root canal therapy. In particular, there was a characteristic stretching vibration peak attributed to ureyl (-NHCONH-) at 3372 cm^{-1}, and an asymmetric stretching vibration peak attributed to carbonyl (C=O) at 1638 cm^{-1} in the FT-IR spectra of the CPU composite gels (IC, IC-B1, and IC-B2, Figure 3B). The peaks indicated that the AT segment was successfully grafted into polymeric chains as the terminal amino group of AT was transformed into a urea group [36]. However, the three characteristic peaks of AT did not appear in the spectra of cured samples (Figure 3B), which indicates that all the terminal amino groups of aniline trimers had been transformed into the urea group. As shown in Figures 3D and S5 (Supplementary Materials), the nanoparticles were uniformly dispersed in the polymeric matrix in SEM images. At the same time, the Ca, Bi, and Zn elements were all uniformly dispersed in the four groups of cured sample matrix, and the Ba element was evenly dispersed in both the IC-B1 and IC-B2 groups.

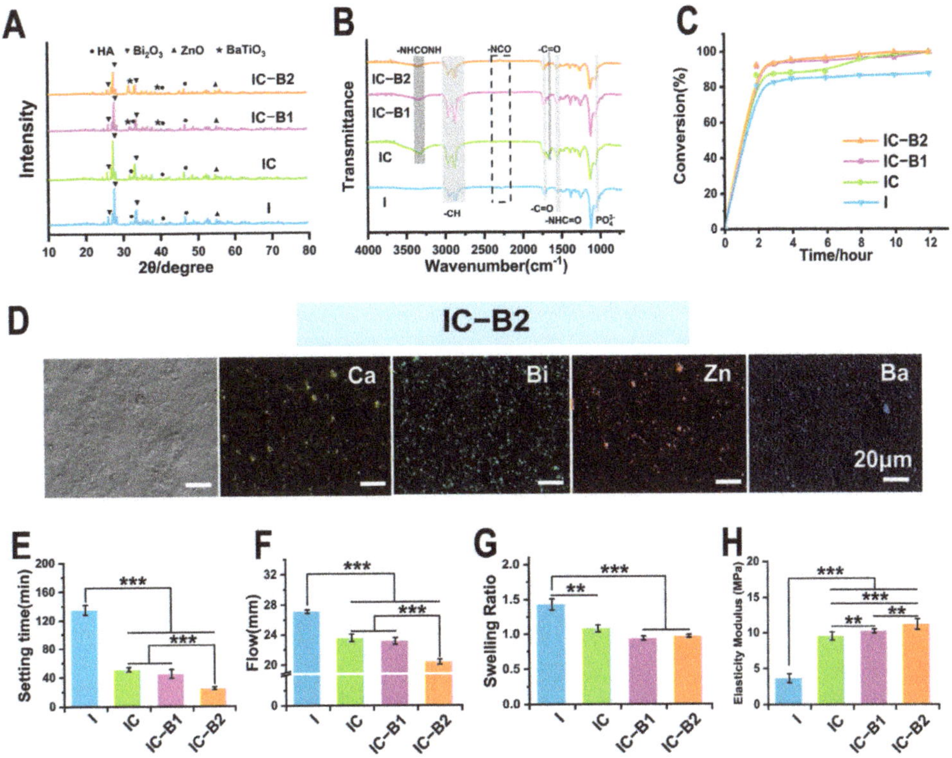

Figure 3. Physicochemical properties of injectable CPU gels for root canal therapy: (**A**) XRD and (**B**) FT-IR spectra of the cured gels. (**C**) NCO conversion degree of the different gels in 12 h. (**D**) SEM image and element distribution of Ca, Bi, Zn, and Ba on the surface of the cured IC-B2. (**E**) Setting time ($n = 3$) and (**F**) flowability ($n = 3$) of the different gels, (**G**) Welling rate ($n = 3$), and (**H**) Elastic modulus ($n = 5$) of cured gels. ** $p < 0.01$ and *** $p < 0.001$.

The anti-washout property of gels is an important factor in successful root canal therapy [37]. As shown in Figure S6, the pristine PU gel of the I group began to disintegrate with some dropped particles appearing after soaking for 30 min. With the time increasing, the disintegration phenomenon of the I group was more serious. In contrast, the samples of the CPU gels (IC, IC-B1, and IC-B2 groups) maintained their original shape until 4 h without any disintegration and particle shedding. The anti-washout results showed that the introduction of AT and n-$BaTiO_3$ improved the anti-elution performance of the CPU gels (IC, IC-B1, and IC-B2 groups).

Setting time and fluidity are two key characteristics of root canal fillings. ISO 6876: 2012 requires the setting time of the filling to be less than 4320 min and the fluidity to be greater than 17 mm [38]. The results in Figure 3E,F showed that the setting time and fluidity of the four groups met the ISO requirements. With the introduction of AT as a chain extender, the setting time of the IC group was significantly reduced to 51 ± 3 min compared with the I group ($p < 0.001$, Figure 3E); with the increasing addition of n-$BaTiO_3$, the setting time of the IC-B1 and IC-B2 groups was further reduced ($p < 0.001$), and the curing time of the IC-B2 group reached 21 ± 2 min. The fluidity of the material also showed a similar trend with the setting time in Figure 3F.

A lot of injectable biomaterials are cured by accelerating the exothermic polymerization reaction through exothermic reactions, wherein a high temperature over 47 °C for 300 s will cause severe thermal damage to tissues [39]. Here, the temperature fluctuation during the curing process was tested and recorded in a range between 34.9 °C and 36.5 °C for all the gel groups (Figure S4C, Supplementary Materials). The narrow temperature range of four PU gels testified that the series of cement could be mildly set in a biological environment without high energy release. After curing for 2 h, the isocyanate groups were diminished (Figure 3C), and the exothermic reaction stabilized (Figure S4B). Among these gels, the isocyanate conversion rate of the IC-B2 group was the highest, reaching 99.99% at 12 h.

Unlike shrinkable root canal filling, polyurethane-based materials can expand in volume after curing, providing effective obstruction within the root canal. However, inorganic particles may influence the extent of this volume expansion [40]. In Figure 3G, the swelling rates of the CPU gels (IC, IC-B1, and IC-B2 groups) were significantly reduced compared with I ($p < 0.01$ and $p < 0.001$). Meanwhile, the gel fractions of three CPU gels (IC, IC-B1, and IC-B2 groups) were higher than that of the I group in Table S1 ($p < 0.05$, Supplementary Materials). The introduction of a chain extender and inorganic particles significantly increased the crosslinking density of IC, IC-B1, and IC-B2 groups after curing. Accordingly, the elastic modulus of samples for IC, IC-B1, and IC-B2 groups have also increased by 2.6 ~ 3 times compared with that of the I group (Figure 3H, $p < 0.01$ and $p < 0.001$). For the IC-B2 group, the maximum modulus was 11.17 ± 0.75 MPa among all the gels, which is far less than the elastic modulus of dentin (16 GPa) [41]. Thus, the stress generated by volumetric expansion would not cause additional damage to the dentin.

2.3. *Piezoelectric Response of Gels Under Cyclic Compressive Load*

Figure 4A illustrates the schematic diagram of the pressurizing device and the principle of the piezoelectric effect under compressive loading. When the sample is subjected to cyclic compressive loading, the positive and negative charge centers of the n-$BaTiO_3$ filler separate, resulting in dipoles that are orderly arranged within the matrix. The action ultimately generates an electric field inside the gel. Based on the selection of cyclic compressive stress parameters, a cyclic compressive stress of 10 N was applied to the cured gels for 30 cycles (half a minute per cycle) optimized in the subsequent experiment (Figure S1).

The UV absorption spectra of the IC, IC-B1, and IC-B2 gels are shown in Figure 4B. The spectral profiles exhibited two well-resolved absorption bands at 311 nm and 600 nm, which were, respectively, attributed to the characteristic π-π^* electronic transition within the aromatic system and the intermolecular charge-transfer exciton transition occurring between the benzenoid and quinonoid moieties [42]. The two peaks did not appear in the spectrum of the I gel. The exciton transition from the benzene ring to the quinone ring, corresponding to the absorption peak at 600 nm, can be attributed to the successful incorporation of the AT fragments into the chain segments of the IC, IC-B1, and IC-B2 gels. This incorporation ameliorated the polymeric structure, thereby enhancing electrical activity [43]. Accordingly, the conductivity of the I gel could not be measured within the detection range of the instrument in Figure 4C. The conductivity of the IC gel reached $3.92 \pm 0.38 \times 10^{-6}$ S/cm. In particular, the conductivity values of the IC-B1 and IC-B2 gels were in the same order of magnitude of $5.33 \pm 0.44 \times 10^{-5}$ S/cm and $8.15 \pm 0.75 \times 10^{-5}$ S/cm, respectively, higher than that of the IC gel ($p < 0.001$). With the increase in n-$BaTiO_3$ content, the conductivity of the IC-B2 gel increased to 1.5 times over than that of IC-B1 ($p < 0.001$).

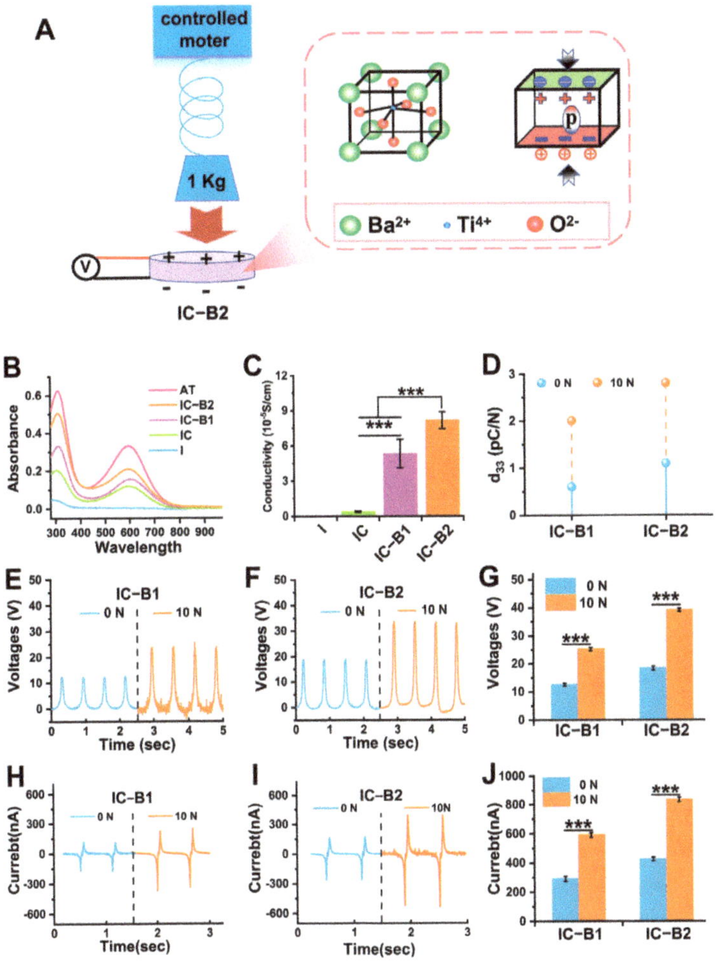

Figure 4. Characterization of electrochemical properties of cured gels under 10 N compression loading for 30 cycles or unloaded condition: (**A**) Schematic diagram of the device for compressive loading: Mechanical stress (10 N, 30 cycles) induces lattice deformation in n-$BaTiO_3$, generating a piezoelectric potential. (**B**) UV-vis spectra of cured gels: The absorption peaks at 311 nm (π-π^* transition) and 600 nm (quinoid exciton transition) confirm the successful integration of AT. (**C**) Conductivity of cured gels: IC-B2 achieved a conductivity of 8.15×10^{-5} S/cm, which is 33.8 times higher than that of non-piezoelectric control (IC group), due to AT/n-$BaTiO_3$ synergy. (**D**) Piezoelectric constant of cured gels (under cyclic loading or no loading): Cyclic loading enhances the piezoelectric response by 3.3× (IC-B1) and 2.5× (IC-B2). (**E–G**) Output voltage spectra of IC-B1 and IC-B2 cured gels and their voltage intensities (under cyclic loading or no loading): Cyclic loading enhances the output voltage by 2.01× (IC-B1) and 2.13× (IC-B2). (**H–J**) Output current spectra of IC-B1 and IC-B2 cured gels and their electrical current intensities (under cyclic loading or no loading): Cyclic loading enhances the output current by 1.9× (IC-B2). ($n = 3$) *** $p < 0.001$.

Under different compressive loading conditions (0 N and 10 N), Figure 4E,F show the representative voltage generated by the cured gels IC-B1 and IC-B2, while Figure 4H,I show the representative current generated by the cured gels IC-B1 and IC-B2. From the quantified results in Figure 4G, without external force loading, the output voltage intensity of the IC-B2 gel was 1.46 times higher than that of IC-B1 (12.56 $\pm$ 0.19 V, $p < 0.001$). After applying a compressive load of 10 N for 30 cycles (half a minute per cycle), the output voltage of IC-B1 and IC-B2 had been significantly improved. Among them, IC-B1 under cyclic loading increased to 2.01 times the output voltage higher than the original without loading, and IC-B2 increased to 2.13 times over the original with the output voltage reaching 39.14 $\pm$ 0.53 V ($p < 0.001$, Figure 4G). The output current intensity of IC-B1 and IC-B2 also showed a similar enhancement trend as output voltage intensity, with IC-B2 exhibiting a 1.9-fold increase under cyclic loading compared to the original under 0 N, reaching 834.34 $\pm$ 16.05 nA ($p < 0.001$, Figure 4J). The piezoelectric constant of IC-B2 was 1.83 times over that of IC-B1 without cyclic loading (Figure 4D). Under the cyclic loading, the piezoelectric constant of IC-B1 increased by 3.33 times than without loading, and the piezoelectric constant of IC-B2 increased by 2.54 times than without loading, reaching 2.8 pC/N (Figure 4D). These electrical characteristics showed that the cyclic loading could significantly improve the electroactivity of the $BaTiO_3$-added composite gels (IC-B1 and IC-B2 groups).

2.4. Antibacterial Properties of Gels Under Cyclic Loading

Both Gram-positive *S. aureus* and Gram-negative *E. coli* were selected as bacterial models to test the broad-spectrum antibacterial effect of the injectable CPU gels (Figure 5A). Figures 5B,C and S7A,B show the inhibition rate of the tested sample against *S. aureus* and *E. coli* directly in a static contact test (SCT). In the static contact experiment with bacteria, the antibacterial ability of the four groups was impaired as the soaking time increased (Figure S7A,B). After 15 days of soaking, the antibacterial rate of all the gels decreased to approximately 90% under unloaded conditions. However, when the axial compressive load of 10 N was applied for 30 cycles, the antibacterial rate of IC-B1 and IC-B2 significantly improved compared to the original unloaded groups from the first day. In contrast, no change was observed in the I and IC groups (Figure 5B,C). Especially, the antibacterial rate of IC-B2 against *S. aureus* and *E. coli* increased to approximately 95% after 15 days of soaking. After 15 days of soaking, the antibacterial rate of the IC-B1 group increased significantly by 2.86%, while that of the IC-B2 group increased significantly by 4.76%, which was 1.66 times higher than that of the IC-B1 group ($p < 0.01$ and $p < 0.001$, Figure 5F). The antibacterial efficacy of the four groups of materials against *E. coli* was similar to that against *S. aureus*. After soaking for 15 days, the antibacterial rate of IC-B2 increased by 5.25% under cyclic loading ($p < 0.001$, Figure 5F).

According to the data obtained from the Dynamic Contact Test (DCT) for *S. aureus* and *E. coli* in Figures 5D,E and S7C,D, the four groups of gels significantly inhibited the proliferation of *S. aureus* and *E. coli* in the first 6 h after soaking for 15 days. The inhibitory effects of the I and IC groups did not change after the application of pressure. The IC-B1 group had a significant inhibitory effect in the first 8 h, while the IC-B2 group effectively inhibited *S. aureus* and *E. coli* for up to 24 h. The DCT curves showed that the mechano-powered CPU gels exhibited a good antibacterial effect. Under the cyclic compressive loading, the antibacterial effect of the IC-B1 and IC-B2 groups after 15 days soaked in PBS was significantly increased to 24 h (from $p < 0.01$ to $p < 0.001$, Figure 5G). It can be postulated that mechano-powered gels could exhibit an excellent and sustained antibacterial effect under the frequent chewing action in the oral cavity.

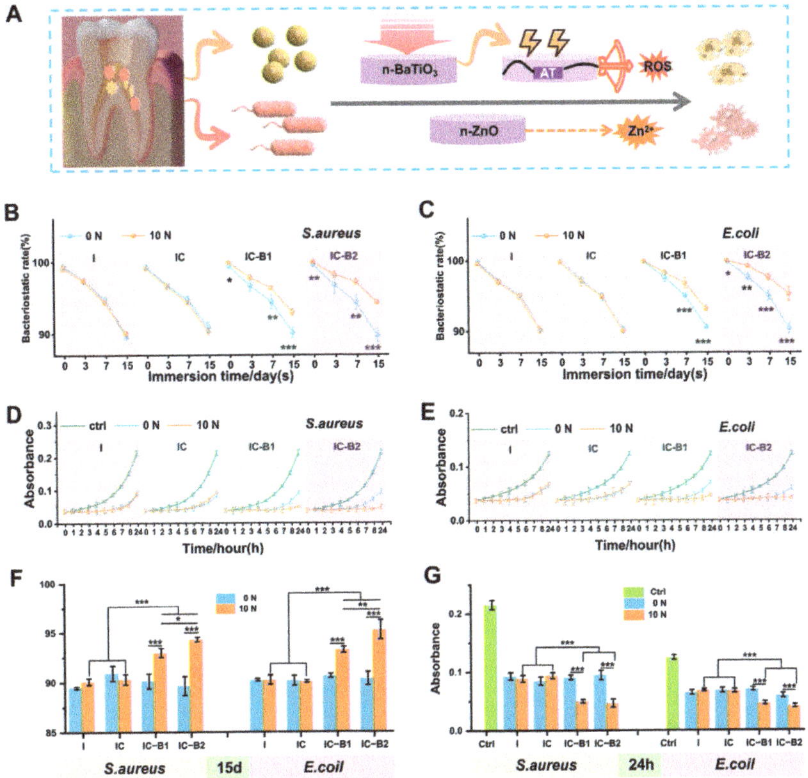

Figure 5. Bacteriostasis properties of cured gels under 10 N compression loading for 30 cycles or unloaded condition: (**A**) Antibacterial diagram of different bacteria. (**B**) Antibacterial rates from the static contact test (SCT) over 15 days against *S. aureus* and (**C**) *E. coli*. (**D**) Dynamic contact test curves (DCT) of the various groups after 15 days of soaking in PBS solution against *S. aureus* and (**E**) *E. coli*. (**F**) Antibacterial rates from the SCT over 15 days against *S. aureus* and *E. coli*. (**G**) After 15 days of soaking in PBS solution, the absorbance of DCT of each group against *S. aureus* and *E. coli* at 24 h. The uncoated wells were used as the blank control group. * $p < 0.05$, ** $p < 0.01$, and *** $p < 0.001$ ($n = 3$).

2.5. The Ability of Gels to Eradicate Bacterial Biofilms Under Cyclic Loading

Bacterial biofilm is a self-protection mechanism that bacteria develop during their process. The formation of biofilm hinders the interaction between bactericidal substances and bacteria, thereby reducing the efficacy of these bactericidal agents [44]. Figure 6A shows a crystal violet staining image of a *S. aureus* biofilm, where the stained crystal violet was dissolved in ethanol and the absorbance measured reflects the level of residual biofilm. Figure 6B presents a semi-quantitative analysis of the residual biofilm level. The efficiency of biofilm eradication for the four gel groups was maintained at 88% under pressure-free conditions. After cyclic compressive loading, the biofilm eradication rate for the IC-B1 group was significantly improved to 92% compared to the original groups without loading ($p < 0.001$), and the biofilm eradication rate for IC-B2 was also significantly improved to 96% ($p < 0.001$). Figure 6C,D show agar plate images of bacterial colony counts in residual biofilms, as well as a quantitative analysis of bacterial viability. In the absence of loading pressure, 15% of the bacteria survived; after cyclic loading, the survival rate decreased to 10% for IC-B1 and 7% for IC-B2 ($p < 0.001$). Figure 6F–I show that the eradication ability of the material against the *E. coli* biofilm was similar to that of the *S. aureus* biofilm. After applying cyclic loading, the eradication rate of the *E. coli* biofilm for IC-B2 increased to 97%, and the survival rate of bacteria within the membrane decreased to 7% ($p < 0.001$).

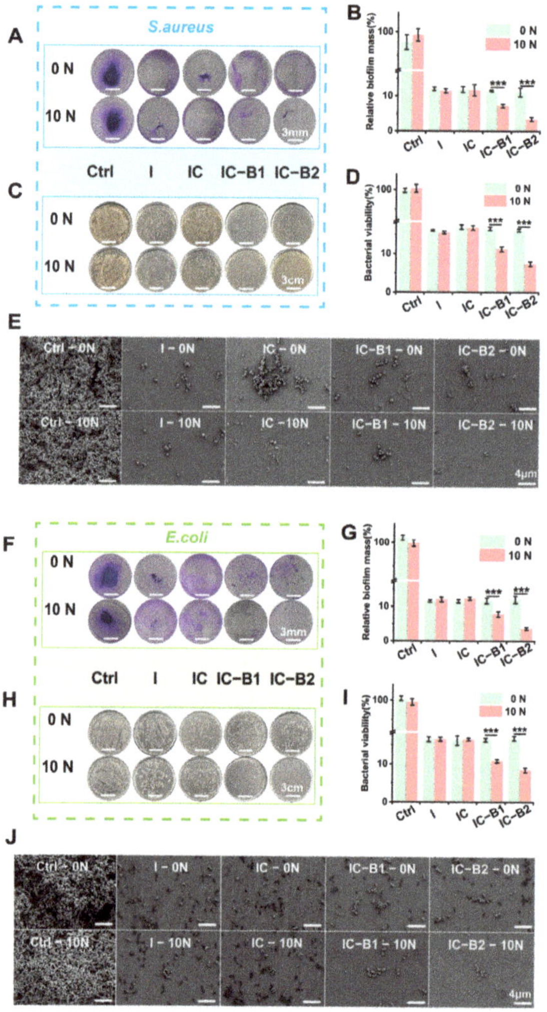

Figure 6. In vitro biofilm eradication and bacterial killing efficiency of cured gels under 10 N compression loading for 30 cycles or unloaded condition: (**A**) Crystal violet staining image of *S. aureus* biofilm (the chromatic intensity of the purple-stained solutions provides a quantitative chromogenic measure of retained biofilm biomass). (**B**) Semi-quantitative analysis of crystal violet staining intensity for residual *S. aureus* biofilm. (**C**) Agar plate count of bacterial viability in *S. aureus* biofilm. (**D**) Quantified results of bacterial viability within the *S. aureus* biofilm. (**E**) SEM images of *S. aureus* residue biofilm (×10,000). (**F**) Crystal violet staining image of *E. coli* biofilm (The chromatic intensity of the purple-stained solutions provides a quantitative chromogenic measure of retained biofilm biomass). (**G**) Semi-quantitative analysis of crystal violet staining intensity for residual *E. coli* biofilm. (**H**) Agar plate count of bacterial viability in *E. coli* biofilm. (**I**) Quantified results of bacterial viability within *E. coli* biofilm (**J**) SEM images of *E. coli* residue biofilm (×10,000). *** $p < 0.001$. ($n = 3$).

Figure 6E,J show the SEM morphology of the residual *S. aureus* and *E. coli* biofilms. The original biofilm was dense and thick; the bacteria in the *S. aureus* biofilm remained typically spherical, while those in the *E. coli* biofilm remained typically rod-shaped. Under cyclic loading, only sparse bacterial biofilms were observed in both the IC-B1 and IC-B2 gels. In Figure 6E,J, the bacterial surface appeared wrinkled and contracted, with some cells losing membrane integrity and leaking cellular contents. In contrast to the two piezoelectric-enhanced gels (IC-B1 and IC-B2), both the I and IC groups still retained a small amount of biofilm regardless of loading or not, even if the chemical fungicide (n-ZnO) and electroactive component (AT) were incorporated in the composite gels (Figure 6E,J).

2.6. Chemical Structure Changes and ROS Generation of Gels Under Cyclic Loading

To investigate the effect of cyclic loading on the composite gels, high-resolution XPS spectra of O 1s were deconvoluted to analyze the oxygen-related bonding states (Figure 7A–D). In the fitted graph, the peak of n-$BaTiO_3$ was observed in both the IC-B1 and IC-B2 gels, confirming the successful incorporation of n-$BaTiO_3$ into the polymeric matrix. High-resolution XPS spectra of gels C-1s were deconvolved to analyze key carbon states. For the C 1s spectra (Figure 7E–J), the peak at 292 eV in the IC gel (with AT as a chain extender) corresponds to the π-π^* transition within the conjugated aromatic system of the benzene ring. Additionally, the increased intensity of the C-N bond peak at 286 eV in the IC gel indicates the successful integration of AT into the polyurethane backbone (red star, Figure 7F), as compared to the control (I gel, Figure 7E). After cyclic loading, the peak intensity of the C-N bond (286 eV, red star) in the C 1s spectra of the IC-B1 gel (Figure 7H) was observed to decrease by 21.78% compared to that in the IC-B1 gel without loading (Figure 7G). Under the same mechanical conditions, the π-π^* peak (292 eV) of IC-B1 remained unchanged regardless of whether loading was applied (black arrow). In particular, the peak intensity corresponding to the C-N bond (286 eV) in IC-B2 decreased by 48.34% after applying cyclic pressures. The π-π^* (292 eV) peak remained stable, confirming the structural integrity of the aromatic system (Figure 7I,J). The changes in the C-N bond related to grafted-AT in PU suggest that IC-B2 is more sensitive to mechanical stress than IC-B1, leading to greater structural change.

During the oxidation process of AT fragments in polyurethane, electrons were produced and transferred to surrounding H_2O and O_2, leading to the formation of ROS such as hydroxyl radicals (•OH) and superoxide anions (•O_2^-). The level of generated ROS is crucial in disrupting bacteria metabolism and biofilm formation [45]. So, the ability and types of ROS generation of the injectable and conductive polyurethane gels were assessed according to a previous study [46].

The colorless TMB could react with •OH to form blue oxidized TMB (oxTMB) with a typical absorption peak at 652 nm. In Figure 8A, no peak has been observed in all the groups without loading. Under cyclic loading, a characteristic peak at 652 nm was observed only in the IC-B1 and IC-B2 groups, whereas other groups showed no detectable peak at this wavelength (Figure 8B,C). The peak observed at 652 nm in both the IC-B1 and IC-B2 groups confirmed the generation of hydroxyl radicals (•OH) under loading conditions. As the number of loading cycles increased, the intensity of peak at 652 nm gradually increased (Figure S1A,B), demonstrating the time-dependent production of •OH for both the IC-B1 and IC-B2 gels. In addition, during the same cycle period, the intensity of the peak at 652 nm of IC-B2 was 129.59% higher than that of IC-B1 under 30 cycles loading, which indicates that IC-B2 gel generated more hydroxyl radicals (•OH) under cyclic loading than that of IC-B1 (Figure S1C). For •O_2^- detection, NBT was reduced to blue formazan, monitoring by its absorption peak at 560 nm [46]. Similarly to the formation of •OH, only IC-B1 and IC-B2 exhibited obvious absorption peaks at 560 nm under cyclic pressure and the peak intensity at 560 nm increased with increasing the loading cycles, confirming the generation of $\bullet O_2^-$ as a time-dependence mode (Figure 8E,F and Figure S1D,E). Under the same cycle of loading conditions, the characteristic peak intensity at 560 nm for IC-B2 was 150.72% higher than that of IC-B1, indicating that IC-B2 generated more superoxide anions than that of IC-B1 after 30 loading cycles (Figure S1F). DPBF is an indicator with a characteristic peak at 420 nm. Upon reacting with $^1O_2$ and $\bullet O_2^-$, yellow DPBF forms endoperoxides and transforms into colorless 1, 2-dibenzoylbenzene, causing the characteristic absorption peak at 420 nm to weaken or disappear [47]. Under cyclic loading, the characteristic peaks of IC-B1 and IC-B2 at 420 nm weaken and disappear, confirming the generation of $^1O_2$ and $\bullet O_2^-$ (Figure 8H,I). As the cycle period increased, the absorption intensity at 420 nm gradually decreased, indicating that the production of $^1O_2$ and $\bullet O_2^-$ is time-dependent (Figure S1G,H). Moreover, within the same cycle period, the absorption peak intensity at 420 nm for the IC-B2 gel decreased by 153.03% more than that of IC-B1, indicating that a high sensitivity to $^1O_2$ and $\bullet O_2^-$ generation in the IC-B2 group compared to IC-B1 (Figure S1I).

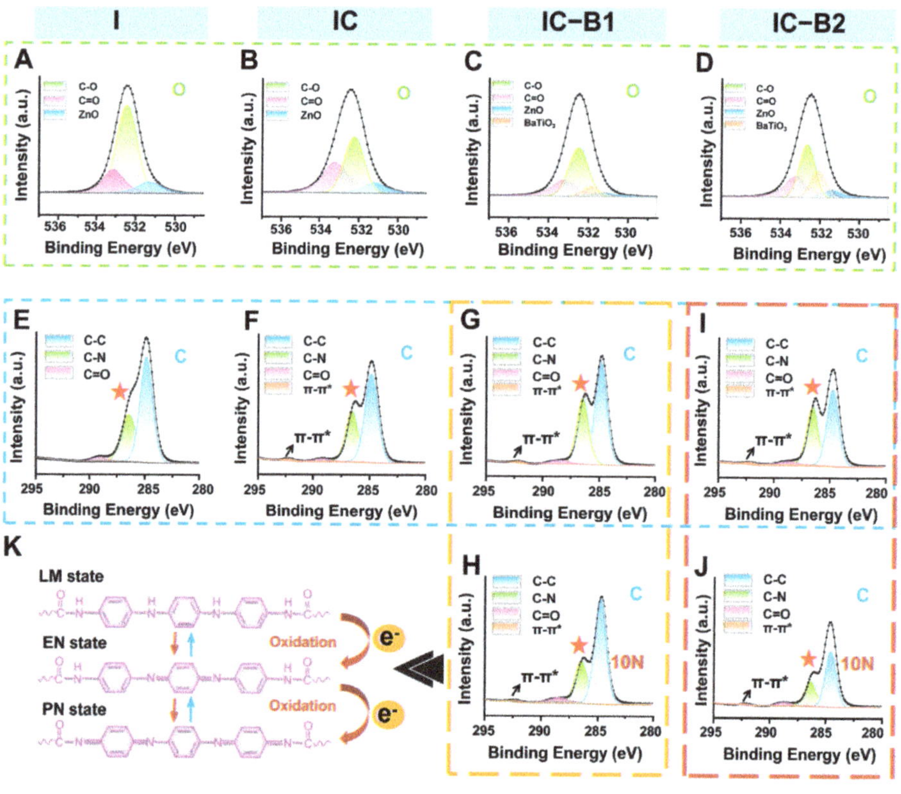

Figure 7. Deconvoluted XPS spectra of cured gels under 10 N compression loading for 30 cycles or unloaded condition: High-resolution XPS spectra of (**A**) O 1s in I gel; (**B**) O 1s in IC gel; (**C**) O 1s in IC-B1 gel; (**D**) O 1s in IC-B2 gel; (**E**) C 1s in I gel; (**F**) C 1s in IC gel; C 1s in IC-B1 gel (**G**) under no load and (**H**) under cyclic loading; C 1s in IC-B2 gel (**I**) under no load and (**J**) under cyclic loading. (**K**) Schematic diagram of chemical structure change in AT fragment in CPU matrix under loading. Red star: C-N bond at 286 eV; black arrow: the π-π^* peak at 292 eV.

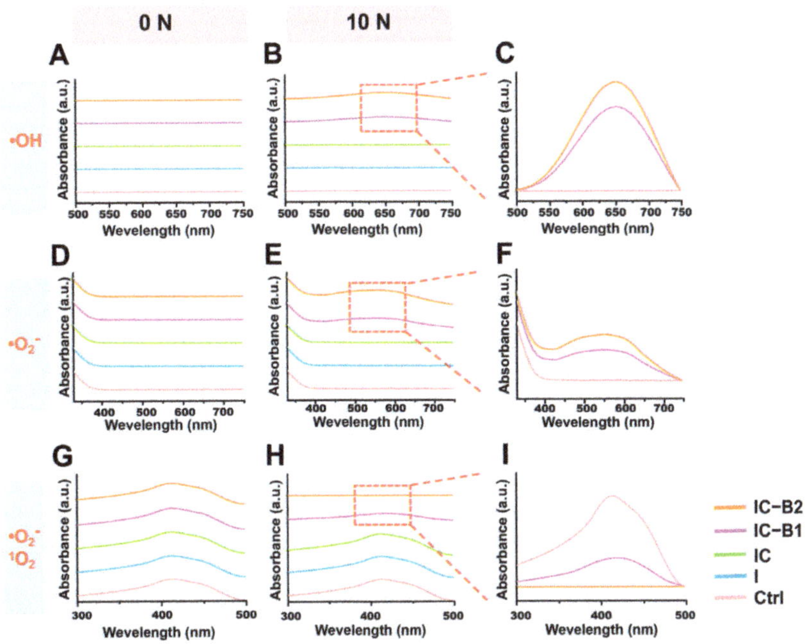

Figure 8. ROS generation performed by UV-vis spectra for gels under 10 N compression loading for 30 cycles or unloaded condition: (**A**) The UV-vis spectra of TMB were used to characterize the ·OH produced by four groups of conductive polyurethanes without pressure. (**B**) The UV-vis spectra of TMB were used to characterize the ·OH produced by four groups of conductive polyurethanes under cyclic pressure. (**C**) The UV-vis spectra of TMB were used to characterize the ·OH produced by IC-B1 and IC-B2 under cyclic pressure. (**D**) The UV-vis spectra of NBT were used to characterize the ·O_2^- produced by four groups of conductive polyurethanes without pressure. (**E**) The UV-vis spectra of NBT were used to characterize the ·O_2^- produced by four groups of conductive polyurethanes under cyclic pressure. (**F**) The UV-vis spectra of NBT were used to characterize the ·O_2^- produced by IC-B1 and IC-B2 under cyclic pressure. (**G**) The Uv-absorbable spectra of DPBF were used to characterize the $^1O_2/O_2^-$ produced by four groups of conductive polyurethanes without pressure. (**H**) The Uv-absorbable spectra of DPBF were used to characterize the $^1O_2/O_2^-$ produced by four groups of conductive polyurethanes under cyclic pressure. (**I**) The Uv-absorbable spectra of DPBF were used to characterize the $^1O_2/O_2^-$ produced by IC-B1 and IC-B2 under cyclic pressure. Ctrl group: No sample was placed in the indicator solution, and other conditions were consistent with those of the experimental group.

2.7. Biocompatibility of Gels Under Cyclic Loading

The proliferation of L929 cells was measured using cck-8. When subjected to a compressive load of 10 N for 30 cycles, the mechano-powered gels (IC-B1 and IC-B2) generated a current that could be transmitted through the wire to the pore plate of the cultured cell (Figure S8A). Figure S8B shows that each group of cells exhibited a tendency to proliferate as the culture time extended. In the absence of loading, the OD value of materials in each group was slightly lower than that in the control group with no significant difference (Figure S8B). The cell viability shows that all the gels had good biocompatibility. According to the aforementioned working mechanism, the IC-B2 group exhibited the highest cell viability, which was significantly higher than that of the original groups without loading from 1 day ($p < 0.01$) to 7 days ($p < 0.001$), while the IC-B1 group significantly proliferated from 4 d to 7 days ($p < 0.05$). Simultaneously, the cell proliferation in groups I and IC remained unchanged regardless of whether pressure loading was applied. The staining results of live/dead cells presented in Figure 9 confirm the biocompatibility of all four gel groups. Specifically, the IC-B1 and IC-B2 gels exhibit enhanced L929 cell proliferation under loading conditions, attributable to the bioelectric currents generated by their piezoelectric activity. Critically, these gels show no cytotoxic effects, as evidenced by high cell viability and minimal cell death in the stained assays.

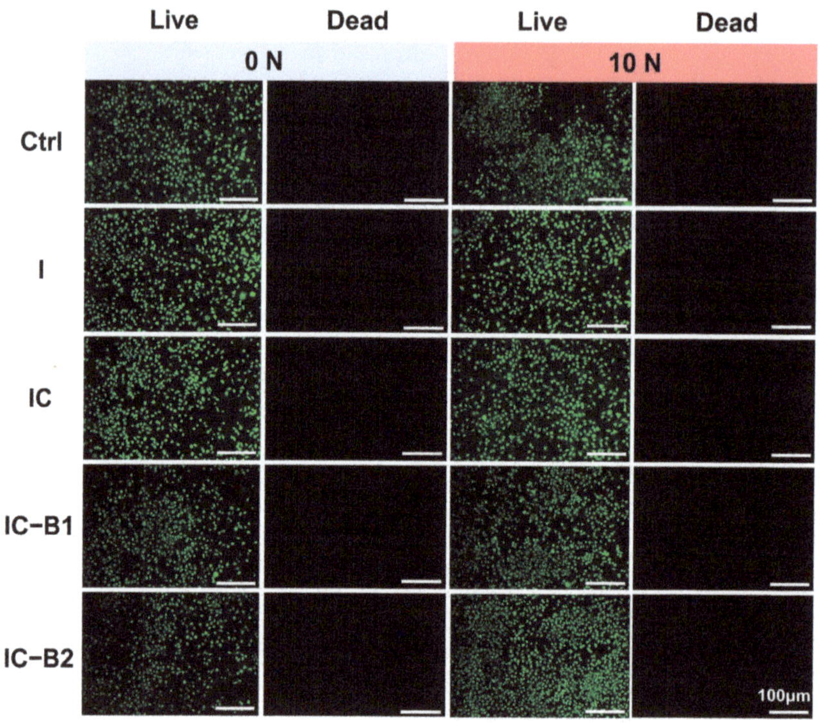

Figure 9. Biocompatibility assessment of the polyurethane gels under mechanical loading. The staining of live/dead cells was performed after culturing L929 cells in the polyurethane gel extract for 7 days under two conditions: (1) 10 N compression loading applied for 30 cycles, or (2) an unloaded condition. In the loaded condition, one group had the culture solution in contact with the solid material under pressure, while the other group remained untreated without additional operations. Green: living cells; red: dead cells).

2.8. Discussion

The frequent chewing behavior in daily life generates significant compressive stress within the oral environment [13]. Under the influence of this stress, an electric field is induced on the surface of piezoelectric materials [25]. The mouth, acting as a natural energy source, can be harnessed to power these piezoelectric materials. The conversion of mechanical energy into electrical effects ensures sustained antibacterial efficacy against recurrent infections in the complex oral environment. To address the limited lifespan of traditional antibacterial agents, a mechano-powered strategy has been developed to achieve prolonged bacteriostasis. This approach utilizes an injectable and conductive polyurethane composite gel for root canal therapy. In this study, piezoelectric n-$BaTiO_3$ and antimicrobial n-ZnO were synergistically combined with conductive AT segments in a polyurethane matrix to create a piezoelectric antibacterial gel. Within this system, an integrated piezoelectric/conductive network was constructed to facilitate efficient charge transport via the mechano-electric coupling effect. Powered by the piezoelectric/conductive matrix, the conductivity of the composite gels was significantly enhanced under cyclic compressive loading (Figure 4). Building upon this piezoelectric enhancement, the antibacterial properties of the gels under pressure-induced activation were further investigated. Focusing on the protection mechanism against bacterial biofilms, the mechano-powered gel was dynamically activated through cyclic loading, leveraging the mechano-electric coupling effect to effectively eradicate biofilms in the oral environment.

2.8.1. Electro-Responsive Behavior of the Conductive Gels Under Cyclic Loading

The human body contains abundant untapped biomechanical energy that could potentially be harvested and transformed into therapeutic electrical signals through specialized energy conversion processes. Everyday human activities provide numerous opportunities for energy harvesting, such as the mechanical energy generated by chewing in the mouth. A promising way is to harvest the "intermittent motion" energy by using piezoelectricity [48]. Piezoelectric substances demonstrate intrinsic electromechanical coupling through spontaneous electrical potential generation upon mechanical deformation [49]. The incorporation of AT segments formed urea bonds with the polyurethane matrix, enhancing crosslinking density and mechanical strength (reduced swelling and higher modulus). Synergy with n-$BaTiO_3$ via hydrogen bonding created a multiscale network, optimizing both mechanical resilience and electrical performance, consistent with prior studies [31,50]. After the AT segment was successfully grafted onto the molecular chain of polyurethane, the conductive polyurethane performed a typical electric response (Figure 4B,C). The CPU polymeric matrix could form a continuous conductive network providing high electro-sensitivity for the mechano-powered gel, efficiently transferring the charges transduced by the piezoelectric effect, and playing a positive role in the mechano-electric coupling effect on the dynamic antibacterial activity of the CPU gels (Figures 5 and 6). Even without mechanical loading, both the IC-B1 and IC-B2 groups demonstrated significantly enhanced conductivity, with values 21.09 and 33.81 times higher than the IC group (without n-$BaTiO_3$) (Figure 4C, $p < 0.001$).

The piezoelectric phenomenon fundamentally originates from atomic lattice rearrangement induced by applied mechanical forces, governed by the material's intrinsic lattice asymmetry [50]. Mechanical strain induces lattice symmetry breaking through charge center displacement in crystalline structures, triggering spontaneous dipole formation via electrostatic imbalance that ultimately drives charge displacement along potential gradients. According to the literature, the average chewing force of human premolars is about 10 N which frequently produces a cyclic pressure in daily life [51]. So, the compressive load of 10 N was applied as the power source for the piezoelectric n-$BaTiO_3$, which was the transducer in the CPU gel (IC-B1 and IC-B2). After applying cyclic compressive stress for 30 cycles to the cured samples, the output voltage and output current of the mechano-powered gels were significantly improved. Among them, the output voltage of the IC-B2 group was increased to 39.14 ± 0.53 V and the current was 834.34 ± 16.05 nA, significantly higher than that of the original gel without loading (Figure 4G,J, $p < 0.001$). In comparison, Kim added 20 wt% n-$BaTiO_3$ to the insulated polyurethane substrates, and the maximum output voltage was only 2.16 V under pressure from the weight of an adult male [52]. Moreover, the piezoelectric constants of IC-B1 and IC-B2 were increased by 3.33 times and 2.54 times, respectively, after applying cyclic stress lasting 30 cycles (Figure 4D). In daily life, the chewing action is more frequent with a long-time working that could excite the dynamic antibacterial activity via the mechano-electric coupling effect of the conductive gels.

Different from the piezoelectric $BaTiO_3$, n-ZnO is a widely used n-type semiconductor contributing to the presence of intrinsic defects, such as oxygen vacancies and zinc interstitials [53]. But, the pure ZnO exhibits limited electron conductivity due to its wide bandgap of approximately 3.37 eV at room temperature. The surface conductivity of n-ZnO enables efficient electron transport through doped heterojunctions [54], thermal excitation [55], and exposure to oxidizing gases (O_2 and NO_2) or reducing gases (H_2 and CO) [56]. In the composite gels, no method or dopants can enhance the charge carrier mobility to introduce free electrons into the conduction band of ZnO particles. Consequently, the n-ZnO component in the I group did not display any electrical properties which were testified by the UV-spectra and conductivity values (Figure 4B,C). Under the same test condition, the conductivity of the IC group also cannot be strengthened under the mechanical loading for the chemical structure of AT conductive molecules, same as the low intrinsic conductivity of the ZnO nanoparticles. For both the I and IC groups (without n-$BaTiO_3$), piezoelectric characteristics could not be measured, regardless of external force loading.

At this point, the cyclic loading induces chemical structural changes in the CPU gels (IC-B1 and IC-B2). The phenomenon is attributed to the piezoelectric field generated by the n-$BaTiO_3$ nanoparticles under mechanical stress. AT has different redox states similar to that of polyaniline, and redox states could be exchanged under different voltages and redox actions [57]. Therefore, the electric field facilitates the oxidation of C-N bonds to C=N bonds within the AT-grafted polyurethane chains (Figure 7K). The higher n-$BaTiO_3$ content in IC-B2 amplifies this effect, as the intensified piezoelectric potential drives greater electron transfer, further reducing C-N bonds and enhancing the conjugation within the polymer backbone [31]. The increased conjugation aligns with the elevated conductivity observed in IC-B2 (Figure 4C), as extended π-electron delocalization promotes charge carrier mobility [50]. The preservation of π-π^* transition peaks (Figure 7F) confirms that the aromatic structure of AT remains intact, ensuring sustained electronic communication across the conductive network [34,35]. Such structural evolution under cyclic loading establishes a synergistic piezoelectric/conductive pathway, where the mechano-electrical coupling not only amplifies charge generation but also stabilizes the redox-active states of AT, similar to polyaniline systems [57]. This dynamic interplay between piezoelectric polarization and conductive polymer redox activity underpins the enhanced electroactivity, enabling efficient charge transduction for prolonged antibacterial efficacy. These findings align with prior studies on redox-active polymers, where structural modifications via electric fields significantly modulate charge transport and interfacial interactions [31,50].

2.8.2. Dynamic Antibacterial Activity of the Gels Under Cyclic Loading

Recent studies have demonstrated that piezoelectric materials, when mechanically or otherwise excited, can produce electric effects such as ROS generation, which contribute to their antibacterial properties [58–60]. The mechano-powered antibacterial mechanism provided a promising approach for developing piezoelectric antibacterial gel in root canal therapy. Root canal fillers are subjected to the stress of frequent tooth occlusion during service, but few design consideration is postulated by using the teeth generating pressure. Therefore, the piezoelectric n-$BaTiO_3$ filler is added as the transducer of the CPU gel (IC-B1 and IC-B2) and the response sensitivity of the gels is highly excited by the bite of the teeth. Based on the natural source of stress, the mechanical energy has been converted into electrical energy, achieving the sustained killing effect of bacteria in a complex oral environment.

The first key finding of the antibacterial activity is that the decreasing trend of SCT and DCT values was not affected by the addition of n-ZnO and AT in the composites, whether cyclic loading was applied or not, even as the soaking time increased (I and IC groups, Figures 5 and S7). The data show that n-ZnO played an antibacterial role in all four gel groups in the absence of pressure. Several investigations have discovered the decreasing size of ZnO nanoparticles can enhance their antibacterial efficacy [61]. Based on our previous study, n-ZnO acts as a chemical fungicide in injectable PU-based root canal filling by releasing Zn^{2+} in comparison to the bacteriostatic rate of commercial AH Plus® and Apexit® Plus which imploded to about 30% at 15 days [24]. Unfortunately, the gradient ions release of n-ZnO from the surface of cured root canal filling will be fading in the dry condition of a root canal, while the ROS generation of ZnO also is weakened without light irradiation in the anaerobic and darkness conditions.

To overcome the short life of chemical antibacterial agents and avoid the overuse of antibiotics, a sustained strategy has been designed by virtue of refreshed power generated by tooth occlusion. But, the antibacterial activity of n-ZnO (I group) was not affected by different compressive conditions (Figure 5), in that n-ZnO as an n-type oxide semiconductor could not transfer the mechanical press to electrical energy (Figure 4). Similar phenomena were observed for the addition of electroactive AT (IC group) shown in Figure 5. Therefore, n-$BaTiO_3$ as a piezoelectric filler has been introduced in the injectable CPU gels to stimulate dynamical antibacterial activation under intermittent loading. Under cyclic loading, the piezoelectric n-$BaTiO_3$ of the mechano-powered gels sustained a significantly strong antibacterial activity (IC-B1 and IC-B2 groups, 95%) for 15 days ($p < 0.001$). The antibacterial way could make full use of the natural stress generated by mouth to play an antibacterial role while avoiding the production of bacterial resistance [62].

Bacterial biofilms are key contributors to oral infections. These complex communities of microorganisms develop on various surfaces—both living (biotic) and non-living (abiotic). Once formed, the bacteria become embedded within a protective matrix, making them more resistant to viral attacks. Bacteria in biofilms are more resistant to antibiotics than their isolated confreres; especially, both living and dead biofilms present a risk for the pharmaceutical product. The community-like microbial lifestyle helps them survive longer under challenging conditions such as nutrient deficiencies, dryness, unfavorable temperatures, exposure to ultraviolet radiation, and pH, as bacterial cells embedded in biofilms are up to 1000 times more resistant [63]. Due to the existence of biofilms, traditional antimicrobial methods are insufficient in dealing with the challenges of harmful bacteria [64]. As shown in Figure 6A–I, the material eradicated approximately 90% of the biofilm under no pressure, with a bacterial survival rate within the membrane of about 15%. After applying cyclic pressure for 30 cycles, the biofilm eradication efficiency of the IC-B1 and IC-B2 groups was significantly enhanced ($p < 0.001$). Notably, the IC-B2 group demonstrated the highest efficacy, reducing the biofilm to approximately 3%, while the bacterial survival rate dropped from 15% to 7%. The exceptional antibacterial performance of the piezoelectric IC-B2 gel meets stringent clinical requirements for root canal therapy, which necessitates >90% biofilm eradication and <10% bacterial survival to prevent reinfection [9,12]. As systematically summarized in Table S3, conventional root canal materials exhibit significant limitations: Epiphany SE provides only transient antibacterial efficacy (<1 week) and lacks detectable biofilm resistance under static conditions [65,66]; Pulp Canal Filling EWT and EndoSequence BC fail to maintain antibacterial activity beyond one week and lack documented biofilm inhibition capabilities [10]. Of particular concern is AH Plus®/Apexit® Plus, which shows a dramatic decline in antibacterial efficacy to ~30% after 15 days under static loading, with no biofilm resistance [24]. While Shah's

$BaTiO_3$ nanoparticles achieve ~85 ± 3.5% biofilm resistance against *S. aureus* under static conditions [59], tour IC-B2 gel achieves a biofilm resistance rate of ~97% against both pathogens under cyclic loading (10 N, 30 cycles), surpassing the clinical threshold (>90% biofilm eradication) [9,12].

The electrochemical inhibition of bacteria is a promising and robust approach currently under investigation for both the degradation and formation of biofilm. High temperatures can destroy biofilms, but the approach is not safe to deeply use inside the delicate interior of a root canal. Studies indicate that the combination of low concentrations of antibacterial agents and direct current (DC) results in a synergistic effect, significantly enhancing biofilm degradation [12]. This study presents a novel piezoelectric-enhanced strategy, integrating mechano-electric coupling effect with chemical antimicrobial activity to enable sustained biofilm eradication through dynamic antibacterial effects (Figure 10). Under cyclic compressive loading, the lattice of piezoelectric nanoparticles (n-$BaTiO_3$) underflows asymmetrical deformation, leading to the displacement of positive and negative charge centers and the subsequent generation of piezoelectric potential [67]. Under cyclic loading, the repeated deformation of n-$BaTiO_3$ produces alternating surface charges, achieving sustained charge transfer [31]. After grafting, the aniline trimer (AT) preserved the complete π-π^* conjugated structure in the polyurethane-based conductive polymer (Figure 7F), synergistically forming a continuous piezoelectric/conductive network with n-$BaTiO_3$. The interconnected network functions as an "electron highway", facilitating efficient carrier migration and establishing a stable internal electric field. Similarly to polyaniline, AT exhibits multiple redox states that can be reversibly interconverted under applied voltage or redox stimulus [57]. Driven by the internal electric field, the benzene rings in AT undergo oxidation (with the partial conversion of C-N to C=N bonds), releasing electrons into the surrounding environment (Figure 7K). These electrons subsequently react with ambient H_2O and O_2 to generate abundant reactive oxygen species (ROS), including hydroxyl radicals (•OH), singlet oxygen ($^1O_2$), and superoxide anions (•O_2^-), of which the characteristic peaks were observed in the UV-vis spectra in Figure 8. The generating ROS exerts oxidative stress to disrupt bacterial cell membranes, degrade biofilm matrix components (polysaccharides and proteins), and damage genetic material, thereby effectively eradicating biofilms and inhibiting bacterial proliferation [45]. Simultaneously, the electric field has enhanced ROS interaction with negatively charged bacterial surfaces through electrostatic attraction (Figure 6E,J), amplifying antibacterial efficacy (Figure 5). Crucially, the dynamic mechanical stimulation from natural oral chewing can sustainably promote ROS generation through the piezoelectric–electrochemical coupling mechanism, enabling the long-term targeted disruption of biofilm regeneration and reactivation. Although ROS was detected under cyclic loading (Figure 8B,F,H), CCK-8 and Live/Dead assays confirmed no detectable cytotoxicity (Figures S8B and 9). This selective antibacterial effect arises from fundamental differences in oxidative stress defense between bacteria and mammalian cells. Unlike mammalian cells, bacteria lack sophisticated ROS-neutralizing systems such as superoxide dismutase (SOD), catalase (CAT), and glutathione peroxidase (GPx) [68,69]. For instance, *S. aureus* and *E. coli* depend on limited enzymatic pathways, such as alkyl hydroperoxide reductase AhpCF [70], making them highly vulnerable to oxidative stress. In contrast, mammalian cells possess robust antioxidant networks capable of efficiently neutralizing moderate ROS levels [69]. The finding aligns with prior studies showing that piezoelectric materials can generate therapeutically useful ROS concentrations without compromising mammalian cell viability [63]. The mechano-electrical coupling, therefore, represents a critical therapeutic strategy that ensures adequate ROS production for effective biofilm eradication while preserving tissue biocompatibility. The self-powered strategy of the CPU gels can smooth the deficiencies of static antimicrobial agents, such as diminishing efficacy over time while avoiding antibiotic resistance linked to conventional approaches [62]. By harnessing the energy of intermittent motion, the mechano-powered CPU system has achieved a dynamic sustained antibacterial activity without external energy input or chemical overuse.

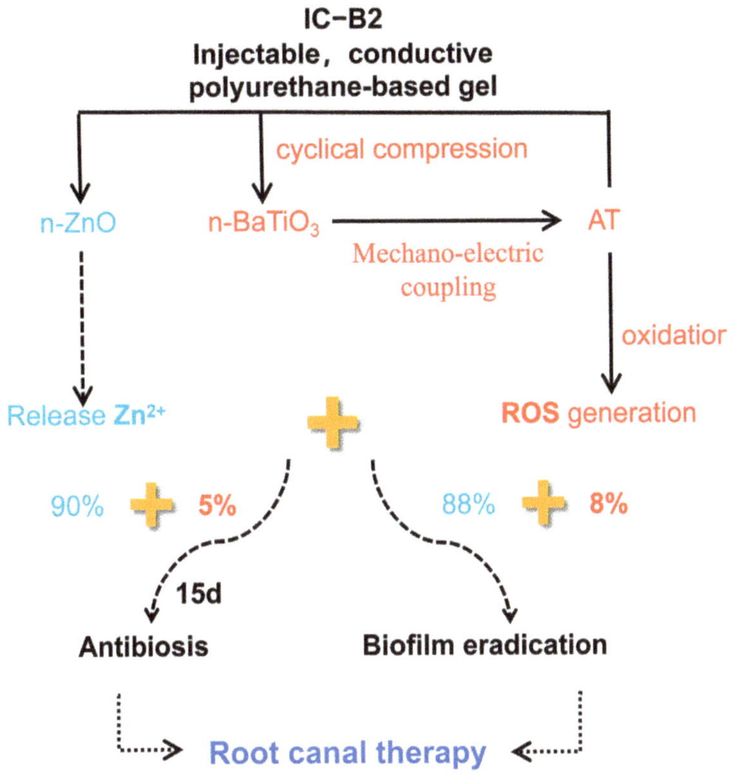

Figure 10. Mechanical and electrical coupling effect combined with chemical antibacterial activity to achieve root canal therapy.

In future studies, we will systematically evaluate different sterilization methods (autoclaving, gamma irradiation, low-temperature plasma) on material properties to identify the optimal clinical protocol. Concurrently, we will explore various approaches to enhance material stability, including encapsulation under an inert gas atmosphere, to prolong shelf life. Special attention will be given to optimizing setting time for clinical operability. Further validation will include performance testing under simulated chewing loads and comprehensive biocompatibility assessment to ensure clinical applicability.

3. Conclusions

A novel mechano-responsive therapeutic approach has been successfully developed through an injectable conductive polyurethane (CPU) composite gel incorporating piezoelectric n-$BaTiO_3$ nanoparticles and functional components. Under cyclic compressive loading, the formation of an interconnected piezoelectric/conductive network within the composite matrix enables efficient charge transport via mechano-electric coupling. This design results in enhanced sensitivity to mechanical stimuli, with intermittent motion being converted into sustained charge transport and ROS generation, synergistically combined with chemical antimicrobial activity for improved bacteriostasis and biofilm eradication. The experimental results confirm exceptional biofilm eradication efficiency (97%) alongside maintained biocompatibility, representing a significant advancement from conventional passive materials to actively responsive therapeutic systems for prolonged infection control in root canal therapy.

4. Materials and Methods

4.1. Materials

The key materials were sourced from commercial suppliers (Shanghai Aladdin (Shanghai, China); Chengdu Kelong (Chengdu, China)): PTMEG (M_n = 2000), Isophorone Diisocyanate (IPDI), nano-zinc oxide (n-ZnO), nano-bismuth oxide (n-Bi_2O_3), nano-barium titanate (n-$BaTiO_3$), polyethylene glycol (PEG_{600}, M_n = 600), stannous salt, trolamine, aniline, and p-phenylenediamine. Nano-hydroxyapatite (n-HA) particles were synthesized via chemical precipitation (method detailed previously) followed by vacuum-drying (108 °C, 2 h) [71]. All the chemicals were analytical grade.

4.2. Synthesis of AT

On the basis of previous studies, amino-terminated aniline trimers (ATs) were synthesized [72]. In brief, after the p-phenylenediamine was dissolved in a solution, ammonium persulfate was added under circulated cooling. When the color of the solution changed, aniline was added till the reaction ended and the crude products were filtered; washed with hydrochloric acid, distilled water, and ammonia water; and dried by a freeze dryer. AT product was then purified by a rotary evaporator and ground into a powder state with an agate mortar. The purification process removes unreacted monomers and solvents, effectively preventing cytotoxicity or inflammatory reactions caused by these impurities, thereby significantly improving the biocompatibility of the material [35].

4.3. Preparation of Injectable Conductive Polyurethane-Based Composite Gels

4.3.1. Component A: Injectable Prepolymer

PTMEG and mannitol were mechanically mixed in a three-neck flask under N_2 and thermally processed at 70 °C for 1 h. Then, n-HA as a calcified promoter with good biocompatibility, Bi_2O_3 as an X-ray inhibitor with radiopacity, and n-ZnO (the hexagonal fibrous zincite structure (Figure S2A) [73]) as a chemical fungicide providing broad-spectrum bacteriostasis were added referring to a previous study [24] and dispersed for 1 h with/without piezoelectric n-$BaTiO_3$ addition (Figure S2B) (gels' composition shown in Table 1) [74]. Stage II comprised the controlled addition of IPDI/AT under a nitrogen atmosphere, followed by pre-polymerization at 75 °C for 2 h, with rigorous maintenance of the isocyanate index at a molar ratio of NCO:OH = 1.5 throughout the process.

Table 1. Composition and abbreviation of the injectable and conductive polyurethane gels.

	I	IC	IC-B1	IC-B2
AT (wt%)	0	2.5	2.5	2.5
$BaTiO_3$ (wt%)	0	0	5	10

Abbreviations of the composite gels: **I**: injectable polyurethane composite (polyurethane/n-HA/Bi_2O_3/n-ZnO at a weight ratio of 13: 2: 4: 1); **IC**: injectable conductive polyurethane composite; **IC-B**: injectable conductive polyurethane composite with the addition of n-$BaTiO_3$.

4.3.2. Component B: Curing Reagent

The curing reagent, referred to as component B, was prepared by mixing triethanolamine, polyethylene glycol (PEG_{600}), and tin salt at a mass ratio of 45:17:1. The mixture was then subjected to ultrasonication for 30 min to ensure thorough homogenization.

4.3.3. Preparation of Gels

Components A and B were combined at a 20:1 volume ratio under ambient temperature, thoroughly mixed to form a homogeneous paste, and subsequently transferred into a cylindrical polytetrafluoroethylene mold (6 mm diameter × 12 mm height). The filled mold was then thermally cured in a controlled-temperature oven maintained at 37 °C. Four groups of polyurethane composite gels were fabricated as preceding steps and the abbreviated names and formulations of each group were shown in Table 1. After the mixtures had been cured in the Teflon mold (Φ 10 mm × 2 mm), we removed them from the mold and dried and sterilized them through low-temperature plasma sterilization for electrical and biological tests.

4.4. Physicochemical Evaluation of Gels for Root Canal Therapy

4.4.1. XRD and FT-IR

The copolymers had been characterized by the X-ray diffraction of Cu Kα radiation (XRD, EMPYERAN, Panalytical, Almelo, The Netherlands). Scans and 2 θ range was from 10 to 80° at a step size of 0.03° under the conditions of 30 kV and 25 mA. The FTIR spectra (Nicolet 6700, PerkinElmer, Waltham, MA, USA) of PU composites were recorded from 650 to 4000 cm^{-1} with 4 cm^{-1} resolution to analyze molecular interactions.

4.4.2. Anti-Washout Property

After components A and B were mixed evenly, the mixture was immediately transferred to a syringe, injected into a Petri dish, and then injected with 10 mL of deionized water at room temperature. The washout resistance of the composites was evaluated according to the method in previous research [75].

4.4.3. Material Morphology and Element Distribution

Morphological analysis was conducted using a JEOL JSM-7500F scanning electron microscope, with coupled energy-dispersive X-ray spectroscopy (EDS) employed of Ca, Bi, Zn, and Ba across the gel surfaces. Prior to imaging, specimens underwent gold sputter-coating to ensure surface conductivity under high-vacuum conditions.

4.4.4. Setting Time and Flowing

The flowability and setting time of the gels were tested in accordance with ISO 6876: 2012 [38]. After the A and B components were mixed, the mixed slurry was injected into the Teflon mold (Φ 5 mm × 3 mm) and then placed in the oven at 37 °C. Setting time was defined as the duration required for the material to achieve structural resistance against Gillmore needle penetration. Each group of samples was measured three times.

After mixing component A and component B for 180 ± 5 s, we used a syringe with a scale to inject (0.05 ± 0.001) ml of gels into a glass plate center (20 g), and then placed a glass plate (20 g) on top of the gels and place a 100 g weight on it. After 10 min, diameters (max/min) were measured with vernier calipers through triplicate testing per sample group.

4.4.5. Degree of Crosslinking

According to the expansion equilibrium method, the degree of crosslinking was evaluated through the characterization of the swelling rate and gel in the polyurethane composite gels. Following the stoichiometric homogenization of components A/B, the reactive mixture was cast into cylindrical PTFE molds (Φ 6 × 12 mm) and cured in the oven at 37 °C, where the original weight of the cured sample was M_1, and then the sample was soaked in tetrahydrofuran alcohol for 24 h. After the soaked sample was removed and the liquid on its surface was wiped clean, the wet weight was weighed as M_2. Finally, the sample was dried in an oven at 50 °C till its weight no longer changed; thus, its dry weight was weighed to M_3. Each group of samples was measured three times. The swelling rate and gel fraction of polyurethane material were calculated by Formulas (1) and (2).

$$Q = (M_2 - M_1)/M_1 \tag{1}$$

$$G = M_3/M_1 \tag{2}$$

where Q was the swelling rate of the polyurethane material, G was the gel fraction of the polyurethane material, M_1 was the original weight of the polyurethane material, M_2 was the mass of the polyurethane material after soaking in tetrahydrofuran for 24 h, and M_3 was the mass of the polyurethane material after drying.

4.4.6. Mechanical Test

Test specimens were fabricated following Section 4.3.3, injected into Teflon molds (Φ 6 mm × 12 mm), and oven-cured at 37 °C. Mechanical characterization utilized an AUTOGRAPH AG-IC 20/50KN tester (Tokyo, Japan) at 1 mm/min displacement until the 50% compression of initial length (25 °C). Compression modulus was determined from linear stress–strain curve slopes. Each group tested five independent samples.

4.5. Electrochemical Properties of Composite Gels Under Compressive Loading

4.5.1. Conductivity and Electrochemical Properties

The electrical conductivity of the cured specimens was measured using a Keithley 6517A 4-probe tester equipped with 1.0 mm-spaced linear probes. Each set of experiments tested three samples. The conductivity was calculated by Formula (3).

$$\delta = 1/\rho \tag{3}$$

where δ represented conductivity and ρ represented the resistance.

The cured samples' electroactivity was analyzed via UV-vis spectroscopy (PerkinElmer Lambda 35) using DMF solution. Spectral acquisition employed the same spectrophotometer configuration.

4.5.2. Piezoelectric Properties Under Cyclic Compressive Loading

A piezoelectric impact test was conducted at 1.0 Hz using a linear motor as the impact source. The voltage output from the PEH devices was measured using a Keithley 6514 electrometer, while the current output was recorded with a low-noise current preamplifier (SR570, Stanford Research Systems, Sunnyvale, CA, USA). A DAQ card (NIPCI-6221, Beijing Altai Technology Company, Beijing, China) acquired and processed the signals. The ability and types of ROS generation of the mechano-powered gels were tested by the colorimetric method of 3,3′,5,5′-tetra- methylbenzidine (TMB) and 1,3-diphenyl isobenzofuran (DPBF) to screen suitable parameters of cyclic compressive loading. The screening process is detailed in Figure S1 (Supplementary Materials). The optimized compressive loading was as follows: 10 N for 30 cycles (half a minute per cycle). The cured samples were divided into two groups: one group was subjected to an axial compressive load of 10 N for 30 cycles, and the other group was not subjected to the compressive load. Each group of samples was measured three times. Under the same loading conditions, ZJ-3A was used to test the piezoelectric constant of the cured material.

4.6. Antibacterial Properties Under Cyclic Compressive Loading

The cured samples were split into two groups. One group was applied a 10 N force for 30 cycles (half a minute per cycle) and the other group was not under compression; thereafter, the same mechanical conditions were loaded in the following biological tests.

Gram-negative *Escherichia coli* (*E. coli*, ATCC 25922) and Gram-positive *Staphylococcus aureus* (*S. aureus*, ATCC 25923) evaluated the antibacterial properties of the conductive polyurethanes gels in the absence of load and under cyclic compressive load. *E. coli* and *S. aureus* were grown in Oxoid broth at 37 °C with oxygen. After centrifugation, the bacteria were transferred to fresh broth. Bacterial suspensions were adjusted to specific optical densities matching known concentrations to prepare inocula. The antimicrobial activity of the polyurethane gels was evaluated using two methods: static exposure conditions following ASTM G21-15 and dynamic exposure conditions based on ASTM E2149-2013a.

4.6.1. Static Contact Test (SCT)

After curing in a Φ 10 mm × 2 mm Teflon mold, the mixtures were removed and ground. After curing, the samples were immersed in 10 mL PBS and oscillated in a 37 °C water bath for 0, 3, 7, and 15 days. All the samples underwent low-temperature plasma sterilization before testing. When the cured sample was placed in the center of the Petri dish, 20 μL of suspended bacterial solution (2×10^6 CFU per mL) was added to the surface of the sample. The sterilized samples were incubated in a humid atmosphere at 37 °C for 2 h to evaporate the bacterial suspension under different loading conditions to ensure direct contact between the bacteria and the sample. Next, the specimen surfaces were washed three times with 5 mL of phosphate-buffered saline (PBS), and the wash solutions were collected. A 1 mL aliquot of the collected wash liquid was diluted in 9 mL PBS, after which 200 μL of this diluted sample was plated onto Brain Heart Infusion (BHI) agar. The resulting colonies were enumerated (C_2) following 48 h of incubation. A sterile, dry polyethylene film (10 mm diameter) served as the negative control (C_1). The bacteriostatic rate was calculated using Equation (4), and the experiment was replicated three times.

$$Bacteriostatic\ rate\ (\%) = (C_1 - C_2)/C_1 \times 100\% \quad (4)$$

4.6.2. Dynamic Contact Test (DCT)

The fresh mixture of different composites was coated on the bottom and side walls of the 96-well plate. The material was immersed in PBS solution for 15 days. Following the extraction of the immersed solution, all the coated 96-well plates were sterilized using low-temperature plasma sterilization. Then, 20 μL of bacterial suspension ($(3-7) \times 10^6$ CFU per mL) was added dropwise to the coated well and cultured under different loading conditions for 2 h at 37 °C: one group was subjected to a cyclic 10 N force (30 cycles, with each cycle lasting 30 s), while the other group remained uncompressed. In the process, the bacterial suspension was evaporated, ensuring that the bacteria were in direct contact with the curing gel surface. Next, 200 μL of culture solution was added to each well. Absorbance readings were taken at 0, 1, 2, 3, 4, 5, 6, 7, 8, and 24 h post-incubation using a microplate reader (PerkinElmer 1420 Multilabel Counter, PerkinElmer, Waltham, MA, USA) at 650 nm. Sterilized, uncoated wells served as the blank control group. Three replicates were measured for each treatment group.

4.7. In Vitro Bacterial Biofilm Culture and Eradication Under Cyclic Compressive Loading

Escherichia coli (*E. coli*, ATCC 25922) and *Staphylococcus aureus* (*S. aureus*, ATCC 25923) strains were cultured in broth medium (Oxoid Ltd., Basingstoke, UK) at 37 °C. Next, 1×10^8 CFU/mL of bacteria were inoculated into 12-well plates and incubated at 37 °C for 48 h to form bacterial biofilms. The unbound bacteria were rinsed with PBS and then the sterilized samples were placed tightly on the biofilm under different loading conditions for 2 h. To quantify residual biofilm biomass, the biofilms were stained with 0.1% crystal violet for 30 min. The stained biofilm was washed with PBS and dispersed in 1.0 mL of 95% ethanol, and the remaining biofilm biomass was quantified by measuring absorbance at 590 nm. The relative biofilm mass was calculated by Formula (5).

$$Relative\ biofilm\ mass\ \% = A_1 / A_2 \times 100\% \quad (5)$$

where $\boldsymbol{A_1}$ and $\boldsymbol{A_2}$ represented the absorbance values of crystal violet solution at 590 nm for the experimental group and control group, respectively.

Bacterial viability was assessed via plate counting after disrupting the treated biofilms. For the morphological analysis, the biofilms were fixed with 2.5% glutaraldehyde (2 h) and examined using a JSM-7500F SEM (Jeol, Tokyo, Japan).

4.8. Chemical Structural Changes and ROS Generation Under Cyclic Compressive Loading

In order to test the piezoelectric effect of conductive polyurethane gels under cyclic compressive loading on AT segments of molecular chain, the cured samples (1 cm × 1 cm × 5 mm) were divided into two groups. One group was subjected to 10 N loading for 30 cycles (half a minute per cycle), and the other group was not under compression. The material's chemical structure was analyzed using X-ray photoelectron spectroscopy (XPS, Thermo Scientific K-Alpha, USA).

Considering biofilm is affected by the reactive oxygen species (ROS), the colorimetric method of TMB and DPBF was used to investigate the ability and types of ROS generation of cured gels under cyclic loading. The generation of $\bullet O_2^-$ was verified using a Nitrotetrazolium Bluechloride (NBT) assay [46]. It was dispersed in sodium citrate buffer (pH 5.0) with TMB as substrate.

The cured samples were placed in the TMB solution under the compression of 10 N loading for 30 cycles (half a minute per cycle), respectively, wherein the samples were soaked in the solution for 30 min without load as a contrast. Then, the UV-vis spectroscopy of the solution was measured by UV-vis spectrophotometer. The material was placed in PBS solution and the corresponding amount of NBT developing agent was added. The same treatment was applied to the material, and the UV-vis spectrum of the solution was determined by UV-vis spectrophotometer. Finally, the DPBF was dispersed in PBS (pH 7.4). The cured samples were placed in the DPBF-PBS solution loading with the pressure of 10 N for 30 cycles (half a minute per cycle), respectively, wherein the samples were soaked in the solution for 30 min without pressure as a contrast. Then, the UV-vis spectroscopy of the solution was measured by UV-vis spectrophotometer.

4.9. *In Vitro Cell Evaluation Under Cyclic Compressive Loading*

4.9.1. Cell Culture and Extraction Solution Preparation

The L929 mouse fibroblast cell line was provided by West China School of Stomatology, Sichuan University. The L929 cells were cultured in DMEM medium (Gibco, USA) with 10% fetal bovine serum (FBS, Gibco) at 37 °C in a 5% CO_2 atmosphere. The cells were subcultured at 80% confluence, and the medium was refreshed every other day.

The preparation of the extracted solution of the different gels was according to ISO 10993-9 [76]. The sterilized samples were immersed in DMEM (Gibcos; Thermo Fisher Scientific, Waltham, MA, USA) medium supplemented with 10% fetal calf serum (FBS, Gibco). The samples were immersed at 0.1 g/mL and stored in a 5% CO_2 atmosphere (37 °C, 24 h). The extract was collected and stored at 4 °C.

4.9.2. Cell Proliferation

Cell proliferation was assayed by Cell Counting Kit-8 (CCK-8, KeyGEN BioTech, Nanjing, China). The L929 cells were seeded in 24-well plates (2×10^4 cells/well) and cultured for 1, 4, and 7 days in a 5% CO_2 atmosphere at 37 °C. The culture medium was replaced every 2 d. The cured material was trimmed into 1 cm × 1 cm × 5 mm squares, which were completely wrapped with conductive glue. Two conductive wires were drawn from the upper and lower sides. The orifice plates were divided into two groups, one of which did not apply additional treatment. The other group inserted conductive wires into the orifice plates and dipped them into the medium. The sterilized samples were then subjected to pressure under different loading conditions. Each well received 900 µL fresh medium and 100 µL CCK-8 solution, followed by 2 h incubation at 37 °C. The solution was transferred to a 96-well plate, and absorbance at 450 nm was measured using a PerkinElmer Wallac Victor 31420 microplate reader. Triplicate samples were tested per group.

4.9.3. Live-Dead Cell Staining Assay

The L929 cells were seeded in 24-well plates (2×10^4 cells/well) and cultured for 1, 4, and 7 days in a 5% CO_2 atmosphere at 37 °C. The cured material was trimmed into 1 cm × 1 cm × 5 mm squares, which were completely wrapped with conductive glue. Two conductive wires were drawn from the upper and lower sides. The orifice plates were divided into two groups, one of which did not apply additional treatment. The other group inserted conductive wires into the orifice plates and dipped them into the medium. The sterilized samples were then subjected to pressure under different loading conditions. Cell viability was assessed using a LIVE/DEAD stain (Invitrogen, Waltham, MA, USA) following the manufacturer's protocol. The stained cells were visualized with a Nikon fluorescence microscope (Tokyo, Japan).

4.10. *Statistical Analysis*

The experimental data are presented as mean ± standard deviation (SD). Statistical evaluations were performed using the SPSS software (version 25.0; LEAD Technologies, Chicago, IL, USA) through one-way ANOVA followed by Tukey's or Dunnett's post hoc tests. Significance thresholds were defined at $p < 0.05$ (statistically significant), $p < 0.01$ (highly significant), and $p < 0.001$ (exceptionally significant), with $p > 0.05$ indicating non-significant differences.

Supplementary Materials: The following supporting information can be downloaded at: https://www.mdpi.com/article/10.3390/gels11050346/s1, Figure S1: ROS generation of gels under 10 N compression loading for 10, 20 and 30 cycles. (A) IC-B1 and (B) IC-B2: The production of

$\cdot$OH was detected by TMB. (C) Peak value of IC-B1 and IC-B2 at 652 nm under cyclical loading for 30 cycles. (D) IC-B1 and (E) IC-B2: The production of $\cdot O_2$- was detected by NBT. (F) Peak value of IC-B1 and IC-B2 at 560 nm under cyclical loading for 30 cycles. (G) IC-B1 and (H) IC-B2: The production of $1O_2/\cdot O_2$- was detected by DPBF. (I) Peak value of IC-B1 and IC-B2 at 420 nm under cyclical loading for 30 cycles; Figure S2: XRD patterns of (A) n-ZnO and (B) n-$BaTiO_3$. (C) Magnified XRD pattern of n-$BaTiO_3$; Figure S3: FT-IR spectra of the AT; Figure S4: Properties of different gels in curing process. (A) The representative FTIR spectra of gel with cure time (2 h and 12 h). (B) Isocyanate conversion calculated by the integrated peak areas of representative FTIR spectra (at 2200–2300 cm^{-1}) during curing of copolymers (2 h). (C) Diagram of temperature variation during curing of copolymers; Figure S5: SEM images of polyurethane gels after cured ($\times$2000) and element distribution of Ca, Bi, Zn, and Ba on the surface of cured samples; Figure S6: Anti-washout property of gels after setting; Figure S7: Antibacterial properties of cured seals in the absence of loading. Antibacterial rates of static contact experiment against (A) *S. aureus* and (B) *E. coli*. Dynamic contact test curves of the various groups after soaking in PBS solution for 15 d against (C) *S. aureus* and (D) *E. coli*. The uncoated wells were set up as the blank control group; Figure S8: Proliferation of L929 cells in polyurethane gels under 10 N compression loading for 30 cycles or unloaded condition. (A) Schematic diagram of experimental apparatus. (B) The OD values of CCK-8 were determined after L929 cells were cultured in polyurethane gels extract for 1, 4, and 7 days (one group was connected to the culture medium through a wire when pressure was applied to the solid material, and the other group was not subjected to other operations). * $p < 0.05$, ** $p < 0.01$, *** $p < 0.001$. (n = 3); Table S1: Gel fraction of polyurethane gels; Table S2: Average molecular weight and polydispersion coefficient of p-I and p-IC prepolymers; Table S3: Comparison of Antibacterial Rates and Biofilm Inhibition Efficiencies Against *E. coli* and *S. aureus* Among Selected Commercial Materials, Literature-Reported Materials, and IC-B2 Gel; Video S1: Video demonstration of the injectable property of the gels.

Author Contributions: B.M.: data curation, writing—original draft, investigation, formal analysis, and methodology. X.L.: investigation. Y.Z. (Yinglong Zhang): investigation. J.Z.: investigation. Q.D.: data curation. Y.L. (Yuping Li): investigation. D.H.: data curation. L.W.: data curation. J.L.: data curation. Y.L. (Yubao Li): project administration and conceptualization. Y.Z. (Yi Zuo): conceptualization, methodology, writing—review and editing, and funding acquisition. All authors have read and agreed to the published version of the manuscript.

Funding: This work was supported by the Sichuan International Science and Technology Innovation Cooperation Project (No. 2024YFHZ0308), and the Fundamental Research Funds for the Central Universities.

Institutional Review Board Statement: Not applicable.

Informed Consent Statement: Not applicable.

Data Availability Statement: All the data supporting the conclusions are presented in the manuscript and the Supplementary Materials.

Conflicts of Interest: The authors declare no competing interests.

References

1. Kilian, M.; Chapple, I.L.C.; Hannig, M.; Marsh, P.D.; Meuric, V.; Pedersen, A.M.L.; Tonetti, M.S.; Wade, W.G.; Zaura, E. The oral microbiome—An update for oral healthcare professionals. *Br. Dent. J.* **2016**, *221*, 657–666. [CrossRef] [PubMed]
2. Deo, P.; Deshmukh, R. Oral microbiome: Unveiling the fundamentals. *J. Oral Maxillofac. Pathol.* **2019**, *23*, 122–128. [CrossRef]
3. Palmer, R.J., Jr. Composition and development of oral bacterial communities. *Periodontology 2000* **2013**, *64*, 20–39. [CrossRef] [PubMed]
4. Sibley, C.D.; Peirano, G.; Church, D.L. Molecular methods for pathogen and microbial community detection and characterization: Current and potential application in diagnostic microbiology. *Infect. Genet. Evol.* **2012**, *12*, 505–521. [CrossRef] [PubMed]
5. Botelho, M.P.J.; Maciel, S.M.; Cerci Neto, A.; Dezan, C.C.; Fernandes, K.B.P.; de Andrade, F.B. Cariogenic Microorganisms and Oral Conditions in Asthmatic Children. *Caries Res.* **2011**, *45*, 386–392. [CrossRef]

6. Kilian, M. The oral microbiome—Friend or foe? *Eur. J. Oral Sci.* **2018**, *126*, 5–12. [CrossRef]
7. Zheng, D.-W.; Deng, W.-W.; Song, W.-F.; Wu, C.-C.; Liu, J.; Hong, S.; Zhuang, Z.-N.; Cheng, H.; Sun, Z.-J.; Zhang, X.-Z. Biomaterial-mediated modulation of oral microbiota synergizes with PD-1 blockade in mice with oral squamous cell carcinoma. *Nat. Biomed. Eng.* **2021**, *6*, 32–43. [CrossRef]
8. Karpiński, T.M. Role of Oral Microbiota in Cancer Development. *Microorganisms* **2019**, *7*, 20. [CrossRef]
9. Long, J.; Kreft, J.U.; Camilleri, J. Antimicrobial and ultrastructural properties of root canal filling materials exposed to bacterial challenge. *J. Dent.* **2020**, *93*, 103283. [CrossRef]
10. Baras, B.H.; Melo, M.A.S.; Thumbigere-Math, V.; Tay, F.R.; Fouad, A.F.; Oates, T.W.; Weir, M.D.; Cheng, L.; Xu, H.H.K. Novel Bioactive and Therapeutic Root Canal Sealers with Antibacterial and Remineralization Properties. *Materials* **2020**, *13*, 1096. [CrossRef]
11. Zhang, N.; Ma, Y.; Weir, M.D.; Xu, H.H.K.; Bai, Y.; Melo, M.A.S. Current Insights into the Modulation of Oral Bacterial Degradation of Dental Polymeric Restorative Materials. *Materials* **2017**, *10*, 507. [CrossRef] [PubMed]
12. Srivastava, A.; Verma, N.; Kumar, V.; Apoorva, P.; Agarwal, V. Biofilm inhibition/eradication: Exploring strategies and confronting challenges in combatting biofilm. *Arch. Microbiol.* **2024**, *206*, 212. [CrossRef] [PubMed]
13. de Abreu, R.A.M.; Pereira, M.D.; Furtado, F.; Prado, G.P.R.; Mestriner, W.; Ferreira, L.M. Masticatory efficiency and bite force in individuals with normal occlusion. *Arch. Oral Biol.* **2014**, *59*, 1065–1074. [CrossRef]
14. Quiudini, P.R.; Pozza, D.H.; Pinto, A.d.S.; de Arruda, M.F.; Guimarães, A.S. Differences in bite force between dolichofacial and brachyfacial individuals: Side of mastication, gender, weight and height. *J. Prosthodont. Res.* **2017**, *61*, 283–289. [CrossRef] [PubMed]
15. van Steenberghe, D.; de Vries, J.H. The development of a maximal clenching force between two antagonistic teeth. *J. Periodontal Res.* **2006**, *13*, 91–97. [CrossRef]
16. Fahlgren, A.; Bostrom, M.P.G.; Yang, X.; Johansson, L.; Edlund, U.; Agholme, F.; Aspenberg, P. Fluid pressure and flow as a cause of bone resorption. *Acta Orthop.* **2010**, *81*, 508–516. [CrossRef]
17. Huo, B.; Lu, X.L.; Costa, K.D.; Xu, Q.; Guo, X.E. An ATP-dependent mechanism mediates intercellular calcium signaling in bone cell network under single cell nanoindentation. *Cell Calcium* **2010**, *47*, 234–241. [CrossRef]
18. Yang, Y.; Yang, Y.; Li, X.; Cui, L.; Fu, M.; Rabie, A.B.; Zhang, D. Functional analysis of core binding factor a1 and its relationship with related genes expressed by human periodontal ligament cells exposed to mechanical stress. *Eur. J. Orthod.* **2010**, *32*, 698–705. [CrossRef]
19. *ISO 7405*; Dentistry—Evaluation of Biocompatibility of Medical Devices Used in Dentistry. Standardization IOf. International Organization for Standardization: Geneva, Switzerland, 2018.
20. Lee, S.J.; Heo, M.; Lee, D.; Han, S.; Moon, J.-H.; Lim, H.-N.; Kwon, I.K. Preparation and characterization of antibacterial orthodontic resin containing silver nanoparticles. *Appl. Surf. Sci.* **2018**, *432*, 317–323. [CrossRef]
21. Popkin, D.L.; Zilka, S.; Dimaano, M.; Fujioka, H.; Rackley, C.; Salata, R.; Griffith, A.; Mukherjee, P.K.; Ghannoum, M.A.; Esper, F. Cetylpyridinium Chloride (CPC) Exhibits Potent, Rapid Activity Against Influenza Viruses in vitro and in vivo. *Pathog. Immun.* **2017**, *2*, 253. [CrossRef]
22. Melo, M.A.S.; Cheng, L.; Zhang, K.; Weir, M.D.; Rodrigues, L.K.A.; Xu, H.H.K. Novel dental adhesives containing nanoparticles of silver and amorphous calcium phosphate. *Dent. Mater.* **2013**, *29*, 199–210. [CrossRef]
23. Das, J.; Ghosh, J.; Manna, P.; Sil, P.C. Taurine provides antioxidant defense against NaF-induced cytotoxicity in murine hepatocytes. *Pathophysiology* **2008**, *15*, 181–190. [CrossRef] [PubMed]
24. Lei, X.; Wang, J.; Chen, J.; Gao, J.; Zhang, J.; Zhao, Q.; Tang, J.; Fang, W.; Li, J.; Li, Y.; et al. The in vitro evaluation of antibacterial efficacy optimized with cellular apoptosis on multi-functional polyurethane sealers for the root canal treatment. *J. Mater. Chem. B* **2021**, *9*, 1370–1383. [CrossRef] [PubMed]
25. Yang, N.; Chen, Y.; Dan, N.; Zheng, X.; Feng, R.; Yu, G.; He, X.; Dan, W.; Wang, Y. Flexible nano-piezoelectric membranes with spontaneous electric field generation for bacteria elimination and wound healing. *J. Mater. Sci.* **2022**, *57*, 19532–19552. [CrossRef]
26. Montoya, C.; Du, Y.; Gianforcaro, A.L.; Orrego, S.; Yang, M.; Lelkes, P.I. On the road to smart biomaterials for bone research: Definitions, concepts, advances, and outlook. *Bone Res.* **2021**, *9*, 12. [CrossRef]
27. Buschanga, P.H.; Hayasakib, H.; Throckmorton, G.S. Quantication of human chewing-cycle kinematics. *Arch. Oral Biol.* **2000**, *45*, 461–474. [CrossRef]
28. Jiang, Y.; Parsonnet, E.; Qualls, A.; Zhao, W.; Susarla, S.; Pesquera, D.; Dasgupta, A.; Acharya, M.; Zhang, H.; Gosavi, T.; et al. Enabling ultra-low-voltage switching in $BaTiO_3$. *Nat. Mater.* **2022**, *21*, 779–785. [CrossRef]
29. Montoya, C.; Jain, A.; Londoño, J.J.; Correa, S.; Lelkes, P.I.; Melo, M.A.; Orrego, S. Multifunctional Dental Composite with Piezoelectric Nanofillers for Combined Antibacterial and Mineralization Effects. *ACS Appl. Mater. Interfaces* **2021**, *13*, 43868–43879. [CrossRef]

30. Wei, J.; Xia, J.; Liu, X.; Ran, P.; Zhang, G.; Wang, C.; Li, X. Hollow-structured $BaTiO_3$ nanoparticles with cerium-regulated defect engineering to promote piezocatalytic antibacterial treatment. *Appl. Catal. B Environ.* **2023**, *328*, 122520. [CrossRef]
31. Zheng, T.; Yu, Y.; Pang, Y.; Zhang, D.; Wang, Y.; Zhao, H.; Zhang, X.; Leng, H.; Yang, X.; Cai, Q. Improving bone regeneration with composites consisting of piezoelectric poly(l-lactide) and piezoelectric calcium/manganese co-doped barium titanate nanofibers. *Compos. Part B Eng.* **2022**, *234*, 109734. [CrossRef]
32. Tang, J.; Zhang, Y.; Fang, W.; Man, Y.; Zhang, J.; Zhao, Q.; Lei, X.; Chen, J.; Li, J.; Li, Y.; et al. Electrophysiological microenvironment and site-specific cell behaviors regulated by fibrous aniline trimer-based polyurethanes in bone progressive regeneration. *Chem. Eng. J.* **2023**, *460*, 141630. [CrossRef]
33. Fang, W.; Sun, F.H.; Tang, J.J.; Zhao, Q.; Chen, J.; Lei, X.Y.; Zhang, J.Z.; Zhang, Y.L.; Zuo, Y.; Li, J.D.; et al. Porous Electroactive and Biodegradable Polyurethane Membrane through Self-Doping Organogel. *Macromol. Rapid Commun.* **2021**, *42*, 2100125. [CrossRef] [PubMed]
34. Chen, J.; Guo, B.; Eyster, T.W.; Ma, P.X. Super Stretchable Electroactive Elastomer Formation Driven by Aniline Trimer Self-Assembly. *Chem. Mater.* **2015**, *27*, 5668–5677. [CrossRef] [PubMed]
35. Huang, H.-Y.; Huang, T.-C.; Lin, J.-C.; Chang, J.-H.; Lee, Y.-T.; Yeh, J.-M. Advanced environmentally friendly coatings prepared from amine-capped aniline trimer-based waterborne electroactive polyurethane. *Mater. Chem. Phys.* **2013**, *137*, 772–780. [CrossRef]
36. Xie, M.; Wang, L.; Ge, J.; Guo, B.; Ma, P.X. Strong Electroactive Biodegradable Shape Memory Polymer Networks Based on Star-Shaped Polylactide and Aniline Trimer for Bone Tissue Engineering. *ACS Appl. Mater. Interfaces* **2015**, *7*, 6772–6781. [CrossRef] [PubMed]
37. Porter, M.L.; Bertó, A.; Primus, C.M.; Watanabe, I. Physical and Chemical Properties of New-generation Endodontic Materials. *J. Endod.* **2010**, *36*, 524–528. [CrossRef]
38. *ISO 6876*; Dentistry—Root Canal Sealing Materials. Standardization IOf. International Organization for Standardization: Geneva, Switzerland, 2012.
39. Radev, B.R.; Kase, J.A.; Askew, M.J.; Weiner, S.D. Potential for thermal damage to articular cartilage by PMMA reconstruction of a bone cavity following tumor excision: A finite element study. *J. Biomech.* **2009**, *42*, 1120–1126. [CrossRef] [PubMed]
40. Wang, J.; Mei, Q.; Lin, L.; Sun, F.; Li, J.; Zou, Q.; Zuo, Y.; Li, Y. A comparison of the characteristics of polyurethane-based sealers including various antimicrobial agents. *RSC Adv.* **2019**, *9*, 7043–7056. [CrossRef]
41. Williams, C.; Loushine, R.J.; Weller, R.N.; Pashley, D.H.; Tay, F.R. A Comparison of Cohesive Strength and Stiffness of Resilon and Gutta-Percha. *J. Endod.* **2006**, *32*, 553–555. [CrossRef]
42. Zhao, X.; Dong, R.; Guo, B.; Ma, P.X. Dopamine-Incorporated Dual Bioactive Electroactive Shape Memory Polyurethane Elastomers with Physiological Shape Recovery Temperature, High Stretchability, and Enhanced C2C12 Myogenic Differentiation. *ACS Appl. Mater. Interfaces* **2017**, *9*, 29595–29611. [CrossRef]
43. Moini, N.; Jahandideh, A.; Shahkarami, F.; Kabiri, K.; Piri, F. Linear and star-shaped π-conjugated oligoanilines: A review on molecular design in syntheses and properties. *Polym. Chem.* **2022**, *13*, 2714–2756. [CrossRef]
44. Landis, R.F.; Li, C.-H.; Gupta, A.; Lee, Y.-W.; Yazdani, M.; Ngernyuang, N.; Altinbasak, I.; Mansoor, S.; Khichi, M.A.S.; Sanyal, A.; et al. Biodegradable Nanocomposite Antimicrobials for the Eradication of Multidrug-Resistant Bacterial Biofilms without Accumulated Resistance. *J. Am. Chem. Soc.* **2018**, *140*, 6176–6182. [CrossRef]
45. Dai, C.; Lin, J.; Li, H.; Shen, Z.; Wang, Y.; Velkov, T.; Shen, J. The Natural Product Curcumin as an Antibacterial Agent: Current Achievements and Problems. *Antioxidants* **2022**, *11*, 459. [CrossRef]
46. Yang, L.; Tian, B.; Xie, Y.; Dong, S.; Yang, M.; Gai, S.; Lin, J. Oxygen-Vacancy-Rich Piezoelectric BiO_{2-x} Nanosheets for Augmented Piezocatalytic, Sonothermal, and Enzymatic Therapies. *Adv. Mater.* **2023**, *35*, 2300648. [CrossRef] [PubMed]
47. Yu, C.; Li, J.; Zang, P.; Feng, L.; Tian, B.; Zhao, R.; Xie, Y.; Wu, L.; Chen, Z.; Yang, P. A Functional "Key" Amplified Piezoelectric Effect Modulates ROS Storm with an Open Source for Multimodal Synergistic Cancer Therapy. *Small* **2024**, *20*, 2401931. [CrossRef] [PubMed]
48. Lay, R.; Deijs, G.S.; Malmström, J. The intrinsic piezoelectric properties of materials—A review with a focus on biological materials. *RSC Adv.* **2021**, *11*, 30657–30673. [CrossRef] [PubMed]
49. Wu, M.; Zhang, Z.; Liu, Z.; Zhang, J.; Zhang, Y.; Ding, Y.; Huang, T.; Xiang, D.; Wang, Z.; Dai, Y.; et al. Piezoelectric nanocomposites for sonodynamic bacterial elimination and wound healing. *Nano Today* **2021**, *37*, 101104. [CrossRef]
50. Mokhtari, F.; Azimi, B.; Salehi, M.; Hashemikia, S.; Danti, S. Recent advances of polymer-based piezoelectric composites for biomedical applications. *J. Mech. Behav. Biomed. Mater.* **2021**, *122*, 104669. [CrossRef]
51. Li, J.; Zhao, X.; Xia, Y.; Qi, X.; Jiang, C.; Xiao, Y.; Jiang, F.; Jiang, X.; Yuan, G. Strontium-Containing Piezoelectric Biofilm Promotes Dentin Tissue Regeneration. *Adv. Mater.* **2024**, *36*, 2313419. [CrossRef]
52. Kim, J.S.; Nam, I.W.; Lee, H.K. Piezoelectric characteristics of urethane composites incorporating piezoelectric nanomaterials. *Compos. Struct.* **2020**, *241*, 112072. [CrossRef]

53. Sirelkhatim, A.; Mahmud, S.; Seeni, A.; Kaus, N.H.M.; Ann, L.C.; Bakhori, S.K.M.; Hasan, H.; Mohamad, D. Review on Zinc Oxide Nanoparticles: Antibacterial Activity and Toxicity Mechanism. *Nano-Micro Lett.* **2015**, *7*, 219–242. [CrossRef]
54. Wang, Z.L. Zinc oxide nanostructures: Growth, properties and applications. *J. Phys. Condens. Matter* **2004**, *16*, R829–R858. [CrossRef]
55. Look, D.C.; Reynolds, D.C.; Sizelove, J.R.; Jones, R.L.; Litton, C.W.; Cantwell, G.; Harsch, W.C. Electrical Properties of Bulk ZnO. *Solid State Commun.* **1998**, *105*, 399–401. [CrossRef]
56. Na, C.W.; Woo, H.-S.; Kim, I.-D.; Lee, J.-H. Selective detection of NO_2 and C_2H_5OH using a Co_3O_4-decorated ZnO nanowire network sensor. *Chem. Commun.* **2011**, *47*, 5148–5150. [CrossRef]
57. Sadykov, T.T.; Massalimov, I.A.; Mustafin, A.G. Synthesis and physico-chemical properties of composites based on polyaniline and nanosized sulfur. *Polym. Bull.* **2023**, *81*, 2757–2776. [CrossRef]
58. Marin, E.; Boschetto, F.; Sunthar, T.P.M.; Zanocco, M.; Ohgitani, E.; Zhu, W.; Pezzotti, G. Antibacterial effects of barium titanate reinforced polyvinyl-siloxane scaffolds. *Int. J. Polym. Mater. Polym. Biomater.* **2020**, *70*, 425–436. [CrossRef]
59. Shah, A.A.; Khan, A.; Dwivedi, S.; Musarrat, J.; Azam, A. Antibacterial and Antibiofilm Activity of Barium Titanate Nanoparticles. *Mater. Lett.* **2018**, *229*, 130–133. [CrossRef]
60. Swain, S.; Padhy, R.N.; Rautray, T.R. Polarized piezoelectric bioceramic composites exhibit antibacterial activity. *Mater. Chem. Phys.* **2020**, *239*, 122002. [CrossRef]
61. Padmavathy, N.; Vijayaraghavan, R. Enhanced bioactivity of ZnO nanoparticles—An antimicrobial study. *Sci. Technol. Adv. Mater.* **2016**, *9*, 035004. [CrossRef]
62. Rumbaugh, K.P.; Sauer, K. Biofilm dispersion. *Nat. Rev. Microbiol.* **2020**, *18*, 571–586. [CrossRef]
63. Gao, X.; Liu, Y.; Li, Y.; Jin, B.; Jiang, P.; Chen, X.; Wei, C.; Sheng, J.; Liu, Y.-N.; Li, J.; et al. Piezoelectric Nanozyme for Dual-Driven Catalytic Eradication of Bacterial Biofilms. *ACS Appl. Mater. Interfaces* **2023**, *15*, 14690–14703. [CrossRef]
64. Wei, W.; Bing, W.; Ren, J.; Qu, X. Near infrared-caged d-amino acids multifunctional assembly for simultaneously eradicating biofilms and bacteria. *Chem. Commun.* **2015**, *51*, 12677–12679. [CrossRef] [PubMed]
65. Faria-Junior, N.B.; Tanomaru-Filho, M.; Berbert, F.L.; Guerreiro-Tanomaru, J.M. Antibiofilm activity, pH and solubility of endodontic sealers. *Int. Endod. J.* **2013**, *46*, 755–762. [CrossRef]
66. Zhang, H.; Shen, Y.; Ruse, N.D.; Haapasalo, M. Antibacterial activity of endodontic sealers by modified direct contact test against Enterococcus faecalis. *J. Endod.* **2009**, *35*, 1051–1055. [CrossRef] [PubMed]
67. Yuan, X.; Shi, J.; Kang, Y.; Dong, J.; Pei, Z.; Ji, X. Piezoelectricity, Pyroelectricity, and Ferroelectricity in Biomaterials and Biomedical Applications. *Adv Mater* **2024**, *36*, e2308726. [CrossRef] [PubMed]
68. Strzyz, P. Synthetic biology: Designer cells tackle diabetes. *Nat. Rev. Mol. Cell Biol.* **2017**, *18*, 69. [CrossRef]
69. Imlay, J.A. The molecular mechanisms and physiological consequences of oxidative stress: Lessons from a model bacterium. *Nat. Rev. Microbiol.* **2013**, *11*, 443–454. [CrossRef]
70. Seaver, L.C.; Imlay, J.A. Alkyl hydroperoxide reductase is the primary scavenger of endogenous hydrogen peroxide in Escherichia coli. *J. Bacteriol.* **2001**, *183*, 7173–7181. [CrossRef]
71. Li, L.; Zuo, Y.; Zou, Q.; Yang, B.; Lin, L.; Li, J.; Li, Y. Hierarchical Structure and Mechanical Improvement of an n-HA/GCO–PU Composite Scaffold for Bone Regeneration. *ACS Appl. Mater. Interfaces* **2015**, *7*, 22618–22629. [CrossRef]
72. Zhang, Y.; Tang, J.; Fang, W.; Zhao, Q.; Lei, X.; Zhang, J.; Chen, J.; Li, Y.; Zuo, Y. Synergetic Effect of Electrical and Topographical Cues in Aniline Trimer-Based Polyurethane Fibrous Scaffolds on Tissue Regeneration. *J. Funct. Biomater.* **2023**, *14*, 185. [CrossRef]
73. Ning, X.; Hao, A.; Cao, Y.; Hu, J.; Xie, J.; Jia, D. Effective promoting piezocatalytic property of zinc oxide for degradation of organic pollutants and insight into piezocatalytic mechanism. *J. Colloid Interface Sci.* **2020**, *577*, 290–299. [CrossRef] [PubMed]
74. Zhang, W.; Feng, Q.; Hosono, E.; Asakura, D.; Miyawaki, J.; Harada, Y. Tetragonal Distortion of a $BaTiO_3/Bi_{0.5}Na_{0.5}TiO_3$ Nanocomposite Responsible for Anomalous Piezoelectric and Ferroelectric Behaviors. *ACS Omega* **2020**, *5*, 22800–22807. [CrossRef] [PubMed]
75. Wu, M.; Tao, B.; Wang, T.; Zhang, Y.; Wei, W.; Wang, C. Fast-setting and anti-washout tricalcium silicate/disodium hydrogen phosphate composite cement for dental application. *Ceram. Int.* **2019**, *45*, 24182–24192. [CrossRef]
76. *ISO 10993*; Biological Evaluation of Medical Devices. Standardization IOf. International Organization for Standardization: Geneva, Switzerland, 2019.

gels

MDPI

Article

Development of Liposome-Based Hydrogel Patches Incorporating Essential Oils of African Plants and Deep Eutectic Solvents

Wanhang Jiang [1], Sara Toufouki [1], Subhan Mahmood [1], Ali Ahmad [1], Alula Yohannes [2], Yang Xiang [3,*] and Shun Yao [1,*]

[1] School of Chemical Engineering, Sichuan University, Chengdu 610065, China; wanhang_jiang@126.com (W.J.); sara_toufouki@126.com (S.T.); engnr.chem.subhan@gmail.com (S.M.); aliahmad@stu.scu.edu.cn (A.A.)
[2] College of Natural Science, Wolkite University, Wolkite City P.O. Box 07, Ethiopia; alulayhnns@gmail.com
[3] Children's Drug Research Institute, Jianmin Group, Wuhan 430052, China
* Correspondence: xiangyangjmgroup@126.com (Y.X.); cusack@scu.edu.cn (S.Y.); Tel./Fax: +86-28-85405221 (S.Y.)

Abstract: A nanoliposome-integrated polymeric hydrogel was developed for the controlled release of essential oils (Argania spinosa, Passiflora edulis). A deep eutectic solvent (DES) composed of betaine and phytic acid enhanced the solubility and stability of essential oils, facilitating uniform encapsulation within nanoliposomes. The hydrogel exhibited a swelling capacity of 100% and retained 51.7% of water after 7 h, ensuring prolonged hydration. Structural analysis confirmed a homogeneous dispersion of nanoliposomes, contributing to the gradual release of bioactive components. Additionally, the hydrogel demonstrated high mechanical strength (7.5 MPa), ensuring flexibility and durability. The polymeric network, formed by acrylamide, sodium alginate, and bentonite, provided a stable and elastic matrix, optimizing water retention and mechanical performance. The controlled diffusion mechanism of the nanoliposomes was validated through in vitro release studies, indicating Fickian-controlled release behavior. These findings highlight the potential of this polymeric hydrogel system as a functional material for skincare formulations, offering enhanced hydration and sustained bioactive delivery.

Keywords: hydrogel; nanoliposomes; essential oils; deep eutectic solvents; skin dressing; multifunctions

1. Introduction

As the largest human organ (surface area >2 m^2 for adults), skin serves as the primary protective barrier. Its critical functions include maintaining homeostasis through thermoregulation, UV radiation protection, and defense against physical/chemical/microbial threats. Skin health represents a critical indicator of personal health [1]. As a prevalent dermatological manifestation, dry skin arises from compromised stratum corneum hydration homeostasis. Clinically characterized by desquamation and fine rhytides, this condition may predispose to premature skin aging through mechanisms involving immune barrier dysfunction and epidermal barrier integrity impairment. As the skin ages, it not only loses its biomechanical strength but also becomes increasingly prone to environmental damage [2], which in turn impairs its ability to heal wounds [2,3]. In addition, the skin also faces various problems such as dull color, dandruff, scars, wrinkles, bedsores, inflammation, etc., which constitute the needs of daily care for the body surface. Combined with growing consumer preferences for natural products, the continuous enhancement of hygiene standards within the cosmetics and health and wellness industries has prompted both domestic and international companies in the field of daily chemical products to intensify their research and development efforts on plant-derived ingredients [4]. Compared to those artificially synthesized components, active natural products are more friendly and safe and often have more comprehensive activities/effects to meet multiple nursing requirements.

In recent years, the use of plant-derived extracts in modern medicine and skincare has gained significant attention, while essential oils (EOs) are widely used in medicine, food, and cosmetics due to their antibacterial, antioxidant, anti-inflammatory, antiviral, moisturizing, and nourishing properties [5]. Among plant-derived oils, nut oils (e.g., argan oil) and fruit seed oils (e.g., passion fruit seed oil) have garnered increasing interest due to their richness in essential fatty acids, vitamins, and antioxidants. From North Africa, argan oil (Argania spinosa, AO) consists primarily of oleic and linoleic acids and is known for its exceptional moisturizing and reparative properties [6]. Meanwhile, passion fruit seed oil (Passiflora edulis, PFSO) from South Africa is rich in vitamins A and C [7], which help strengthen the skin barrier and provide soothing and antioxidant effects. However, the high volatility, instability, and penetration limitations of these essential oils hinder their broader application in skincare formulations [8]. To address these limitations, advanced nano-delivery systems, such as nano-liposome technology, have been explored to enhance their stability and bioavailability. Up to now, the EOs of rosemary, lemon eucalyptus, artemisia annua, and evening primrose multiflora [9] have been successfully encapsulated, whose compatibility and activities were significantly enhanced. These systems help reduce the exposure of active ingredients to environmental factors (e.g., water, oxygen, and light), minimize evaporation losses, and improve transdermal efficiency [10]. However, the strategy of encapsulating essential oils in nanoliposomes alone faces a series of challenges such as low water content, poor storage stability, organic solvent residue, and limited biocompatibility [11], etc. Besides that, the combinational use of EOs is rarely explored, which can enable them to complement each other's strengths and weaknesses and play a synergistic role.

Hydrophilic gels, also known as hydrogels, are three-dimensional polymer networks capable of retaining large amounts of water [12], which have exhibited reversible physicochemical changes in response to external stimuli such as pH, temperature, ionic strength, enzymatic activity, and electric fields [12]. Depending on various applications, they can be formulated into various physical forms, including slabs, microparticles, nanoparticles, coatings, patches, and films. Compared to conventional formulations such as gels, creams, and emulsions for skincare, hydrogels exhibit superior capabilities in maintaining sustained therapeutic concentrations of active ingredients at targeted sites [13], enabling precise topical delivery with enhanced percutaneous penetration efficacy [12]. Hydrogels demonstrate superior moisture retention capabilities compared to cream while also allowing for unrestricted cellular migration. Additionally, their 3D polymeric structure offers enhanced tissue protection, in contrast to the lipid-based barrier properties characteristic of conventional creams [8]. Especially, a paper-free, sprayable skin mask was successfully developed, composed of a thermogel formed from nicotinamide-loaded biodegradable amphiphilic block copolymers [10]. Nevertheless, the solubility of essential oils in hydrogels is limited, which restricts the further development of their applications. As an effective green medium, deep eutectic solvents (DESs) may offer a promising solution to address this limitation [14].

As a class of low-melting-point mixtures composed of hydrogen bond donors and acceptors, DESs have recently garnered attention due to their designability, multifunction, and biocompatibility. In pharmaceutical applications, DESs exhibit exceptional drug solubilization capabilities, enhancing the solubility of poorly water-soluble active pharmaceutical ingredients (APIs) by 6–2200-fold compared to conventional solvents such as DMSO while also mitigating toxicity concerns [15]. Their tunable physicochemical properties, governed by the component ratios and hydrogen-bonding networks, enable the tailored synthesis of biodegradable polymers (e.g., poly(diol-co-citrate)) under mild conditions and the stabilization of thermolabile APIs [16]. DESs also advance functional nanomaterials by optimizing biocompatibility and drug-loading efficiency in graphene/DNA hybrids [17] and facilitating the green synthesis of antimicrobial metal–organic frameworks [18]. However, there are several aspects worthy of research and improvement, such as structural instability under aqueous conditions (where >50% water disrupts supramolecular organization) [19], as well as the effective use of those DESs with multifunctions (e.g., as active component, solvent, penetration enhancer, film-forming aid simultaneously, etc.) [20]. The existing challenges highlight the need for innovative delivery systems to harness the advantages of DESs. Our study introduces liposome-based hydrogel patches integrating DESs and African plant essential oils, using liposomal encapsulation to stabilize DESs against hydration-induced degradation, and synergizing with essential oils to enhance transdermal permeation.

As two famous representatives of African EOs, AO and PFSO were combined to be applied in this study for the first time, which were further solubilized with a DES composed of betaine and phytic acid and then encapsulated within nanoliposomes to enhance their stability and activities. The designed system aimed to facilitate their use and skin absorption and prolong their effectiveness, thereby providing new strategies and scientific foundations for the future development of skincare formulations. Guided by this strategy, the preparation of argan oil–passion fruit oil mixtures and their basic properties as well as key bioactivities were first investigated, and, then, the method was developed for the preparation of nanoliposomes encapsulating essential oils and DESs. After necessary characterizations, the obtained nanoliposomes were used in the preparation of hydrogel patches as designed. Finally, the comprehensive measurements and study on releasing behaviors of the hydrogel were performed. The incorporation of nanoliposomes was expected to effectively enhance the stability of the bioactive components within the hydrogels and improve their skin permeability and bioavailability. This whole work aimed to provide a useful strategy and scientific basis for advancing EO-involved skincare systems with multifunctions and good properties.

2. Results and Discussion

2.1. Spectral and Chromatographic Analysis of DES

The prepared betaine-phytic acid (1:3) is a yellow brown uniformly transparent DES [21], chosen for its optimal balance between enhancing skin permeability, stability, and controlled release. This DES also offers excellent biocompatibility, minimal skin irritation, and synergistic effects from the antioxidant properties of phytic acid and the skin-conditioning benefits of betaine, making it ideal for transdermal applications. Due to the lack of more effective UV-absorbing groups, it mainly produces strong absorbance around 200 nm (see Figure 1a), which provides a basis for the selection of detection wavelength and mobile phase in the following liquid chromatography quantification. Compared to the FT-IR spectra of its H-bond acceptor and donor as two components (phytic acid: the stretching vibration of O–H around 3350 cm^{-1}, the stretching vibration of P=O at 1646 cm^{-1}, the stretching vibration of P–O at 1125 cm^{-1}, and the stretching vibration of C–O at 1143 cm^{-1}; betaine: the stretching vibration of O–H around 3399 cm^{-1}, the stretching vibration of C=O at 1615 cm^{-1}, the bending vibration of C–H at 1478 cm^{-1}, and the stretching vibration of C–N at 1053 cm^{-1}; see Figures S1 and S2 in Supplemental Materials), Figure 1b indicates that the stretching vibration of P=O was shifted to 1636 cm^{-1} for the effect of intermolecular H-bonding, and the stretching vibration of C–N was shifted to 1058 cm^{-1} due to the electronegativity of anions together with the induction effect. In addition, the signals of phytic acid were obviously much stronger than those of betaine due to the high proportion of the former. The above signal attribution was based on previous studies [22,23].

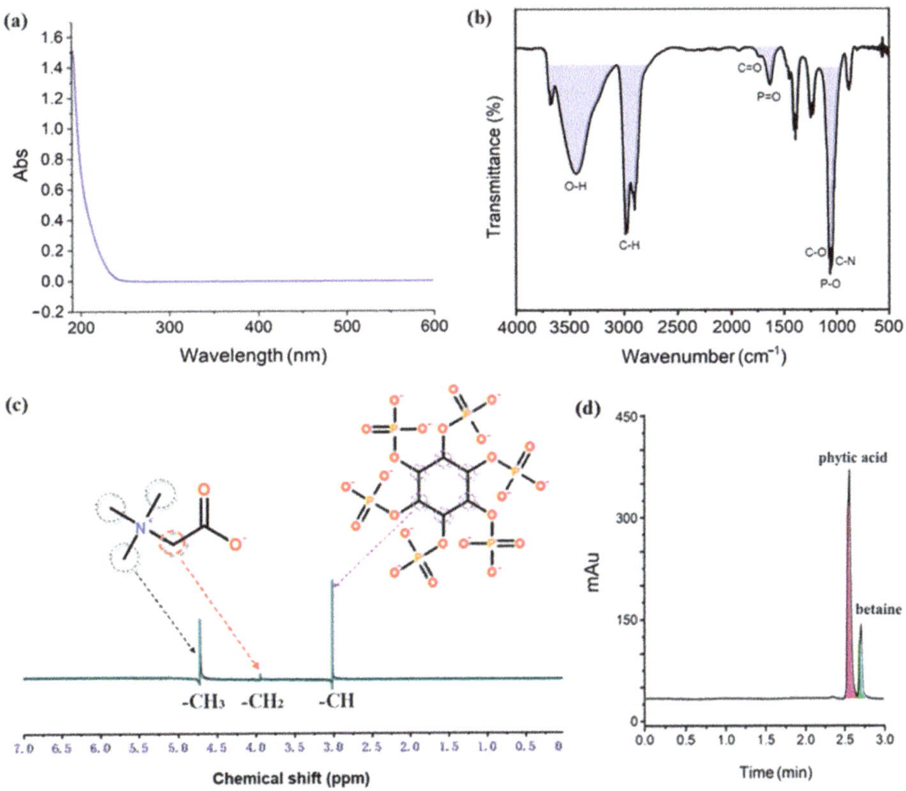

Figure 1. UV spectrum (**a**), FT-IR spectrum (**b**), 1H-NMR spectrum (**c**), and HPLC chromatogram (**d**) of the prepared DES.

For the ^{1}H-NMR spectrum (400 MHz, D_2O) of betaine-phytic acid (1:3) in Figure 1c, all the proton signals of the H-bond acceptor and donor, except the active hydrogen, appeared as expected, and no new signals were found. It proved that no chemical reactions occurred between them, and the DES formation only relied on their intermolecular interactions. In detail, the highest signal at 4.70 ppm belonged to $-CH_3$ groups connected to nitrogen cation in betaine, the shortest peak at 3.93 ppm was contributed by $-CH_2$ connected to nitrogen cation in betaine, and the medium peak at 3.03 ppm resulted from –CH groups in the ring of phytic acid. Due to not being key sites of intermolecular interactions, their chemical shift change was not significant.

Finally, the high-performance liquid chromatographic conditions for the prepared DES were explored. It was found that phytic acid and betaine were eluted successively instead of just one chromatographic peak appearing (see Figure 1d). It also proved that the formation of DES did not represent the synthesis of a new compound. Moreover, the flow rate had little effect on their peak resolution, so it was not taken as the object of investigation. The conditions were first attempted when the flow rate was 1.0 mL/min, the column temperature was 30 °C, and the mobile phase was a mixture of UP water and acetonitrile. The result indicated that, as the proportion of water in the mobile phase increased from 90%, the peaks corresponding to phytic acid and betaine were gradually separated. When the proportion of water reached 97%, the resolution of phytic acid and betaine was the best, and they could achieve baseline separation. When the mobile phase was changed to the mixture of UP water and methanol, the resolution became poorer compared to that resulting from the mobile phase of water–acetonitrile in the same proportion; the corresponding peaks of phytic acid and betaine partially overlapped. Besides that, methanol has strong end absorbance around 200 nm. Therefore, acetonitrile was finally chosen to replace methanol in the mobile phase. When the flow rate was set to 1.0 mL/min and the mobile phase was UP water–acetonitrile (97:3), the resolution of chromatographic peaks of phytic acid and betaine improved with increasing column temperature from 20 °C to 35 °C. The resolution improvement was not significant at higher temperatures, so the column temperature was set to 35 °C. As a result, the standard curves of betaine and phytic acid were $y = 1.6621x + 0.0156$ ($R^2 = 0.9990$) and $y = 1.1847x + 0.0015$ ($R^2 = 0.9991$), where y was the peak area, and x was the DES concentration (mg/mL).

2.2. *Basic Properties and Activities of EO Mixtures*

2.2.1. Basic Properties

It should be noted that the following research can not only provide basic data for the direct use of these essential oils but also is closely related to their preparation into nanoliposomes. As shown in Figure 2A, the appearance of all the samples containing blends of AO and PFSO in different proportions exhibited slight differences. Notably, the argan oil appeared more yellowish in color compared to the passion fruit seed oil. Moreover, the pH results in Figure 2B accord with the fact that most natural essential oils are acidic with a pH value between 5 and 8. The normal pH environment of the skin is usually between 4.5 and 5.5, which is a slightly acidic condition. The weakly acidic pH can help maintain the health of the skin. The preservation of a healthy stratum corneum (SC) and the integrity of the epidermal barrier are primarily dependent on the acidic pH of the skin surface. The homeostasis of the epidermal barrier, the cohesion, and integrity of the SC, the facilitation of proteolytic processes involved in desquamation, the metabolism of extracellular lipids, and the microbial colonization of the SC are all influenced by the acidity of the skin's outermost layer. Both exogenous and endogenous factors contribute to variations in skin pH, which in turn affects the overall function of skin [24].

The density of natural essential oils is usually between 0.78 and 0.84 g/cm^3, and there are also some special essential oils with densities higher than this range (especially those from fruits and kernels), such as cinnamon, cloves, rosehip, and sassafras oils, whose densities are close to or exceed 1 g/cm^3 (density of water). As shown in Figure 2C, the density of the oil mixtures at various ratios is generally in the range of 0.9–1.0 g/cm^3, which makes this type of essential oil prone to emulsification with water. At the same time, the density values of AO, PFSO, and their mixtures do not exhibit significant differences. The density of commercial argan oil was found to be consistent with the value reported by Taribak et al. [25]. Further analysis revealed that an increase in the concentration of passion fruit oil led to a rise in the overall density of the mixture.

In skincare products, viscosity is a critical factor that influences the texture, consistency, and overall sensory experience of the formulation. It affects the ease of application, absorption, and, ultimately, the efficacy of the product. An ideal gel formulation should strike a balance between sufficient fluidity to facilitate smooth application to the target area and adequate viscosity to ensure the maintenance of an effective concentration at the application site over time [25]. At around 20 °C, the apparent viscosity range of common plant EOs is roughly between 50 and 150 mPa·s, and some essential oils have lower viscosity (e.g., 35.4 mPa·s of coconut oil [26]). Here, the apparent viscosity of AO, PFSO, and their mixtures was evaluated, with the results presented in Figure 2D. The apparent viscosity values increased with higher concentrations of PFSO, which is the most viscous component among the samples tested. Specifically, PFSO exhibited an apparent viscosity of 220 mPa·s, significantly higher than that of argan oil (83 mPa·s). For the mixture at the 2:1 ratio, the apparent viscosity decreased to 153 mPa·s, while, for the 1:2 ratio, the apparent viscosity increased as the proportion of argan oil rose. These findings clearly demonstrate that AO has a lower apparent viscosity and that the overall apparent viscosity of the mixture depends on the relative proportions of the two EOs. Dissolving and diluting them with DES is beneficial for uniform dispersion and subsequent processing.

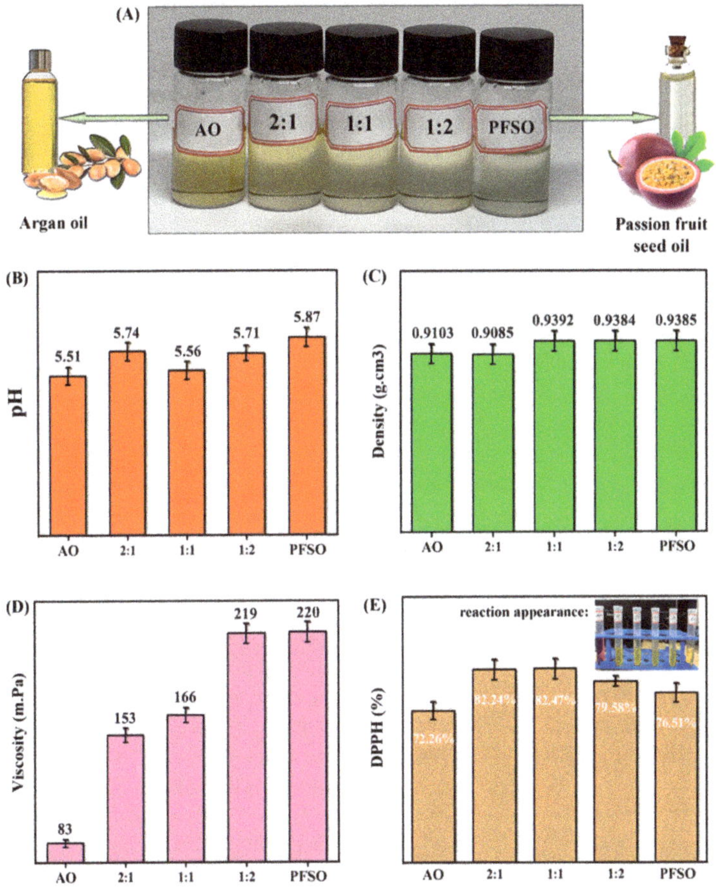

Figure 2. (**A**) Appearance of the AO-PFSO mixtures; (**B**) pH of the oil mixtures at different ratios; (**C**) densities of the oil mixtures at different ratios at 25 °C; (**D**) apparent viscosity values of the oil mixtures at different ratios at 25 °C; (**E**) percentage of free radical scavenging capacity of the oil mixtures. The inset shows the decolorization of the oil samples from purple to yellow in the DPPH assay, indicating the degree of antioxidant reduction.

2.2.2. Typical Bioactivities

The natural essential oils are well recognized for their antioxidant activity, which can help the skin resist the invasion of free radicals, delay skin aging, and promote skin repair and regeneration. Consequently, plant EOs are often regarded as a promising source for the development of novel bioactive components with potential applications in medicine, pharmaceuticals, cosmetics, and other industries [27]. In particular, argan oil is a rich source of essential polyunsaturated fatty acids, including high levels of oleic and linoleic acids. It is also notably abundant in tocopherols and polyphenols, both of which exhibit potent antioxidant properties [6]. Here, the antioxidant activity of mixtures of AO and PFEO is shown in Figure 2E, which was assessed using the stable free radical DPPH. Antioxidants, such as those found in these oils, play a crucial role in protecting the skin from damage and mitigating the skin aging process. It can be found that all the samples demonstrated the ability to reduce DPPH to yellow-colored diphenylpicrylhydrazine, with varying degrees of effectiveness. The extent of discoloration (from purple to yellow) correlates with antioxidant activity, with the strongest antioxidant activities (82.47% and 82.24%) observed in the ratios of 1:1 and 2:1, respectively, which were higher than those of the two essential oils alone. This reflects a certain degree of synergistic effect. For comparison, almond oil exhibited good antioxidant activity (e.g., 73.13%) as a component commonly used in skincare formulations [28], which was also noteworthy in relation to the EOs tested in this study.

Calcium is essential for regulating skin function, playing a significant role in the epidermal barrier [29,30]. Chemical exfoliation with strong calcium chelation activity can effectively rejuvenate skin cells and reduce enlarged conspicuous facial pores, which will make skin radiant and tender. Figure 3A,B show the actual status of the EO samples before and after titration. As indicated by the titration results in Figure 3C (details in Table S1 of Supplemental Materials), the mass of residual calcium present in each sample varies. Passion fruit oil exhibits the highest concentration of calcium ions, followed by the 1:2 mixed EOs, argan oil, the 2:1 mixed EOs, and, finally, the 1:1 mixed EOs, which have the lowest residual concentration of Ca^{2+}. It indicated that the keratinization process of skin results could be blocked by virtue of the combination of 1:1 mixed EOs with calcium in pores; thus, the skin pores could be minimized temporarily. The increasing proportion of both AO and PFSO in the equivalent mixtures will lead to a higher concentration of calcium ions, weakening the performance of softening the skin.

Some studies have shown that EOs can be incorporated into skincare products to block UV radiation. In particular, oily substances are effective at forming a long-lasting sunscreen layer on the skin. Additionally, the moisturizing properties of oils help protect the skin from dryness caused by environmental factors such as wind and sun exposure. The sun protection factor (SPF) of the EOs samples was evaluated spectrophotometrically in vitro under UV-B light wavelengths (290–320 nm). Ultraviolet (UV) radiation, a component of the electromagnetic spectrum, is classified into three types: UV-A (320–400 nm), UV-B (290–320 nm), and UV-C (200–290 nm). UV-C radiation is effectively filtered by the stratospheric ozone layer and does not reach the Earth's surface. In contrast, UV-A and UV-B rays penetrate the ozone layer and reach the surface, with the atmosphere filtering a greater proportion of UV-A radiation compared to UV-B. UV-B radiation, spanning from 290 to 320 nm, is one of the types of ultraviolet rays that can reach the Earth's surface, although much of it is attenuated by the atmosphere. These rays play a significant role in the synthesis of melanin and the induction of sunburn. Therefore, protection from UV-B radiation is critical for preserving skin integrity and preventing damage [31–33].

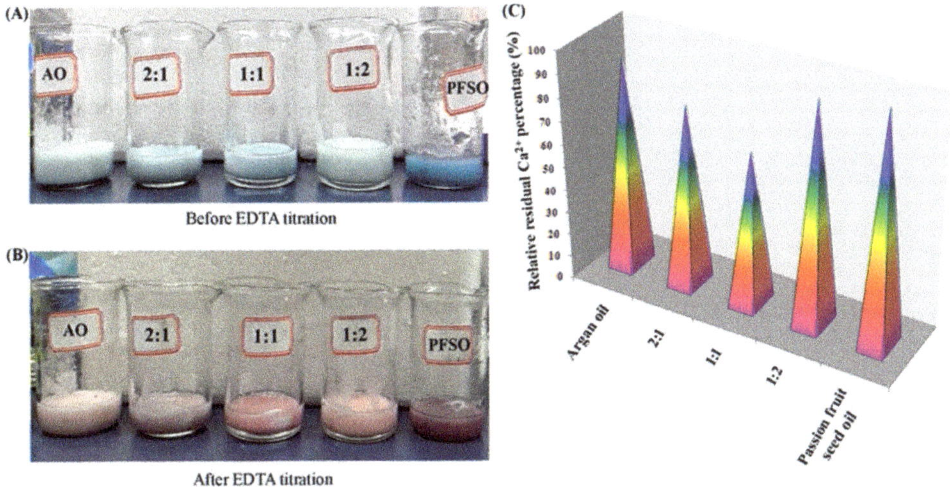

Figure 3. (**A**) Photographs of oil samples before and (**B**) after EDTA titration, from left to right: AO, 2:1 ratio, 1:1 ratio, 1:2 ratio, and PFSO. (**C**) Mass of calcium ions present in the oil samples. Note: the maximum level is treated as 100% to calculate the relative residual Ca^{2+} percentage.

Here, the ability of the oil samples to absorb UV radiation in the 280–320 nm range was assessed, and the sunscreen's effectiveness was quantified using the sun protection factor (SPF). The values of EE(λ) × I(λ) [21] were summarized in Table S2 of Supplemental Materials, and Table 1 presents the SPF values for argan oil, passion fruit oil, and their mixtures, which were calculated using the same Mansur equation as Formula (3). The SPF values for the oil samples ranged from 1.43 to 1.68. The highest SPF value was observed for PFSO (1.68), followed by the AO-PFSO mixture at the 2:1 ratio (1.62), the 1:1 ratio (1.51), AO (1.50), and the 1:2 ratio (1.43), which are higher than that of eucalyptus (0.39), lavender (1.42), oregano (0.74), palmarosa (0.41), tea tree (1.04), and Japanese Yin Yang (0.59) EOs frequently used in skincare products [21]. These results indicate that PFSO demonstrated the strongest UV protection, with the EO mixtures showing a moderate sun protection effect. The difference between them is not so significant. Adding PFSO to AO is also beneficial in preventing skin color from darkening under light exposure when the latter is applied separately. Based on all the above results about the EO properties and activities, the AO-PFSO mixture at the 1:1 ratio was selected for the following preparation of nanoliposomes.

Table 1. SPF values of all the samples.

Sample	Λ (nm)	Abs 1	Abs 2	Abs 3	Aver Abs	SD Abs	CF × EE(λ) × I(λ) × Ab (λ)	Total SPF	*p*
AO	290	0.240	0.248	0.248	0.245	0.005	0.037	1.503	<0.01
	295	0.242	0.240	0.243	0.242	0.002	0.197		
	300	0.181	0.179	0.194	0.185	0.008	0.531		
	305	0.141	0.151	0.152	0.148	0.006	0.485		
	310	0.096	0.105	0.094	0.098	0.006	0.183		
	315	0.065	0.075	0.068	0.069	0.005	0.058		
	320	0.057	0.072	0.069	0.066	0.008	0.012		

Table 1. *Cont.*

Sample	Λ (nm)	Abs 1	Abs 2	Abs 3	Aver Abs	SD Abs	CF × EE(λ) × I(λ) × Ab (λ)	Total SPF	p
	290	0.272	0.269	0.264	0.268	0.004	0.040		
	295	0.232	0.235	0.236	0.234	0.002	0.191		
	300	0.180	0.190	0.191	0.187	0.006	0.537		
AO:PFSO = 2:1	305	0.167	0.157	0.166	0.163	0.006	0.535	1.622	<0.01
	310	0.119	0.109	0.113	0.114	0.005	0.212		
	315	0.113	0.101	0.109	0.108	0.006	0.090		
	320	0.081	0.089	0.085	0.085	0.004	0.015		
	290	0.268	0.271	0.278	0.272	0.005	0.041		
	295	0.233	0.231	0.234	0.233	0.002	0.190		
	300	0.166	0.175	0.163	0.168	0.006	0.483		
AO:PFSO = 1:1	305	0.135	0.158	0.153	0.149	0.121	0.487	1.511	<0.01
	310	0.110	0.113	0.107	0.110	0.003	0.205		
	315	0.098	0.109	0.111	0.106	0.007	0.089		
	320	0.084	0.087	0.088	0.086	0.002	0.016		
	290	0.296	0.305	0.299	0.300	0.005	0.045		
	295	0.237	0.249	0.237	0.241	0.007	0.197		
	300	0.153	0.148	0.155	0.152	0.004	0.437		
AO:PFSO = 1:2	305	0.132	0.134	0.134	0.133	0.001	0.437	1.430	<0.01
	310	0.102	0.114	0.111	0.109	0.006	0.203		
	315	0.116	0.106	0.117	0.113	0.006	0.095		
	320	0.096	0.093	0.089	0.093	0.004	0.017		
	290	0.313	0.308	0.314	0.312	0.003	0.047		
	295	0.273	0.274	0.279	0.275	0.003	0.225		
	300	0.172	0.172	0.188	0.177	0.009	0.510		
PFSO	305	0.153	0.158	0.159	0.157	0.003	0.514	1.679	<0.01
	310	0.128	0.137	0.134	0.133	0.005	0.248		
	315	0.126	0.139	0.136	0.134	0.007	0.112		
	320	0.128	0.132	0.134	0.131	0.003	0.024		

2.3. Characterization of Structure and Properties of Nanoliposomes

The morphology of the prepared nanoliposomes (S1–S4) was first analyzed using optical microscopy, with representative images of the nanoliposomes and their microscopic structures presented in Figure 4A. The morphology of the nanoliposome formulations prepared via the physical dispersion method exhibited variations in both size and shape. Notably, not all the nanoliposomes displayed a spherical morphology; some were non-homogeneous and exhibited characteristics of multilamellar vesicles. In contrast, the S3 and S4 nanoliposomes exhibited relatively more uniform, unilamellar vesicles with a predominantly spherical shape and a more consistent size distribution compared to S1 and S2 nanoliposomes. These morphological differences can be attributed to the preparation method and the specific composition of the formulation. As a phospholipid with both hydrophilic and hydrophobic properties, lecithin plays a crucial role in the formation of these vesicles. Its amphiphilic nature allows lecithin to form stable lipid bilayers, which are essential for encapsulating hydrophobic compounds like essential oils. As the lecithin concentration in the formulation increases, the stability and encapsulation efficiency of the nanoliposomes improve, resulting in more uniform, unilamellar vesicles as observed in the S3 and S4 formulations. These findings contrast with those reported by other studies on nanoliposomes prepared using similar protocols. For instance, Khatibi et al. observed that nanoliposomes encapsulating *Zataria multiflora* Boiss. essential oil were larger and exhibited irregular shapes when they were prepared via the ethanol injection method [6]. Furthermore, the physical appearance of the formulations tended to acquire a more yellowish hue as the lecithin content in the formulation increased. Lecithin, which contains phospholipids, naturally imparts a yellow color, and, as its concentration in the mixtures rises, the intensity of the yellowish coloration becomes more pronounced.

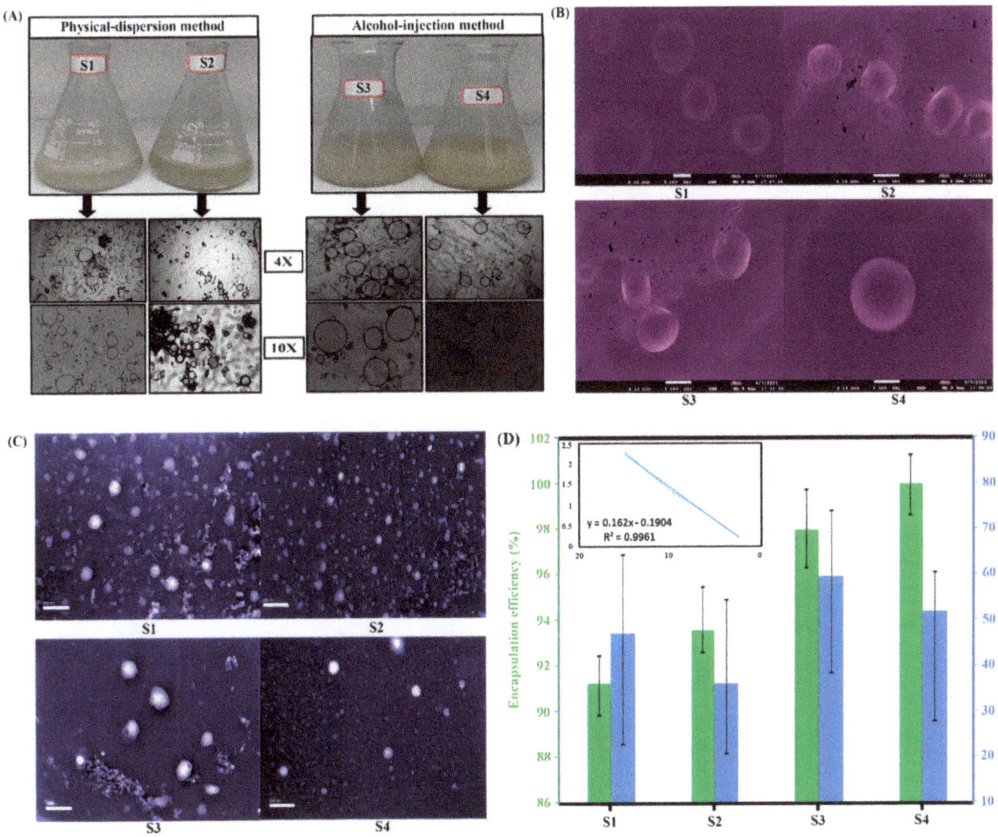

Figure 4. (**A**) Appearance of the prepared nanoliposomes and their optical microscope images in 40× and 100× magnification; (**B**) SEM and (**C**) TEM images of the prepared nanoliposomes as well as (**D**) average particle size and their encapsulation efficiency (inserted with concentration curve of Eos + DES).

Furthermore, Figure 4B and Figure 4C present SEM and TEM images of nanoliposomes (S1–S4) prepared using the two methods, respectively. Obviously, the majority of the particles exhibit a spherical shape, and various samples have various particle size distribution. S1 and S2 nanoliposomes have many small particles and even nonspherical particles. Additionally, the morphology of these nanoliposomes was further examined using TEM, revealing spherical forms with smooth surfaces as shown in the image of the sample of S4 nanoliposome. When comparing these findings with those previously published using the ethanol injection method, the results align well. For instance, current studies reported similar homogenous and spherical nanoliposomes [34–36], as the morphology of encapsulated particles, including size and shape, can be characterized using electron microscopy, with SEM detailing surface features and TEM revealing internal structures. The above results demonstrated that the preparation method employed for the S4 formulation was effective in producing high-quality nanoliposomes. By using ImageJ software (available at https://imagej.net/ij/download.html, accessed on 6 April 2025) to analyze the nanoparticle size distribution, the average particle size of the S4 sample was approximately obtained as 51.84 nm. Shanbnam [37] et al. reported that the use of propylene glycol led to the formation of nanoliposomes with the smallest mean particle size (92.03 nm) and a spherical morphology, making it the preferred co-surfactant. In comparison, the nanoliposomes synthesized in this study exhibited a significantly smaller particle size. Based on the comprehensive morphological evaluation, the nanoliposome based on S4 formulation was considered the most suitable for further studies and potential applications.

Encapsulation efficiency measurement is a crucial step in the characterization of liposomes [38]. To assess the incorporation rate of the active ingredients (EO mixture–DES system) into the lipid phase, UV spectroscopy was employed, which involved measuring the concentration of them in the supernatant following ultracentrifugation. A calibration curve at 280 nm was first established to evaluate the encapsulation efficiency according to the maximum absorbance wavelength of the EO mixture–DES system. To develop this curve, a series of working solutions of EOs and DES at varying concentrations were prepared using ethanol as the solvent. Figure 4D presents the calibration curve obtained and the encapsulation efficiency values for the nanoliposomes prepared using the physical dispersion and ethanol injection methods. The encapsulation efficiency of formulations S1–S4 was found to be showing a gradual upward trend, with values of 91.26%, 93.58%, 97.94%, and 99.95%, respectively. This suggests that variations in lecithin concentration have a certain impact on the encapsulation efficiency of the nanoliposomes (S2 > S1, S4 > S3). Consequently, it can be concluded that the lecithin content in the preparation of nanoliposomes is related to their encapsulation efficiency. When compared to the physical dispersion method, the nanoliposomes prepared via the ethanol injection method exhibited higher encapsulation efficiency, and the former also required a longer preparation time and involved more steps. Furthermore, all the tested formulations demonstrated exceptionally high encapsulation efficiencies above 90%. This is likely attributed to the hydrophobic nature of the encapsulated EOs, which exhibit a low affinity for the aqueous phase. This is further supported by the protective role of the hydrophobic phospholipid component in encapsulating and stabilizing the hydrophobic compounds. As a comparison, Isailović et al. [25] prepared liposomes encapsulating resveratrol using both the thin-film method and the pro-liposome method, achieving encapsulation efficiencies of 92.0% and 97.4%, respectively. Similarly, a study on the encapsulation of ibuprofen and flurbiprofen in nanoliposomes reported high encapsulation efficiencies of 54%, comparable to those observed in the present study [39]. On the basis of the above results, the S4 formulation was chosen for the preparation of the following hydrogel patches.

2.4. *Characterizations and Measurements on Nanoliposome-Loaded Hydrogel Patches*

2.4.1. Appearance and Microstructure

Visual inspection provides valuable insights into the quality and uniformity of hydrogel patches. The physical inspection of the prepared combination patches was conducted visually, with the results presented in Figure 5A,B. The hydrogel patches were colorless and transparent, with no visible air bubbles or cracks on the surface, which makes the users watch the skin status clearly. Furthermore, they exhibit a homogeneous composition. Upon direct application to the human skin, the patches adhere easily and do not leave any residue upon removal. Surface morphology in the optical microscope field of view (see Figure 5C) was also level and smooth, and there were no obvious cracks, grooves, or foreign matter. Based on these observations, the prepared hydrogel patches demonstrate satisfactory quality and are deemed suitable for further testing and characterization.

The microstructure of the hydrogel was further examined using SEM after the samples were dried in a vacuum. To enhance image clarity, the tested samples were coated with gold. As shown in Figure 5D, the originally smooth surface gradually reveals its interconnected network skeleton with an increase in the magnification. Similarly, Basu et al. synthesized an injectable hydrogel with sustained drug release by leveraging noncovalent interactions between DNA and nanosilicates. The incorporation of nanosilicates resulted in a more compact hydrogel network with reduced pore size, enhancing physical crosslinking and contributing to slower drug release [40]. Notably, the prepared patches contained numerous nanoliposomes encapsulating the essential oils (AO and PFSO). The size of the hydrogel patches was standardized to 1.5 cm × 1.5 cm, and the thickness of the hydrogel patches was measured at three different locations, yielding an average thickness of approximately 0.84 mm, which falls within the range of moderately thick hydrogel patches. This thickness was deemed suitable for various applications, offering an optimal balance between mechanical strength, durability, and active ingredient release.

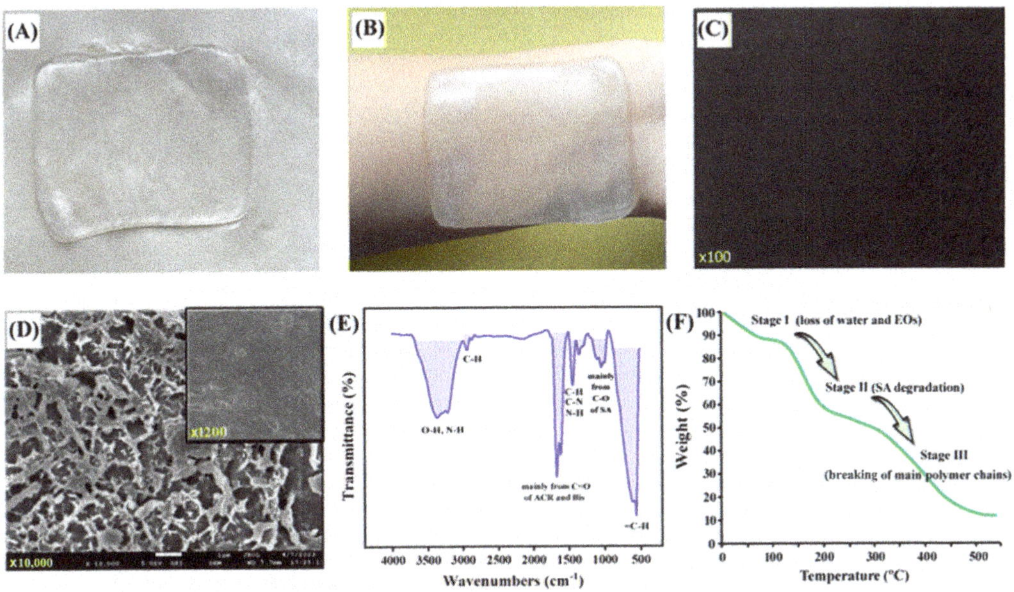

Figure 5. (**A**) The photo of the nanoliposome hydrogel patch; (**B**) the appearance of the prepared hydrogel attached on the human skin; (**C**) surface morphology in the optical microscope field of view (×100); (**D**) SEM images at different magnification (×1200 and ×10,000); (**E**) FT-IR ATR spectrum and (**F**) TGA results.

2.4.2. FT-IR and TGA

Here, FT-IR spectroscopy was also applied to identify functional groups present in the prepared hydrogel patches. As shown in Figure 5E, the peak at 3334 cm^{-1} corresponded to the stretching vibration of the hydroxyl and amino groups (–OH, –NH), indicating the presence of water, polysaccharides, and nitrogen-containing components in hydrogel patches. Moreover, the peak at 2918 cm^{-1} corresponded to asymmetric C-H stretching. The peak at 1658 cm^{-1} resulted from the C=O groups mainly in ACR and Bis, and the peaks in the range of 1450–1300 cm^{-1} were ascribed to C–N stretching vibration and the bending vibration of C–H and N–H. The broad peaks around 1020 cm^{-1} mainly belonged to the C–O stretching in SA. Finally, 550 cm^{-1} corresponded to the wavenumber of =C–H, respectively.

These spectra were similar to previous reports. On the other hand, the weight losses of the hydrogel in different temperature regions are generally associated with a splitting of the main chain and thermal decomposition of the polymer [41,42]. When compared with the TGA results in the current study [43], the three weight loss stages of hydrogel patches were determined as shown in Figure 5F, which accorded with the main components in the whole system. The water and EOs were easily lost in the sample first, and, then, it was followed by the breaking of segments of mannuronic acid and glucuronic acid in SA in the temperature range of 100–225 °C. The degradation in the third stage (above 300 °C) accounts for the breaking of the main polymer chains.

2.4.3. Swelling and Mechanical Properties

Fluid diffusion and swelling ratio variations are influenced by factors such as crosslink density, pore size, molecular weight, gel composition, and filler content. Swelling occurs as the polymer network occupies the available free volume within the sample until it reaches its maximum expansion, at which point equilibrium is achieved. Due to the flexibility of the polymer network, this expansion continues until osmotic pressures balance [44]. The swelling behavior of the nanoliposomal hydrogel patches developed in this study was investigated in a neutral medium (pH = 7.4) over a period of 1 to 7 h at 37 °C. Phosphate-buffered solution (PBS) was selected as the swelling medium, as it closely mimics the osmolality of physiological conditions [45]. The swelling ratio, which is defined as the percentage increase in volume relative to the dry hydrogel, reached its maximum value after 7 h, demonstrating the hydrogel capacity to absorb the swelling medium. In this study, the maximum swelling ratio observed was approximately 100%, meaning that the hydrogel's volume increased by a factor of two after 7 h of exposure to the PBS. This behavior is consistent with the hydrogel's intended application, as it ensures sufficient hydration without excessive swelling, which could compromise the stability and mechanical properties of the hydrogel. The results of the swelling capacity are presented in Figure 6A,B, showing the time-dependent swelling behavior and confirming the ability of the hydrogel to maintain its structural integrity over time. Besides that, it can be found from the comparison in Figure 6A that, because the liposome content in the whole system was limited, the swelling property did not show a great difference, so there was no more comparison on the basic properties between them.

In practical applications, it is essential to examine the water retention behavior of highly absorbent materials. In this study, the water retention capacity of the nanoliposomal hydrogel patch was assessed over time intervals ranging from 1 to 7 h. The results are shown in Figure 6C. After the first hour, the hydrogel lost 6.4% of its weight due to water absorption. By the second hour, the weight loss increased to 20.5%, and, after the third hour, it reached 27.7%. The hydrogel continued to lose water over time, with 35% loss after 4 h, 39.3% after 5 h, and 48.9% after 6 h. Interestingly, the hydrogel began to retain more water after the sixth hour, with only a slight increase in water loss between the sixth and seventh hours. After 7 h, the hydrogel had lost 51.7% of its weight in water. Gravimetric measurements are considered the gold standard for the measurement of water content [46]. This method involves determining the change in the weight of a sample before and after the removal of water. The difference in weight is attributed to the water content. Therefore, the weight loss measured during this test is attributed solely to water evaporation from the hydrogel matrix, thus directly reflecting its water retention capacity.

The prepared nanoliposomal hydrogel patches were found to be highly stretchable and flexible. When force was applied to break the hydrogel patch, it exhibited significant stretchability, and the tensiometer stretched it (length: 240 mm) under 5 N tension without breaking it. Measured with an electronic universal material tester, the stress–strain curves at room temperature and 37 °C are shown in Figure 6D. It can be found that the maximum stress was nearly 7.5 MPa, which would be reduced by 8% after remaining at 37 °C for 7 h. Similarly, Atefen et al. investigated the effect of curcumin liposome nanocarriers on the drug release behavior and mechanical properties of PVA/PEG hydrogels. Optimal nanoliposomes, prepared via thin-film hydration, were incorporated into the hydrogels, which were physically cross-linked through freeze–thaw cycles. Their study reported a maximum tensile stress below 1 MPa [47]. Furthermore, upon repeatedly folding the patch at the same spot on the skin more than 100 times, no cracks or defects were observed. These observations underscore the remarkable flexibility and durability of the hydrogel patches.

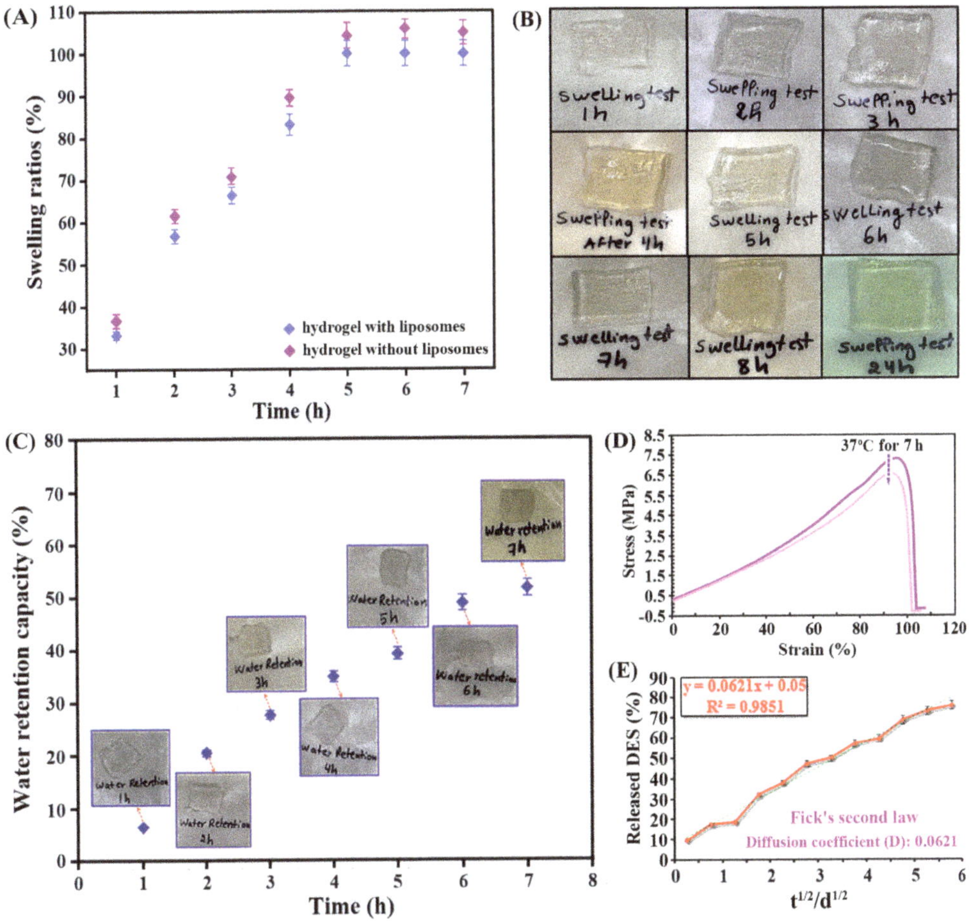

Figure 6. (**A**) Swelling ratio (the hydrogel sample without liposomes as control); (**B**) photographs of the prepared nanoliposome–hydrogel patch as a function of time (1–7 h) in PBS at 37 °C; (**C**) water retention capacity of the prepared nanoliposome–hydrogel patch as a function of time (1–7 h) at 37 °C; (**D**) stress–strain curves at room temperature and 37 °C; (**E**) DES release behavior in in vitro transdermal testing.

2.4.4. Skin Adaptability and DES Release

When formulating skincare products, it is crucial to consider their pH effect on the skin. Regulating skin pH is an essential and controllable factor that can enhance skin penetration while also ensuring the stability, efficacy, and safety of products for direct skin application [34]. The skin serves as the first line of defense against external aggressions, and its microbiome and structural integrity are delicately maintained by various factors. Alterations in skin pH can lead to pathogenic effects, including compromised skin barrier function and increased bacterial colonization [34]. By using an intelligent skin tester, the pH of the hydrogel patch was measured at 5.57 at room temperature. Given that the normal pH range for skin is typically between 4.5 and 5.5, the measured pH of 5.57 is considered to fall within an acceptable range. Also, by using an intelligent skin tester, the water content of the skin increased from 48% to 54% after contact for half an hour, and oil content of the skin increased from 27% to 35%, indicating that the hydrogel patch exerts a hydrating effect on the skin. No significant changes were observed in the skin color and appearance, suggesting that the patch did not induce irritation and is likely safe for use. Due to the fact that the relevant components are commonly used in transdermal systems and no additional substances with obvious irritation or sensitization were used, no significant skin adverse reactions were found within 12 h (this patch was designed to be changed daily and used during sleep). These findings support the safety and efficacy of the hydrogel patch in promoting skin hydration without causing any adverse effects. Finally, removability can also reflect skin adaptability of the patches; if the adhesion is too strong, it may cause skin surface damage during the removal process. After the comparison with a commercial hydrogel patch through the investigation in Section 4.8.4, the difference in adhesion was not great (commercial hydrogel patch: 18.3 ± 0.5 g; our hydrogel patch: 15.8 ± 0.3 g). If it aimed to make peeling easier, a small volume of water was suggested to be dropped on it for its swelling property.

As introduced in Section 4.8.7, the two kinds of African essential oils have complex chemical components, so it is not easy to choose a well-recognized single compound for quantitative analysis in the study of in vitro releasing behaviors. Therefore, the DES of betaine–phytic acid (1:3) was chosen as the marker of a released object for investigation to demonstrate that the relevant components can be released, which was proven for its bioactivities and biocompatibility for skin in a previous study [21]. Plotting the percentage of DES release against the square root of release time, the release curve of DES is shown in Figure 6E. It can be seen that there is an approximate linear relationship between the two parameters, which conforms to Fick's second law. Therefore, the release of the DES from hydrogel patches is mainly controlled by the diffusion mechanism, which is in accordance with the releasing behaviors of perillyl alcohol from its related hydrogel and emulsion [48]. Moreover, the permeability coefficient was determined based on the Franz diffusion cell as 1.12×10^{-3} cm/s. In the above mechanism, active molecules diffused through the polymer barrier in a concentration gradient, were then desorbed from the polymer, and diffused into body fluids or media.

3. Conclusions

This work presents a groundbreaking nano-liposome hydrogel patch engineered for advanced skincare applications, integrating argan oil, passion fruit oil, and a phytic acid–betaine deep eutectic solvent (DES). In our study, the hydrogel patches are designed to be "ready-to-use" immediately after production, following a vacuum drying step. This drying process ensures that the patches maintain their physicochemical integrity and sterility, making them suitable for further application in skincare. The hydrogel demonstrated exceptional hydration properties, achieving a swelling ratio of 100% and sustaining 51.7% moisture retention over 7 h, highlighting its efficacy in prolonged skin protection. The 100% swelling ratio indicates that the hydrogel can efficiently absorb and retain moisture, ensuring optimal hydration without excessive expansion that could compromise its mechanical integrity. This behavior is crucial for skincare applications, where long-lasting hydration is necessary for improving skin texture and appearance. Formulated with biocompatible polymers, specifically acrylamide and sodium alginate, combined with bentonite for structural reinforcement, the material, as an advanced platform designed for skincare, demonstrated robust mechanical properties, evidenced by a tensile strength of 7.5 MPa, as well as excellent flexibility. These characteristics ensure the system's durability and comfort during prolonged direct contact with body skin. Notably, the encapsulation of bioactive oils within nanoliposomes significantly improved colloidal stability, transdermal permeability, and enabled sustained release kinetics. Collectively, these features not only promote the efficient and targeted transdermal delivery of hydrophobic actives but also markedly reduce issues such as burst release or rapid degradation. Therefore, the synergistic use of DESs and natural products within a polymer matrix holds strong potential for versatile applications in the controlled delivery of essential oils, skincare, and wearable bioactive interfaces, offering enhanced bioavailability, minimized biological disruption, lower skin irritation potential, and improved human compliance through its tailored release profiles and superior material performance.

4. Materials and Methods

4.1. Reagents and Materials

A variety of chemicals were procured from Kelong Chemicals Co., Ltd. (Chengdu, China), including ethanol (99.7%), 2,2-diphenyl-1-picrylhydrazyl (DPPH, 96%), ethylenediaminetetraacetic acid (EDTA, AR), sodium hydroxide (NaOH, AR), trichloromethane (99%), methanol (99%), sodium chloride (99.5%), calcium chloride (96%), N,N,N′,N′-Tetramethyl (96%), acrylamide (99%) and ethylenediamine. Furthermore, lecithin (soybean) (R-D) was obtained from Macklin Biochemical Co., Ltd. (Shanghai, China), while cholesterol (99%) was sourced from Rhawn Reagent Co., Ltd. (Shanghai, China). Betaine (98%), bentonite (R-D), sodium alginate (R-D), ammonium peroxydisulfate (98.5%), N,N-methylene-bis-acrylamide (99%), and phytic acid (97%) were provided by Aladdin Biochemical Technology Co., Ltd. (Shanghai, China), which also provided Eriochrome Black T (96%). Among above reagents, sodium alginate, bentonite, and acrylamide were basic ingredients for the hydrogel formulation. All the chemical reagents were employed in their original form without any further modifications. Ultrapure (UP) water was used as the experimental water and prepared by using a water purification system from Millipore Co., Ltd. (Bedford, MA, USA). Argan oil (AO) was sourced from Rahiq lhayat in Morocco, and passion fruit seed oil (PFSO) of South Africa was obtained from El Aroma in Germany (both > 90%), respectively.

4.2. Instruments

The main experimental instruments used in this study include a TU-1810 ultraviolet (UV) spectrometer from Purkinje General Instrument Co., Ltd. (Beijing, China) and a JSM-7001F scanning electron microscope (SEM) from JEOL Co., Ltd. (Tokyo, Japan). Additional instruments include the N-117M optical microscope (×40–100, Novel Optics, Ningbo, China), the H-600 negative-stain transmission electron microscope (TEM, Hitachi, Tokyo, Japan), and a KS-400KDE ultrasonicator (Jielimei Ultrasonic Inc., Kunshan, China). The proton nuclear magnetic resonance (1H NMR) spectrum of the diluted samples with deuterated solvent was recorded at room temperature using a JNM-ECZ400s/L1 spectrometer (400 MHz, JEOL, Tokyo, Japan). A Spectrum Two L1600300 Spectrometer (Perkin Elmer, Waltham, MA, USA) was applied to record FT-IR in the range of 4000–450 cm^{-1}. Finally, a ZP-5 HNA tensile testing machine (±0.001 N; Fuma Electrical Instruments Inc., Dongguan, China), 5967 electronic universal material tester (Instron Corporation, Boston, MA, USA), and HTG-2 microcomputer differential thermal balance (Hengjiu Scientific Instrument Factory, Beijing, China) were employed for hydrogel characterization. Skin detection was completed with RBX-916 type intelligent skin tester (Real Bubee Medical Instrument Inc., Ningbo, China).

4.3. Preparation, Identification and Quantitation of DES

The DES was prepared by heating the betaine and phytic acid in specific molar ratio (betaine–phytic acid = 1:3) at 60 °C in water bath (±0.1 °C) until a clear liquid was formed [21]. Then, the mixture was taken out from the bath, and it was kept under vacuum in 50 °C for 24 h before use, which was further confirmed by UV, IR, and nuclear magnetic resonance (NMR) spectroscopy. In order to complete its purity and quantitative analysis, the UltiMate3000 high-performance liquid chromatography (DIONEX Co., Sunnyvale, CA, USA) was used according to the following conditions: 4.6 × 250 mm C_{18} column (5 μm, Welch Materials, West Haven, CT, USA) as stationary phase, the mixture of acetonitrile and water (3:97, *V*/*V*) as mobile phase 35 °C as column temperature, 1 mL/min as flow rate, 20 μL as injected sample volume, and 200 nm as the detection wavelength according to the UV full wavelength scanning.

4.4. Preparation of Argan Oil-Passion Fruit Oil Mixtures and Their Basic Properties

AO and PFSO were prepared in different volume ratios (2:1, 1:1, 1:2) using simple mixing method followed by stirring (300 rpm) for 30 min to ensure the thorough dispersion of the two EOs. The oil mixtures were kept at room temperature for the following characterization on pH, density, and apparent viscosity.

4.4.1. Determination on pH of the Oil Mixtures

The chemical activity of hydrogen ions in solution was first measured by pH (hydrogen potential). The pH determines whether a sample is basic or acidic, and it is closely associated with its potential influence on skin. After inserting the pH-meter electrode directly into the EO mixtures, the ionic acidity of the product to be analyzed was characterized after the data were stable. This experiment was repeated three times to obtain the results.

4.4.2. Determination on Density of the Oil Mixtures

The density of oil mixtures was determined by the ratio between the 10 mL of oil mixtures and the mass of the same volume of UP water taken at the same temperature. This experiment was repeated three times to obtain the results.

The density of each sample was calculated by using Formula (1) [49]:

$$\rho_M = \rho_w \frac{(m_M - m_p)}{(m_w - m_p)} \tag{1}$$

where ρ_M is density of the oil mixtures, g/cm^3; ρ_w is the density of water at 25 °C, which is 0.9970 g/cm^3; m_p is the weight of the empty pycnometer, g; m_w is the weight of pycnometer filled with water, and m_M is the weight of pycnometer filled with the oil mixtures, g.

4.4.3. Determination on Apparent Viscosity of the Oil Mixtures

All the measurements were performed using a STM-IV type viscometer (Meiyu Instrumental Inc., Shanghai, China) at 25 °C by inserting the spindle of viscometer directly into the sample. The mark on the mixing blade shaft was level with the liquid surface, and its rotational speed was 200 rpm. The measurements were performed in triplicate, and the average values were calculated. This experiment was repeated three times to obtain the results.

4.5. Key Bioactivities of Argan Oil–Passion Fruit Seed Oil Mixtures

4.5.1. Diphenyl-1-Picrylhydrazyl Antioxidant Activity

The antioxidant activity (AA%) was determined using 2,2′-diphenyl-1-picrylhydrazyl (DPPH) method [50]. Briefly, 200 μL of different ratios of the essential oil mixtures was mixed with 2.8 mL of ethanol and then mixed thoroughly; after that, 2 mL of 0.004% ethanoic DPPH• solution was added. The mixtures were well shaken and kept at room temperature in the dark for 30 min. The absorbance of each mixture was measured at 517 nm using UV–Vis spectrophotometer. Ethanol was used as a negative control, and the radical scavenging activity (RSA) was calculated as a percentage of DPPH• discoloration by using Formula (2).

$$AA\% = \frac{(A_c - A_s)}{A_c} \times 100 \quad (2)$$

where A_c is the control reaction absorbance that contains all the tested reagents except the essential oils; A_s is the absorbance of tested samples. This experiment was repeated three times to obtain the results.

4.5.2. Ca^{2+}-Binding Activity (Antikeratotic Activity)

Based on current report [51], complexometric volumetric titrations with the ethylenediamine tetra-acetic acid (EDTA) were used to measure the quantity of calcium in oil samples. For 5 mL sample of the oil, a 5 mol/L NaOH solution was added until the mixture became strongly alkaline (around pH 12). Besides that, 2 drops of 1% (*W*/*V*) Eriochrome Black T indicator solution were further added into each sample, and they were titrated with 0.01 mol/LEDTA. When the system color was changed from pink to blue, the titration was complete. The residual concentration of calcium was determined after the titration to evaluate the Ca^{2+}-binding activity of each sample. This experiment was repeated three times to obtain the results.

4.5.3. Sun Factor Protection Activity

The measurement of the sun factor protection (SFP) activity was used to assess the photoprotective impact of oil mixtures with different compositions, and the samples were prepared by weighing 0.1 g of the tested oil mixture. The precisely weighed oils were mixed with 10 mL of ethanol under the radiation of 100 W ultrasound waves until homogenization, and the homogenous mixture was tested three times by using the spectrophotometer at 5 nm intervals within the wavelength range of 290–320 nm. The obtained values were calculated using the Mansur equation as Formula (3) [14–17].

$$SPF = CF\sum_{290}^{320} EE(\lambda)I(\lambda)Abs(\lambda) \quad (3)$$

where EE(λ) is the erythemal effect spectrum, I(λ) is the solar intensity spectrum, and Abs is the absorbance of essential oils and their mixtures. CF is the correction factor (=10). The value of EE(λ) × I(λ) is constant determined by Sayre et al. [18]. This experiment was repeated three times to obtain the results.

4.6. Preparation and Characterization of Nanoliposomes Encapsulating Two EOs and DES

4.6.1. Preparation of Essential Oils and DES-Encapsulated Nanoliposomes

In this section, the nanoliposomes were prepared by physical solid dispersion and ethanol injection methods to encapsulate the mixture of two essential oils, which are two commonly used methods in current studies. By comparing the appearance, encapsulation performance, energy consumption, and operational time required for each method, it was aimed to explore the most efficient way to prepare hydrogel patch nanoliposomes. For simple comparison, Figure 7 includes the main procedures of two methods; at the same time, Table 2 provides the details of ingredients for the two preparation methods.

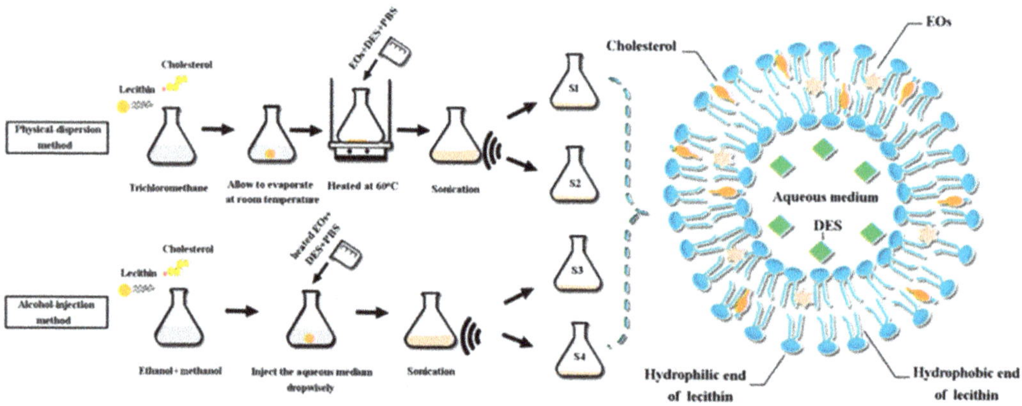

Figure 7. Diagram of preparation process for EO-encapsulated nanoliposomes in two ways.

Table 2. Ingredients of essential oil-DES nanoliposomes.

Ingredients	Physical-Dispersion Method		Alcohol-Injection Method	
	S1	S2	S3	S4
Cholesterol (mg)	100	100	100	100
Lecithin (mg)	100	200	100	200
EOs (g)	1	1	1	1
DES (g)	1	1	1	1
Ethanol (mL)	-	-	7	7
Trichloromethane (mL)	5	5	-	-
Methanol (mL)	-	-	3	3
PBS (mL)	50	50	50	50

(1) Physical-dispersion method

Nanoliposomes were prepared by solid dispersion method described previously by Shivhare [52]; different quantities of lecithin were used, while the cholesterol quantity was kept constant. The lipids were dissolved in 5 mL of trichloromethane and then transferred to a flat-bottom conical flask, then, they were evaporated at room temperature for 1 day without disturbing them to form a lipid layer. The oil mixture and DES were mixed with 50 mL of PBS buffer (pH 7.4) and ultrasonicated (500 W) to form a uniform emulsion, which acted as an aqueous medium. The flask containing the lipid film was tilted to one side, and the aqueous medium was introduced down the side of the flask and slowly returned to its upright position. The flask was then placed on the water bath and maintained at 60 °C for 3 h to ensure complete hydration of the lipid film. The formulations were subjected to ultrasonication (500 W) to reduce the size of the formed nanoliposomes. Finally, the formed nanoliposome suspensions were marked as S1 and S2 and were stored at 4 °C.

(2) Alcohol-injection method

Differently, S3 and S4 were prepared by alcohol-injection method [53], and the lipids were dissolved in ethanol and methanol. The aqueous medium was made by mixing oil mixture and the DES in 50 mL of PBS as performed previously, which was heated to 60 °C in the water bath. The heated aqueous medium was then introduced dropwisely into the lipid solution, triggering the spontaneous formation of nanoliposomes through interfacial self-assembly at the aqueous-lipid phase boundary. The nanoliposome suspensions were subjected to ultrasonication (500 W) to reduce the size of the formed nanoliposomes and stored at 4 °C.

4.6.2. Characterization of Essential Oils and DES-Encapsulated Nanoliposomes

The following studies were based on current reports [54,55]. Firstly, a slide was prepared by placing a droplet of the prepared nanoliposome suspension on a glass slide, and cover slip was placed over it. And this slide was viewed under optical microscope at 40× and 100× magnification. Photographs were taken to the prepared slides using digital camera. Then, the structure of the diluted nanoliposome suspension was observed by Negative-stain transmission electron microscope (TEM). Briefly, a droplet of nanoliposome suspension was placed onto copper grids, and the excess was removed using filter paper after 3 min. The TEM was then used to evaluate negatively stained nanoliposomal sample at different magnification. Moreover, the SEM was further used to observe the appearance and morphology of nanoliposomes at high magnitude. The diluted samples were made by mixing 100 μL of nanoliposome suspension with 900 μL of PBS (pH 7.4), and, then, placing one droplet of the produced samples on a clean glass sheet, the sample was then placed on aluminum stubs and finally coated with gold. Finally, the efficacy of encapsulation was determined using UV spectroscopy. Briefly, after the maturation of nanoliposomes, the mixture was further centrifuged (15,000 rpm, 20 min). The supernatant was separated, and its absorbance was measured at 280 nm as maximum absorbance wavelength (see Figure S3). The encapsulation efficacy (EE, %) was calculated as Formula (4).

$$EE\ (\%) = \frac{Total(EO_1 + DES_1) - (EO_2 + DES_2)}{Total(EO_1 + DES_1)} \times 100, \quad (4)$$

where EO_1 refers to the concentration of argan oil and passion fruit oil used to prepare the nanoliposomes; DES_1 refers to the concentration of DES used to prepare the nanoliposomes; EO_2 refers to the concentration of argan oil and passion fruit oil in the supernatant; and DES_2 refers to the concentration of DES in the supernatant. This experiment was repeated three times to obtain the results.

4.7. Preparation of Hydrogel Patches Containing the Formed Nanoliposomes

The hydrogel was prepared by mixing sodium alginate (SA) with UP water (1:50, W/W) under magnetic stirring for 30 min to ensure complete dissolution. Then, bentonite was dispersed in ultrapure water at a mass ratio of 1:100 and subsequently incorporated into the sodium alginate solution. The mixture was stirred under magnetic agitation for another 30 min in a 40 °C water bath (±0.1 °C). Next, 2 mL of the previously prepared EOs-DES nanoliposome was added into the mixture of sodium alginate and bentonite. Following this, 15 g of acrylamide (ACR) was added and mixed thoroughly, and 0.01 g of N,N′-methylene-bis-acrylamide (Bis) in 10 mL of UP wat was added to the solution. After thorough mixing, ammonium persulfate (APS, 10 mg/mL) and N,N,N′,N′-tetramethylethylenediamine (TEMED, 0.76 mg/mL) were added at a mass ratio of 1:2 to initiate crosslinking and gelation. The reaction proceeded at 60.0 ± 0.1 °C for 0.5–1 h to ensure uniform crosslinking and gelation. The viscous product was immediately transferred into the clean mold, cooled to room temperature for structural stabilization and finally demolded. Post-treatment procedures included vacuum drying (50 °C, 3 h), UV sterilization (60 min, 254 nm), sectioning into standardized specimens, and vacuum-sealed storage to preserve physicochemical integrity and sterility, successively. Post-treatment, the hydrogel patches were ready for immediate use in actual applications. For specific experiments in the following contents (such as swelling tests), the hydrogel patches needed to be immersed in phosphate-buffered saline (PBS) for up to 24 h to observe their swelling behavior and achieve the desired hydration state.

4.8. Characterization and Testing on the Prepared Hydrogel Patches

The prepared hydrogel patches were first evaluated through visual inspection for color, clarity, flexibility, and smoothness. Then, the following measurements and testing were performed for their main characteristics and properties. Each of the following experiments was repeated three times to obtain the results.

4.8.1. Thickness Measurement

The thickness of hydrogel patches was determined using digital caliper (±0.01 cm) at three different places of each patch; the patch was placed between the two jaws of the calipers, which were gently closed until the patch was lightly compressed. Then, the thickness values on the digital display were recorded, and the mean value (n = 3) was calculated as final data.

4.8.2. SEM Observation

The morphology of the hydrogel patches containing nanoliposomes was examined using SEM. Prior to observation, the hydrogel patches were vacuum-dried for 10 min, cut to reveal the internal structure, and then coated with gold by sputtering after drying. SEM images were observed at magnifications of 1200× and 6000×, respectively. It is essential for analyzing the detailed microstructure of materials at high resolution, enabling the visualization of nanoliposome distribution, size, and interaction with the hydrogel matrix.

4.8.3. FT-IR Spectral and Thermogravimetric Analysis (TGA)

FT-IR spectrophotometer analysis was performed on the prepared dried hydrogel patches. The attenuated total reflection (ATR) mode of operation was used with the spectrophotometer, which can directly analyze solid samples without preparing transparent disks, simplify the sample processing process, and directly reveal the microstructure of the sample surface. The analysis was carried out in the wavenumber range from 4000 cm^{-1} to 500 cm^{-1}. In order to confirm the basic composition of the hydrogel patches, TGA was carried out, and the sample was heated from 35 °C under an atmosphere of air at a constant temperature rate of 10 °C/h.

4.8.4. Tensile, Folding Strength, and Removability Analysis

To determine the tensile strength of hydrogel patches, they were mounted in the tensile testing machine and subjected to a controlled stress until the hydrogel patch broke. The folding endurance of the patches was tested by manually folding one patch repeatedly at the same spot until it either broke or developed visible cracks, which was assessed to indicate good patch quality [53]. This is crucial to determine whether the sample can withstand folding during applications. The value of folding endurance was determined by how many times the patches could be folded in the same location without breaking. Finally, the universal material tester was applied to obtain the tensile stress–strain curves at room temperature and 37 °C. Finally, the adhesion of patches with the size of 1.2 × 1.2 cm^2 was measured according to General Principles in the fourth part of Chinese Pharmacopoeia (2020 edition), which reflected their removability, and the commercial E-106 type hydrogel patch (Haiyang Biotechnology Co., Ltd., Qingdao, China) was used as control.

4.8.5. Swelling Testing

The swelling ratio of hydrogels was determined using a general gravimetric method [56]. Three sets of disk specimens with a diameter of 1 cm were made from the dry-weighed samples (recorded as m_d) and then immersed and allowed to soak in PBS (pH 7.4) for 24 h at 37 °C. The samples were removed after the swelling equilibrium was reached, the extra moisture was wiped from the surface of the membrane with filter paper, and, then, they were weighed (recorded as m_s) again. The tests were performed three times, and the average results were noted. The swelling ratio (%) of the samples was calculated using Formula (5) as the following [57]:

$$\text{Swelling ratio } (\%) = \frac{m_s - m_d}{m_d} \times 100 \quad (5)$$

4.8.6. Water Retention Testing

The capacity of water retention was determined according to the following steps [48]: the specimens were submerged in UP water until they reached equilibrium, and the excess water was removed from the surface; the specimen weight was measured and recorded as W_{eq}. Then, the wet hydrogel samples were then stored in an incubator at 37 °C. Then, they were weighed at periodic intervals (recorded as W_t), and the mean values were obtained. The water retention ratio of hydrogel patches was calculated by using Formula (6) as below [23]:

$$\text{Water retention ratio } (\%) = \frac{W_t}{W_{eq}} \times 100 \quad (6)$$

where W_{eq} stands for the initial weight (t = 0 min), and W_t is the weight of hydrogels in different times from 1 h to 8 h.

4.8.7. Releasing of Loaded Component from the Hydrogel Patch

According to the previous design for the whole system, the components encapsulated in nanoliposomes can be released from the hydrogel patch and play their skincare role; that is, the hydrogel patch can also act as the carrier for delivering bioactive ingredients. Due to the fact that the two kinds of African essential oils belong to a complex system composed of multiple components, it is not easy to choose a well-recognized single compound for quantitative analysis in the study of releasing behaviors. Therefore, the DES of betaine–phytic acid (1:3) encapsulated together with EOs was chosen as the released object for investigation to demonstrate that the relevant components can be released. Due to being located at a deeper position in liposomes than EOs, it was treated as a marker to monitor the release behavior and overall diffusion the encapsulated components. Based on the developed method using Franz diffusion cell [51], a medical polymethylsiloxane (PDMS) film (BioExcellence® International Tech. Co., Ltd., Beijing, China; thickness: 0.5 mm, area: 1.5 × 1.5 cm^2) was first soaked in 95% ethanol for 30 min and dried naturally for use.

The effective diffusion area was 1.33 cm^2, and the cell jacket was kept at 32 ± 0.5 °C. The volume of the receiving cell was 15 mL, 0.9% physiological saline (Biosharp Biotechnology Co., Ltd., Hefei, China) was used as the receiving medium, and, then, the permeation experiments were carried out as before. The released amount of DES was detected with the HPLC conditions in Section 4.3. The permeability coefficient was calculated according to previous report [58].

Supplementary Materials: The following supporting information can be downloaded at https://www.mdpi.com/article/10.3390/gels11050364/s1, Figure S1: FT-IR spectra of betaine (KBr disk); Figure S2: FT-IR spectra of phytic acid (KBr disk); Figure S3: UV spectra of encapsulated system; Table S1: The experimental details for the test on Ca^{2+} combined with oil samples; and Table S2: The values of EE(λ) × I(λ).

Author Contributions: Conceptualization, Y.X.; methodology, W.J.; software, S.M.; validation, A.A.; formal analysis, S.T.; investigation, W.J.; data curation, Y.X.; writing—original draft preparation, W.J.; writing—review and editing, A.Y.; visualization, S.T.; supervision, S.Y.; project administration, S.Y.; funding acquisition, S.Y. All authors have read and agreed to the published version of the manuscript.

Funding: This work is supported by the Fundamental Research Funds for the Central Universities/Sichuan University, Luzhou Science and Technology Innovation Platform Construction Project (2022CDLZ-20).

Institutional Review Board Statement: Not applicable.

Informed Consent Statement: Not applicable.

Data Availability Statement: All the data are contained within this article and the Supplementary Materials.

Acknowledgments: The authors acknowledge the technical support from the Center of Engineering Experimental Teaching, School of Chemical Engineering, Sichuan University, as well as the FT-IR and SEM characterization analysis by Lin Xiang and Yanpin Huang, respectively.

Conflicts of Interest: The authors declare no conflicts of interest.

References

1. Stacy, A.; Belkaid, Y. Microbial guardians of skin health. *Science* **2019**, *363*, 227–228. [CrossRef] [PubMed]
2. Balin, A.K.; Pratt, L.A. Physiological consequences of human-skin aging. *Cutis* **1989**, *43*, 431–436. [PubMed]
3. Gilchrest, B.A. Skin aging and photoaging—An overview. *J. Am. Acad. Dermatol.* **1989**, *21*, 610–613. [CrossRef]
4. Meng, H.; Li, J.; Dong, Y.; He, Y. Poly traditional Chinese medicine formulation prepared with skin moisturizing properties. *Dermatol. Ther.* **2020**, *33*, e14105. [CrossRef]
5. Khatibi, S.A.; Misaghi, A.; Akhondzadeh Basti, A. Effect of nanoliposomes containing Zataria multiflora Boiss. essential oil on gene expression of Shiga toxin 2 in Escherichia coli O_{157}:H_7. *J. Appl. Microbiol.* **2018**, *124*, 389–397. [CrossRef] [PubMed]
6. Goik, U.; Goik, T.; Zaleska, I. The properties and application of argan oil in cosmetology. *Eur. J. Lipid Sci. Technol.* **2019**, *121*, 1800313. [CrossRef]
7. Vieira Lopes, R.d.V.; Zamian, J.R.; Resck, I.S.; Sales, M.J.A.; Santos, M.L.; Cunha, F.R. Physicochemical and rheological properties of passion fruit oil and its polyol. *Eur. J. Lipid Sci. Technol.* **2010**, *112*, 1253–1262. [CrossRef]
8. El Asbahani, A.; Miladi, K.; Badri, W.; Sala, M.; Addi, E.H.; Casabianca, H.; El Mousadik, A. Essential oils: From extraction to encapsulation. *Int. J. Pharmaceut.* **2015**, *483*, 220–243. [CrossRef]
9. Aarabi, M.H. Preparation of nanoliposomes containing rosmarinus offi cinalis L essential oil; a comparative study. *Biosci. Biotechnol. Res. Commun.* **2017**, *10*, 105–110. [CrossRef]
10. Cimino, C.; Maurel, O.M.; Musumeci, T.; Bonaccorso, A.; Drago, F.; Souto, E.M.B.; Pignatello, R. Essential oils: Pharmaceutical applications and encapsulation strategies into lipid-based delivery systems. *Pharmaceutics* **2021**, *13*, 327. [CrossRef]
11. Jiang, M.; Ying, W. Research progress on delivery systems of plant essential oils and applications in fruit and vegetable preservation. *Food Sci.* **2024**, *45*, 293–305.

12. Ahmed, E.M. Hydrogel: Preparation, characterization, and applications: A review. *J. Adv. Res.* **2015**, *6*, 105–121. [CrossRef] [PubMed]
13. Zainal, S.H.; Mohd, N.H.; Suhaili, N.; Anuar, F.H.; Lazim, A.M.; Othaman, R. Technology, preparation of cellulose-based hydrogel: A review. *J. Mater. Res. Technol.* **2021**, *10*, 935–952. [CrossRef]
14. Devi, M.; Moral, R.; Thakuria, S.; Mitra, A.; Paul, S. Hydrophobic deep eutectic solvents as greener substitutes for conventional extraction media: Examples and techniques. *ACS Omega* **2023**, *8*, 9702–9728. [CrossRef]
15. Zhuo, Y.; Cheng, H.L. Ionic liquids in pharmaceutical and biomedical applications: A review. *Pharmaceutics* **2024**, *16*, 151. [CrossRef]
16. Prabhakaran, M.P.; Ghasemi-Mobarakeh, L.; Jin, G.; Ramakrishna, S. Electrospun conducting polymer nanofibers and electrical stimulation of nerve stem cells. *J. Biosci. Bioeng.* **2011**, *112*, 501–507. [CrossRef]
17. Hayyan, M.; Abo-Hamad, A.; Alsaadi, M.A.; Hashim, M.A. Functionalization of graphene using deep eutectic solvents. *Nanoscale Res. Lett.* **2015**, *10*, 324. [CrossRef]
18. Huang, W.; Xiao, Y.; Shi, X. Construction of electrospun organic/inorganic hybrid nanofibers for drug delivery and tissue engineering applications. *Adv. Fiber Mater.* **2019**, *1*, 32–45. [CrossRef]
19. Shekaari, H.; Zafarani-Moattar, M.T. Effect of choline chloride/ethylene glycol or glycerol as deep eutectic solvents on the solubility and thermodynamic properties of acetaminophen. *J. Mol. Liq.* **2018**, *249*, 1222–1235. [CrossRef]
20. Palmelund, H.; Andersson, M.P.; Asgreen, C.; Rantanen, J.; Lobmann, K. Tailor-made solvents for pharmaceutical use? Experimental and computational approach for determining solubility in deep eutectic solvents (DES). *Int. J. Pharm. X* **2019**, *1*, 100034. [CrossRef]
21. Chen, C.; Yan, W.; Jiang, W.; Zhu, C.; Yao, S. Transdermal release behaviors of bioactive deep eutectic solvents as natural skin care and mechanism. *J. Mol. Liq.* **2022**, *367*, 120412. [CrossRef]
22. Dave, G.; Modi, H. FT-IR method for estimation of phytic acid content during bread-making process. *J. Food Meas. Charact.* **2018**, *12*, 2202–2208. [CrossRef]
23. Kitamura, Y.; Iwasaki, T. Standard infrared absorption spectrum of betaine and optimal conditions for its measurement. *J. Food Hyg. Soc. Jpn.* **2006**, *47*, 232–236. [CrossRef]
24. Mijaljica, D.; Townley, J.P.; Spada, F.; Harrison, I.P. The heterogeneity and complexity of skin surface lipids in human skin health and disease. *Prog. Lipid Res.* **2024**, *93*, 101264. [CrossRef] [PubMed]
25. Lee, S.H.; Chow, P.S. Developing eco-friendly skin care formulations with microemulsions of essential oil. *Cosmetics* **2022**, *9*, 30. [CrossRef]
26. Maghazechi, A.; Mohammadi Nafchi, A.; Tan, T.C.; Easa, A.M. Rheological characterization and fouling deposition behavior of coconut cream emulsion at heat processing temperature range. *Food Sci. Nutr.* **2022**, *10*, 3801–3813. [CrossRef]
27. Amorati, R.; Foti, M.C. Antioxidant activity of essential oils. *J. Agric. Food Chem.* **2013**, *61*, 10835–10847. [CrossRef] [PubMed]
28. Csakvari, A.C.; Lupitu, A. Fatty acids profile and antioxidant activity of almond oils obtained from six Romanian varieties. *Farmacia* **2019**, *67*, 882–887. [CrossRef]
29. Lee, S.E.; Lee, S.H. Skin barrier and calcium. *Ann. Dermatol.* **2018**, *30*, 265. [CrossRef]
30. Parrado, C.; Mercado-Saenz, S. Environmental stressors on skin aging. Mechanistic insights. *Front. Pharmacol.* **2019**, *10*, 759. [CrossRef]
31. Ebrahimzadeh, M.A.; Enayatifard, R.; Khalili, M.; Ghaffarloo, M. Correlation between sun protection factor and antioxidant activity, phenol and flavonoid contents of some medicinal plants. *Iran. J. Pharm. Res.* **2014**, *13*, 1041.
32. Mbanga, L.; Mulenga, M. Determination of sun protection factor (SPF) of some body creams and lotions marketed in Kinshasa by ultraviolet spectrophotometry. *Int. J. Pharm. Res.* **2014**, *1*, 7–13.
33. Cefali, L.; Ataide, J.; Moriel, P.; Foglio, M.A.; Mazzola, P.G. Plant-based active photoprotectants for sunscreens. *Int. J. Advan. Res. Chem. Sci.* **2016**, *38*, 346–353. [CrossRef] [PubMed]
34. Vella, F.M.; Cautela, D. Determination of antioxidant activity and sun protection factor of commercial essential oils. *Biol. Life. Sci. Forum* **2021**, *6*, 96.
35. Barbosa, A.I.; Lima, S.A.C. Evaluating the skin interactions and permeation of alginate/fucoidan hydrogels per se and associated with different essential oils. *Pharmaceutics* **2023**, *15*, 190. [CrossRef] [PubMed]
36. Bian, J.; Girotti, J.; Fan, Y.; Levy, E.S.; Zang, N.; Sethuraman, V.; Kou, P.; Zhang, K.; Gruenhagen, J.; Lin, J. Fast and versatile analysis of liposome encapsulation efficiency by nanoParticle exclusion chromatography. *J. Chromatogr. A* **2022**, *1662*, 462688. [CrossRef] [PubMed]
37. Nielsen, S.S. Complexometric determination of calcium. In *Food Analysis Laboratory Manual*; Springer: Boston, MA, USA, 2009; pp. 61–67. [CrossRef]

38. Emami, S.; Ahmadi, M.; Nasiraie, L.R.; Shahidi, S.A.; Jafarizadeh-Malmiri, H. Cinnamon extract and its essential oil nanoliposomes–preparation, characterization and bactericidal activity assessment. *Biologia* **2022**, *77*, 3015–3025. [CrossRef]
39. Baek, Y.; Jeong, E.W.; Lee, H.G. Encapsulation of resveratrol within size-controlled nanoliposomes: Impact on solubility, stability, cellular permeability, and oral bioavailability. *Colloid. Surf. B* **2023**, *224*, 113205. [CrossRef]
40. Chen, L.; Gan, L. Preparation and Drug Carrying Characteristics of Flurbiprofen Liposomes. *Tongji Daxue Xuebao/J. Tongji Univ.* **2011**, *39*, 1079–1083.
41. Wang, Q.; Zhang, Y.; Wang, M.; Pan, G. Nano-crosslinked dynamic hydrogels for biomedical applications. *Mater. Today Bio.* **2023**, *20*, 100640. [CrossRef]
42. El-Zawawy, W.K. Preparation of hydrogel from green polymer. *Polym. Adv. Technol.* **2005**, *16*, 48–54. [CrossRef]
43. Yu, Z.; Xia, Y. FT-IR spectrum analysis of the component content of acrylamide copolymers. *Spec. Petrochem.* **2012**, *29*, 79–82.
44. Samanta, H.S.; Ray, S.K. Synthesis, characterization, swelling and drug release behavior of semi-interpenetrating network hydrogels of sodium alginate and polyacrylamide. *Carbohyd. Polym.* **2014**, *99*, 666–678. [CrossRef] [PubMed]
45. Killion, J.A.; Geever, L.M.; Devine, D.M.; Higginbotham, C.L. Fabrication and in vitro biological evaluation of photopolymerisable hydroxyapatite hydrogel composites for bone regeneration. *J. Biomater. Appl.* **2014**, *28*, 1274–1283. [CrossRef] [PubMed]
46. Han, L.; Wang, M.; Li, P.; Gan, D.; Yan, L.; Xu, J. Mussel-inspired tissue-adhesive hydrogel based on the polydopamine–chondroitin sulfate complex for growth-factor-free cartilage regeneration. *ACS Appl. Mater. Interfaces* **2018**, *10*, 28015–28026. [CrossRef]
47. Hamdan, A.M.E. Design, Formulation and characterization of liposomal preparation of voriconazole (VRC). *J. Pharmaceut. Biomed. Sci.* **2015**, 5.
48. Miastkowska, M.; Konieczna, M.; Lason, E. The release of perillyl alcohol from the different kind of vehicles. *Curr. Pharm. Biotechnol.* **2018**, *19*, 573–580. [CrossRef]
49. Zhang, T.; Hu, J.; Tang, S. Densities and surface tensions of ionic liquids/sulfuric acid binary mixtures. *Chin. J. Chem. Eng.* **2018**, *26*, 1513–1521. [CrossRef]
50. Olszowy, M.; Dawidowicz, A.L. Essential oils as antioxidants: Their evaluation by DPPH, ABTS, FRAP, CUPRAC, and β-carotene bleaching methods. *Monatsh. Chem.* **2016**, *147*, 2083–2091. [CrossRef]
51. Era, B.; Floris, S. Anti-aging potential of extracts from *Washingtonia filifera* seeds. *Plants* **2021**, *10*, 151. [CrossRef]
52. Schreml, S.; Kemper, M. Skin pH in the elderly and appropriate skin care. *EMJ Dermatol.* **2014**, *2*, 86–94. [CrossRef]
53. Taribak, C.; Casas, L.; Elfadli, Z.; Metni, R.E. Quality of cosmetic argan oil extracted by supercritical fluid extraction from *Argania spinosa* L. *J. Chem.* **2013**, *2013*, 408194. [CrossRef]
54. Lin, L.; Chen, W.; Li, C.; Cui, H. Enhancing stability of *Eucalyptus citriodora* essential oil by solid nanoliposomes encapsulation. *Ind. Crops Prod.* **2019**, *140*, 111615. [CrossRef]
55. Isailović, B.D.; Kostić, I.T. Resveratrol loaded liposomes produced by different techniques. *Food Sci. Emerg.* **2013**, *19*, 181–189. [CrossRef]
56. Shivhare, U.; Ambulkar, D.; Altamimi, M.A.; Hussain, A. Formulation and evaluation of pentoxifylline liposome formulation. *Dig. J. Nanomater. Bios.* **2009**, *4*, 857–862.
57. Fattahian, A.; Fazlara, A.; Maktabi, S.; Pourmahdi, M.; Bavarsad, N. The effects of chitosan containing nano-capsulated *Cuminum cyminum* essential oil on the shelf-life of veal in modified atmosphere packaging. *J. Food Meas. Charact.* **2022**, *16*, 920–933. [CrossRef]
58. Biswas, G.; Chakraborty, S.; Ghosh, N.; Majee, S. Fabrication of bucco-matrix tablets of amoxicillin trihydrate on the basis of release and permeation kinetics. *J. Appl. Pharm. Sci.* **2015**, *5*, 48–52. [CrossRef]

gels

MDPI

Article

Phytochemical Characterization, Bioactivities, and Nanoparticle-Based Topical Gel Formulation Development from Four *Mitragyna speciosa* Varieties

Pimporn Anantaworasakul [1], Weeraya Preedalikit [2], Phunsuk Anantaworasakul [3], Sudarshan Singh [1,4], Aekkhaluck Intharuksa [1], Warunya Arunotayanun [5], Mingkwan Na Takuathung [6,7], Songwut Yotsawimonwat [1] and Chuda Chittasupho [1,*]

1 Faculty of Pharmacy, Chiang Mai University, Chiang Mai 50200, Thailand; pimporn.a@cmu.ac.th (P.A.); sudarshan.s@cmu.ac.th (S.S.); aekkhaluck.int@cmu.ac.th (A.I.); songwut.y@cmu.ac.th (S.Y.)
2 Department of Cosmetic Sciences, School of Pharmaceutical Sciences, University of Phayao, Phayao 56000, Thailand; weeraya.pr@up.ac.th
3 Regional Medical Sciences Center 2 Phitsanulok, Phitsanulok 65000, Thailand; phunsuk.a@dmsc.mail.go.th
4 Office of Research Administration, Chiang Mai University, Chiang Mai 50200, Thailand
5 Kanchanabhishek Institute of Medical and Public Health Technology, Faculty of Public Health and Allied Health Science, Praboromarajchanok Institute, Nonthaburi 11150, Thailand; warunya@kmpht.ac.th
6 Department of Pharmacology, Faculty of Medicine, Chiang Mai University, Chiang Mai 50200, Thailand; mingkwan.n@cmu.ac.th
7 Clinical Research Center for Food and Herbal Product Trials and Development (CR-FAH), Faculty of Medicine, Chiang Mai University, Chiang Mai 50200, Thailand
* Correspondence: chuda.c@cmu.ac.th

Abstract

Mitragyna speciosa (kratom) is a traditional medicinal plant rich in bioactive alkaloids and phenolics, known for their antioxidant and anti-aging properties. This study aimed to develop nanoparticle-based topical gels from ethanolic extracts of four kratom varieties, including Kan Daeng (KD), Hang Kang (HK), Tai Bai-yao (KY), and Kan Keaw (KG). Kratom NPs were prepared using a solvent displacement method. The resulting nanoparticles (NPs) exhibited sizes of 201.9–256.2 nm, polydispersity indices (PDI) below 0.3, and a zeta potential between −22.6 and −29.6 mV. The phytochemical analysis revealed that KG and KY extracts contained the highest total phenolic content (TPC) and total flavonoid content (TFC), which were mostly retained after NP formulation. The HPLC analysis confirmed HK as the richest source of mitragynine (9.97 ± 0.10% w/w), while NP formulations displayed slightly reduced levels. Antioxidant activities assessed by DPPH, ABTS, and FRAP assays revealed enhanced radical scavenging in nanoparticle formulations, with IC_{50} values ranging from 151.23 to 199.87 μg/mL (DPPH) and 207.37 to 272.83 μg/mL (ABTS). All formulations exhibited a significant inhibition of collagenase (80.56 ± 1.60 to 97.23 ± 0.29%), elastase (45.46 ± 6.53 to 52.19 ± 1.20%), and hyaluronidase (83.23 ± 2.34 to 91.67 ± 3.56%), with nanoparticle forms showing superior enzyme inhibition. Notably, nanoparticle formulations exhibited superior inhibitory effects compared to crude extracts. HaCaT cytotoxicity tests confirmed high biocompatibility (IC_{50} > 700 μg/mL), especially for KD and KG NPs. The NP-loaded gels demonstrated acceptable physicochemical stability after heating/cooling cycle testing, with pH (7.27 to 7.88), viscosity (10.719 to 12.602 Pa·s), and favorable visual and textural properties. In summary, KG and KY cultivars emerged as the most promising cosmeceutical candidates due to their superior phytochemical content, antioxidant capacity, enzyme-inhibitory activities, and formulation performance. These findings support the potential use of KG NP and KY NP-loaded gels as multifunctional cosmeceutical agents for antioxidant protection, anti-aging, and skin rejuvenation.

Keywords: topical gel; poloxamer micelles; antioxidant activity; anti-aging activity; cosmeceutical formulation

1. Introduction

Skin aging is a multifactorial biological process influenced by intrinsic factors such as genetics and hormonal changes, as well as extrinsic factors like ultraviolet radiation, pollution, and oxidative stress [1]. One of the primary contributors to this process is the overproduction of reactive oxygen species (ROS), which leads to oxidative damage and the activation of matrix-degrading enzymes such as collagenase, elastase, and hyaluronidase [2,3]. One of the primary contributors to this process is the overproduction of reactive oxygen species (ROS), which can originate from sources such as UV radiation, environmental pollution, and mitochondrial dysfunction, resulting in oxidative damage and the activation of matrix-degrading enzymes. These enzymes include collagenase, which targets collagen fibers; elastase, which degrades elastin; and hyaluronidase, which breaks down hyaluronic acid, all of which contribute to the deterioration of skin structure and function [4]. These enzymes degrade extracellular matrix (ECM) components, resulting in skin wrinkling, loss of elasticity, and dryness. Consequently, the development of topical agents with antioxidant properties and the ability to inhibit collagenase, elastase, and hyaluronidase activity has gained significant attention in dermatological and cosmeceutical research.

Kratom (*Mitragyna speciosa* (Korth.) Havil.), a tropical medicinal plant native to Southeast Asia, has long been used in traditional medicine for its analgesic, anti-inflammatory, hypolipidemic, and energetic effects [5–8]. Recent studies have reported that kratom leaves contain a rich variety of phytochemicals, including polyphenols, flavonoids, and alkaloids such as mitragynine, which exhibit potent antioxidant and anti-inflammatory activities [9]. Given its antioxidant and enzyme-inhibitory effects, these properties suggest that kratom holds promise as a natural ingredient in anti-aging skincare products by helping to protect against oxidative stress and extracellular matrix degradation in the skin. However, its practical application in topical formulations is limited due to poor water solubility, physicochemical instability, and limited skin dermal permeability [10].

Nanoparticle-based delivery systems have emerged as a promising strategy to overcome the limitations of conventional plant extract formulations. Formulating plant-derived bioactive compounds into nanoparticulate form can enhance their solubility, stability, and bioavailability [11,12]. Among various nanocarriers, poloxamer-based micelles are particularly attractive due to their excellent biocompatibility and capacity to improve the dermal delivery of hydrophobic phytoconstituents. Poloxamers, also known as Pluronics, consist of hydrophilic polyethylene oxide (PEO) and hydrophobic polypropylene oxide (PPO) blocks arranged in a triblock configuration (PEO–PPO–PEO). In aqueous environments, these molecules self-assemble into micelles with a hydrophobic PPO core and a hydrophilic PEO corona, enabling the encapsulation of poorly water-soluble compounds [13]. In addition to the solubilization capacity, poloxamer micelles exhibit thermoresponsive behavior, forming gels upon temperature elevation due to micelle packing and network formation, which is an advantageous feature for controlled drug release and enhanced skin adhesion [14]. Furthermore, the presence of the PEO corona provides steric stabilization, reducing particle aggregation and contributing to the long-term colloidal stability of the formulation [15]. These properties make poloxamer-based micelles particularly suitable for the delivery of hydrophobic phytoconstituents in topical applications. In this study, poloxamer-based micellar nanoparticles were employed to encapsulate various kratom extracts (KD, HK, KY, and KG), with the aim of improving their aqueous solubility, enhancing biological activities, and thereby increasing their cosmeceutical potential. We hypothesize that kratom extracts from different cultivars, when encapsulated into poloxamer-based micellar NPs, will retain or enhance their antioxidant and enzyme-inhibitory activities, while improving biocompatibility and formulation stability. This nanoparticle-based delivery system is expected to provide a multifunctional platform suitable for cosmeceutical skincare applications.

2. Results and Discussion

2.1. Characterization of Kratom Nanoparticles

2.1.1. Particle Size and Polydispersity Index (PDI)

The dynamic light scattering (DLS) analysis confirmed that all kratom nanoparticle (NP) formulations achieved nanoscale dimensions, with mean particle sizes ranging from 201.90 ± 0.69 nm to 256.20 ± 2.68 nm. Among the four formulations, KG NP exhibited the smallest particle size (201.90 ± 0.69 nm), followed by KY NP (221.00 ± 0.60 nm), HK NP (247.70 ± 12.20 nm), and KD NP (256.20 ± 2.68 nm). The blank nanoparticle system displayed a notably smaller particle size of 113.50 ± 0.87 nm, likely due to the absence of encapsulated phytochemicals.

All formulations demonstrated PDI values below 0.3, reflecting narrow and uniform size distributions, which are considered optimal for pharmaceutical nanoparticle systems. The lowest PDI was observed in KY NP (0.097 ± 0.013), followed by KG NP (0.183 ± 0.007), HK NP (0.188 ± 0.030), and KD NP (0.262 ± 0.006). The blank formulation had a PDI of 0.211 ± 0.010. These results meet the accepted criteria for monodisperse nanoparticles, where particle size uniformity is critical for reproducible skin permeation and consistent therapeutic performance. Differences in size and PDI may reflect intrinsic phytochemical variations among kratom varieties.

2.1.2. Zeta Potential Values

Zeta potential measurements revealed that all kratom NPs formulations possessed negative surface charges, ranging from −22.60 ± 1.47 mV to −29.57 ± 3.23 mV. KY NP demonstrated the most negative zeta potential (−29.57 ± 3.23 mV), followed closely by HK NP (−29.40 ± 3.05 mV), KD NP (−26.20 ± 0.38 mV), and KG NP (−22.60 ± 1.47 mV). The blank formulation exhibited a ZP of −23.30 ± 3.32 mV. Typically, NPs with zeta potential values greater than ±30 mV are often considered highly stable due to strong electrostatic repulsion. NPs within the −20 to −30 mV range may still maintain colloidal stability, particularly when formulated with steric stabilizers. In this study, the use of a poloxamer likely provided additional steric hindrance via its hydrophilic polyethylene oxide chains, forming a protective hydration shell that prevented particle aggregation. This dual stabilization mechanism—electrostatic and steric—is consistent with the previous findings where poloxamer coatings contributed to the stability of nanoparticles, even with moderately negative zeta potentials [16]. Together, the observed zeta potential values and the presence of steric stabilizers suggest that all prepared nanoparticles possess adequate colloidal stability for topical delivery.

Total Phenolic Content

The total phenolic content (TPC) of each extract and its corresponding NPs formulation was quantified and expressed in micrograms of gallic acid equivalents per microgram of extract (μg GAE/μg extract) (Figure 1). Among the crude extracts, KG exhibited the highest TPC (0.1616 ± 0.0164 μg GAE/μg extract), followed by KY (0.1576 ± 0.0089), HK (0.1537 ± 0.0192), and KD (0.1138 ± 0.0094). A similar trend was observed in NP formulations, with the highest TPC observed in KG NP (0.1529 ± 0.0102), followed by KY NP (0.1439 ± 0.0137), HK NP (0.1426 ± 0.0079), and KD NP (0.1084 ± 0.0154). The statistical analysis revealed that KD had a significantly lower TPC compared to HK, KY, and KG, as well as significantly lower than KG NP.

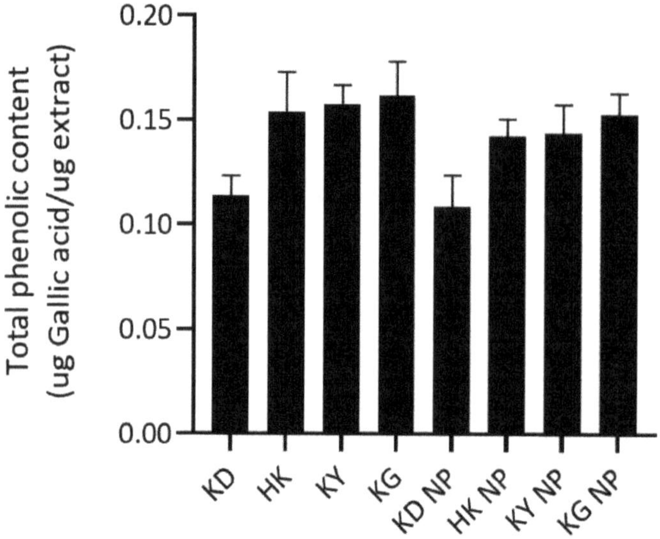

Figure 1. Total phenolic content (TPC) of kratom leaf extracts and their corresponding NP formulations. Data are presented as mean ± SD (n = 3). No statistically significant differences were observed between each extract and its corresponding nanoparticle formulation ($p > 0.05$).

These findings highlight the superior phenolic content of KG and KY varieties, supporting their traditional use and potential antioxidant efficacy. In contrast, the lower TPC in KD suggests a reduced contribution to phenolic-dependent antioxidant activity. Notably, no statistically significant differences in TPC were observed between each crude extract and its corresponding NP formulations ($p > 0.05$). This suggests that a poloxamer-based encapsulation system effectively preserves the phenolic constituents without causing degradation or loss. It also indicates that the phenolic compounds were either not significantly retained within the micelle core or remained fully accessible or reactive with the Folin–Ciocalteu reagent following encapsulation.

Total Flavonoid Content

The total flavonoid content (TFC), quantified using both EGCG and quercetin equivalents, was assessed for each kratom extract and its corresponding poloxamer-based NP formulation (Figure 2). Among the crude extracts, KG exhibited the highest TFC, with values of 1.9539 ± 0.0413 µg EGCG equivalent/µg extract and 1.0370 ± 0.0322 µg quercetin equivalent/µg extract, followed by KY and HK. In contrast, KD showed the lowest TFC, with values of 1.1199 ± 0.0524 and 0.6153 ± 0.0431 EGCG and quercetin equivalents per µg extract, respectively. The NP formulations mirrored these trends: KG NP and KY NP retained the highest TFC values, while KD NP consistently showed the lowest. The statistical analysis indicated no significant differences in TFC between each extract and its corresponding nanoparticle formulation ($p > 0.05$), suggesting that the NP preparation process effectively preserved the flavonoid content of the extracts.

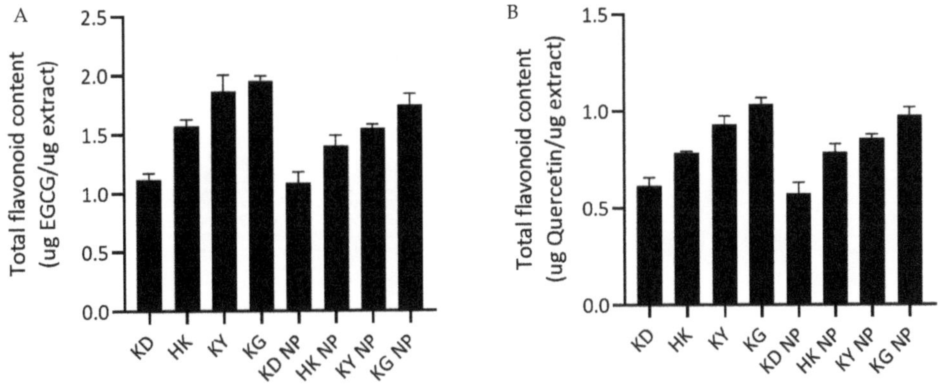

Figure 2. Total flavonoid content (TFC) of kratom leaf extracts and their corresponding NP formulations. (**A**) TFC expressed as epigallocatechin gallate equivalents (µg EGCG/µg extract). (**B**) TFC expressed as quercetin equivalents (µg quercetin/µg extract). Data are shown as mean ± SD (n = 3). No statistically significant differences were observed between each extract and its corresponding NP formulation ($p > 0.05$).

Quantification of Mitragynine in Extracts and Nanoparticle Formulations

High-performance liquid chromatography (HPLC) was used to quantify mitragynine in various kratom extracts and their corresponding NP formulations. Under the selected chromatographic conditions, the retention time (RT) of mitragynine was consistently observed at approximately 5.2–5.3 min. Representative chromatograms included samples from crude extracts (KD, HK, KY, and KG), NP formulations (KD NP, HK NP, KY NP, and KG NP), a blank nanoparticle formulation, and a mitragynine standard (40 ppm) (Figure 3). Each sample showed a sharp peak corresponding to mitragynine. The absence of a corresponding peak in the blank nanoparticle formulation confirmed the specificity of the method and the absence of mitragynine contamination from excipients.

The tailing observed in the chromatograms of nanoparticle (NP) formulations may be attributed to the interaction of the mitragynine with the poloxamer, which can cause delayed elution and result in peak asymmetry [17].

The mitragynine content in the crude extracts and corresponding NP formulations was quantified and expressed as both % *w*/*w* and mg/g of sample. Among the crude extracts, HK exhibited the highest mitragynine content (9.97 ± 0.10% *w*/*w*; 98.39 ± 1.05 mg/g), followed by KG (9.64 ± 0.24%, 96.41 ± 0.24 mg/g), KY (7.65 ± 0.13%, 76.52 ± 0.10 mg/g), and KD (7.48 ± 0.18%, 74.82 ± 0.14 mg/g). Among the NPs, KG NP retained the highest mitragynine content (8.20 ± 0.17% *w*/*w*; 81.97 ± 0.17 mg/g), followed by KY NP (6.27 ± 0.22%; 62.75 ± 0.22 mg/g), HK NP (5.82 ± 0.15%; 58.25 ± 0.15 mg/g), and KD NP (5.15 ± 0.24%; 51.51 ± 0.24 mg/g).

The calibration curve for the quantification method demonstrated excellent linearity over the concentration range of 10–100 µg/mL, with a correlation coefficient (R^2) of 0.9987. The linear regression equation was y = 30,629x + 166,812. The method exhibited good accuracy, with low residuals across all concentrations. The residual standard deviation was calculated to be 39,008.53, which was subsequently used to determine the sensitivity of the method. The limit of detection (LOD) and limit of quantification (LOQ) were calculated based on the ICH Q2 (R1) guideline as 4.20 µg/mL and 12.74 µg/mL, respectively, indicating the method's suitability for detecting and quantifying the analyte within the working range.

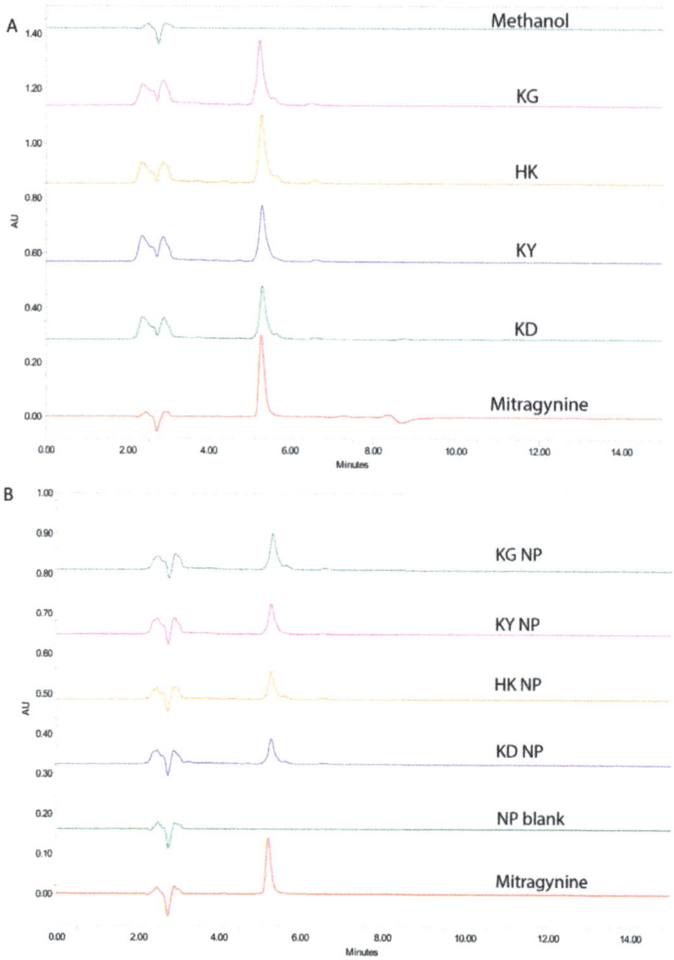

Figure 3. Representative HPLC chromatograms of mitragynine in kratom samples (**A**), crude extracts (KD, HK, KY, KG), and mitragynine standard (40 ppm), and (**B**) corresponding NP formulations and blank formulation (NP blank). Mitragynine was consistently detected at a retention time of 5.2–5.3 min.

The precision analysis displayed low relative standard deviation (RSD) values for all NP samples, indicating excellent analytical repeatability: KG NP (0.20%), HK NP (0.26%), KY NP (0.34%), and KD NP (0.46%). All values remained within acceptable limits for bioanalytical methods. These findings confirm that HK and KG extracts contain the highest levels of mitragynine, supporting their potential as potent sources of bioactive alkaloids. Although the mitragynine content decreased following nanoparticle formulation, KG NP and KY NP still retained relatively high concentrations, suggesting good formulation efficiency and compound stability during the encapsulation process.

The encapsulation efficiency (%EE) of mitragynine in the nanoparticle formulations varied depending on the kratom source. The KD NP formulation showed an average %EE of 79.25 ± 0.49%, while KY NP exhibited a significantly higher encapsulation efficiency of 94.34 ± 0.36%. In contrast, HK NP presented the lowest %EE among the four, with an average of 67.76 ± 0.66%. Notably, the KG NP formulation demonstrated the highest %EE at 97.78 ± 0.13%, indicating a highly efficient encapsulation. These results suggest that the phytochemical profile of different kratom sources may influence the encapsulation capacity of the nanoparticles.

DPPH Radical Scavenging Activity

The antioxidant activities of kratom leaf extracts and their corresponding NP formulations were evaluated using the DPPH radical scavenging assay. As illustrated in Figure 4A,B, all samples exhibited dose-dependent scavenging activity, though with varying degrees of potency. Among the crude extracts, KY showed the highest DPPH radical scavenging activity, with an IC_{50} of 152.47 ± 3.26 μg/mL, followed closely by KD (157.90 ± 1.65 μg/mL) and HK (167.60 ± 1.91 μg/mL). KG demonstrated the lowest antioxidant activity among the crude samples, with an IC_{50} of 197.70 ± 4.30 μg/mL. For the NP formulations, HK NP exhibited the strongest antioxidant capacity, with an IC_{50} of 151.23 ± 3.55 μg/mL, slightly outperforming its crude counterpart. KG NP and KY NP followed, with IC_{50} values of 160.17 ± 3.76 μg/mL and 163.97 ± 0.61 μg/mL, respectively. In contrast, KD NP showed a slightly reduced activity compared to its extract form, with an IC_{50} of 199.87 ± 2.18 μg/mL. This underperformance may be attributed to the specific phytochemical composition of the KD strain, which contained the lowest total phenolic and flavonoid content among the varieties studied. As expected, the positive control, ascorbic acid, demonstrated the strongest antioxidant activity, with an IC_{50} of 34.99 ± 1.47 μg/mL. In comparison, mitragynine—the major alkaloid in kratom—showed a relatively weak radical scavenging activity, with an IC_{50} of 602.60 ± 17.11 μg/mL, which was higher than that of all kratom samples tested.

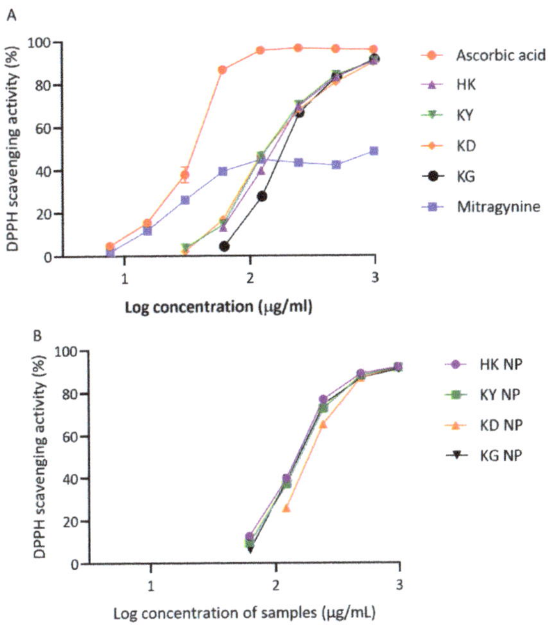

Figure 4. DPPH radical scavenging activity of kratom extracts and their NP formulations. (**A**) Dose-dependent scavenging activity of crude extracts (HK, KY, KD, and KG) compared to ascorbic acid (a positive control). (**B**) Scavenging activity of NP formulations (HK NP, KY NP, KD NP, and KG NP). Data are shown as mean ± SD (n = 3).

ABTS Radical Scavenging Activity

The ABTS assay further confirmed the antioxidant potential of both kratom leaf extracts and their corresponding NP formulations (Figure 5A,B). In this assay, the IC_{50} values of the crude extracts followed a slightly different pattern compared to the DPPH results. KG showed the strongest activity among the extracts (201.67 ± 5.05 μg/mL), followed by KY (217.93 ± 5.53 μg/mL) and HK (229.77 ± 22.01 μg/mL). KD exhibited the weakest ABTS scavenging capacity, with an IC_{50} of 304.23 ± 38.52 μg/mL. The NP formulations generally retained or slightly enhanced antioxidant activity compared to their corresponding crude extracts. HK NP again showed the strongest activity among the NP samples (207.37 ± 28.26 μg/mL). KG NP and KY NP followed, with IC_{50} values of 216.40 ± 11.40 μg/mL and 255.63 ± 12.58 μg/mL, respectively. Interestingly, KD NP showed improved activity compared to its crude extract, with a lower IC_{50} of 272.83 ± 28.84 μg/mL As expected, ascorbic acid remained the most effective ABTS scavenger, with an IC_{50} of 83.16 ± 6.27 μg/mL. In contrast, mitragynine—the major alkaloid in kratom—showed minimal ABTS scavenging activity, with a markedly higher IC_{50} value of 2022.00 ± 108.53 μg/mL, confirming that it is not a major contributor to kratom's antioxidant effect.

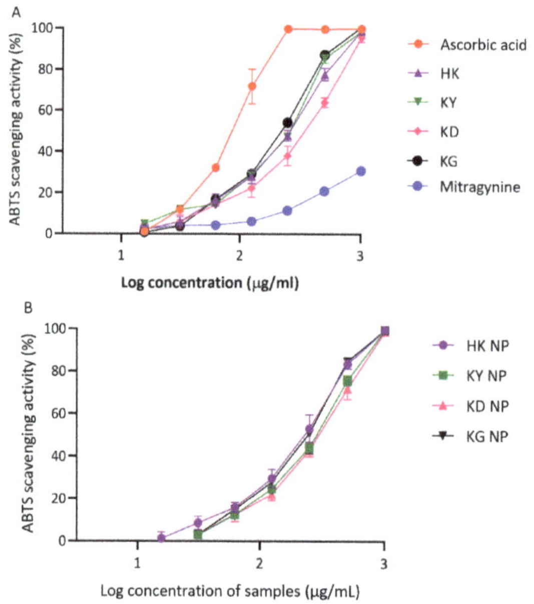

Figure 5. ABTS radical scavenging activity of kratom extracts and their NP formulations. (**A**) Antioxidant activity of crude extracts (HK, KY, KD, and KG) compared with ascorbic acid (positive control). (**B**) Antioxidant activity of NP formulations (HK NP, KY NP, KD NP, and KG NP). Results are presented as percentage inhibition of ABTS versus log concentration (μg/mL). Data are expressed as mean ± SD (n = 3).

Overall, all kratom extracts possessed a moderate antioxidant capacity in both DPPH and ABTS assays. Among the extracts, KY and HK showed slightly stronger activities than KD and KG, consistent with their relatively higher total phenolic and flavonoid contents. The incorporation of the extracts into NP formulations generally preserved or modestly improved the antioxidant capacity, particularly in the ABTS assay. Notably, HK NP demonstrated the lowest IC_{50} values among NP formulations in both assays, suggesting the effective retention and stability of antioxidant constituents during formulation. Interestingly, mitragynine exhibited the weakest antioxidant activity, reinforcing that the observed effects are likely attributable to other phenolic or flavonoid constituents in kratom rather than this primary alkaloid. These findings support the hypothesis that kratom's antioxidant potential is primarily driven by its phenolic and flavonoid compounds rather than by mitragynine. Ascorbic acid, used as a positive control, consistently demonstrated superior radical scavenging performance, underscoring the moderate nature of kratom's antioxidant capacity.

Ferric Reducing Antioxidant Power (FRAP) Activity

The reducing power of kratom leaf extracts and their corresponding NP formulations was assessed using the ferric reducing antioxidant power (FRAP) assay, with ascorbic acid serving as a positive control. Results are expressed in terms of Fe(II) equivalents (μM), with higher values indicating a greater ferric reducing capacity. As shown in Figure 6A, ascorbic acid demonstrated the strongest reducing ability in a dose-dependent manner, reaching 1063.06 μM Fe(II) equivalents at 1000 μg/mL. Among the crude extracts, KY exhibited the highest FRAP value (810.55 μM), followed by KG (752.48 μM), HK (708.18 μM), and KD (582.61 μM), all measured at 1000 μg/mL. These results indicate that all kratom leaf extracts possess a notable ferric reducing activity, with KY and KG showing the strongest effects. In the NP formulations (Figure 6B), a similar trend of dose-dependent increase in FRAP activity was observed. At 1000 μg/mL, KG NP demonstrated the highest Fe(II) equivalent value (642.55 μM), followed by KY NP (619.12 μM), HK NP (608.65 μM), and KD NP (470.49 μM). These results suggest that the ferric reducing power was generally preserved after nanoparticle formulation, although slightly reduced compared to the crude extracts.

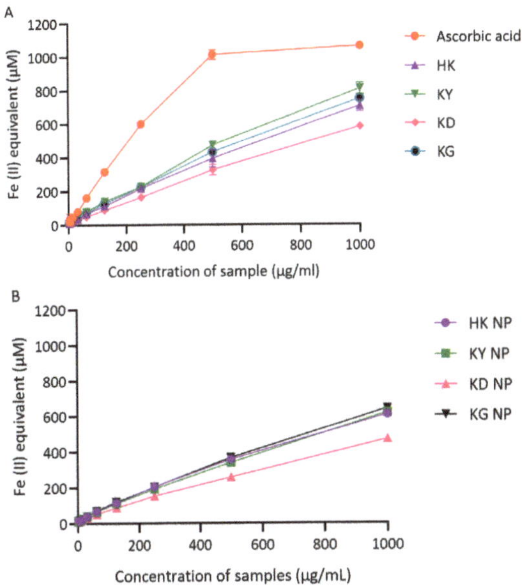

Figure 6. Ferric reducing power (FRAP) of kratom extracts and their corresponding NP formulations. (**A**) Reducing capacity of crude extracts (HK, KY, KD, and KG) compared with ascorbic acid (positive control), expressed as Fe(II) equivalents (μM). (**B**) Reducing capacity of NP formulations (KD NP, HK NP, KY NP, and KG NP). Results are shown as mean ± SD (n = 3).

Correlation Between Phytochemical Content and Antioxidant Activities

A Pearson correlation analysis was conducted to evaluate the relationships between phytochemical content (TPC and TFC, measured as EGCG and quercetin equivalents) and IC_{50} values of crude kratom extracts (KD, HK, KY, KG) and their corresponding NP formulations (KD NP, HK NP, KY NP, KG NP) (Table 1).

Table 1. Comparative Pearson correlation analysis between phytochemical content and the IC_{50} values of kratom extracts and nanoparticle formulations.

Comparison	Kratom Extract		Kratom NPs	
	Pearson r	*p*-Value	Pearson r	*p*-Value
TPC vs. DPPH	0.5802	0.4198	−0.7984	0.2016
TPC vs. ABTS	−0.9755	0.0245	−0.9308	0.0692
TPC vs. FRAP	0.8492	0.1508	0.8566	0.1434
TFC (EGCG) vs. DPPH	0.6272	0.3728	−0.7891	0.2109
TFC (EGCG) vs. ABTS	−0.9856	0.0144	−0.8581	0.1419
TFC (EGCG) vs. FRAP	0.8713	0.1287	0.9106	0.0894
TFC (quercetin) vs. DPPH	0.6274	0.3726	−0.7889	0.2111
TFC (quercetin) vs. ABTS	−0.9856	0.0144	−0.8577	0.1423
TFC (quercetin) vs. FRAP	0.8712	0.1288	0.9108	0.0892

Among the extract samples, a strong and statistically significant negative correlation was observed between TPC and ABTS IC_{50} ($r = -0.9755$, $p = 0.0245$), indicating that phenolic compounds play a key role in enhancing free radical scavenging. Similarly, TFC, whether measured as EGCG equivalents ($r = -0.9856$, $p = 0.0144$) or quercetin equivalents ($r = -0.9856$, $p = 0.0144$), also showed a strong and statistically significant negative correlation with ABTS IC_{50} values. These findings highlight flavonoids as potent contributors to hydrogen atom donation and radical neutralization. In contrast, correlations between the phytochemical content and DPPH IC_{50} values were weaker and not statistically significant. TPC showed a moderate positive correlation ($r = 0.5802$, $p = 0.4198$), while TFC (EGCG and quercetin) yielded similar trends ($r \approx 0.63$, $p > 0.37$). The FRAP assay results demonstrated strong positive correlations between reducing power and both TPC ($r = 0.8492$, $p = 0.1508$) and TFC ($r \approx 0.87$, $p \approx 0.13$). These findings suggest that both phenolic and flavonoid compounds contribute meaningfully to the electron donation involved in ferric ion reduction.

In the NP formulations, similar trends were observed. TPC exhibited a strong negative correlation with ABTS IC_{50} ($r = -0.9308$, $p = 0.0692$), suggesting that the phenolic content retained in the NP formulations continues to contribute to the antioxidant capacity. Likewise, TFC (EGCG and quercetin) also showed strong negative correlations with ABTS IC_{50} ($r = -0.86$), supporting the role of flavonoids in radical scavenging even after nanoencapsulation. For the FRAP assay, all phytochemical parameters demonstrated strong positive correlations: TPC ($r = 0.8566$), TFC-EGCG ($r = 0.9106$), and TFC-quercetin ($r = 0.9108$), with *p*-values ranging from around 0.09 to 0.14. Although these correlations did not reach statistical significance, the trends clearly reflect that nanoparticle incorporation preserved the reducing potential associated with phenolic and flavonoid compounds. Correlations with DPPH were again weaker and not statistically significant ($r = -0.79$ for both TPC and TFC), reinforcing the notion that this assay may be less responsive to subtle phytochemical variations in both crude and nanoparticle-formulated systems.

The antioxidant performance of kratom extracts and their NP formulations can be largely attributed to their phenolic content. Strong positive correlations between total phenolic content (TPC) and both FRAP and ABTS assays suggest that phenolic compounds are the primary contributors to antioxidant activity, particularly in hydrophilic systems where electron donation and radical scavenging predominate. These assays, which operate in aqueous environments, effectively reflect the redox potential of polyphenols commonly present in kratom. In contrast, the DPPH assay, conducted in a less polar (absolute ethanol) medium, showed no significant correlation with TPC. This discrepancy may stem from the specific reactivity of certain phenolic compounds, which may be less efficient at quenching DPPH radicals or hindered by limited solubility or matrix effects in non-aqueous environments [18]. Therefore, while TPC is a strong predictor of antioxidant performance in ABTS and FRAP assays, it does not reliably predict DPPH scavenging under these conditions. Moreover, the consistently low correlation between total flavonoid content (TFC) and antioxidant activity across all assays indicates that flavonoids, although present, may not be the dominant bioactive constituents in these extracts. The reduced antioxidant capacity observed in most NP formulations may be due to partial encapsulation, which limits the accessibility of free phenolic compounds or from dilution effects. Notably, KG NP retained a relatively high TPC and antioxidant activity compared to other NP formulations, indicating possible differences in formulation efficiency or improved stability of active compounds.

Collagenase Inhibitory Activity

Collagenase, a zinc-dependent multidomain enzyme, plays a key role in collagen degradation, contributing to skin wrinkling and the aging process. Therefore, inhibiting collagenase activity represents a strategic approach to delaying collagen breakdown and maintaining skin integrity [19]. This study evaluated the collagenase inhibitory activity of kratom leaf extracts and their corresponding NP formulations, using EGCG as a positive control and mitragynine as a bioactive marker for comparison.

As shown in Figure 7, all tested samples demonstrated varying degrees of collagenase inhibition. EGCG demonstrated the highest inhibitory activity (80.8 ± 3.9%), while mitragynine showed a moderate inhibitory activity (69.9 ± 3.7%), which was significantly lower than that of EGCG but notably higher than the HK and KY extracts. Interestingly, KD and KG showed the most pronounced collagenase inhibition (70.7 ± 1.0 and 71.6 ± 0.8%, respectively) among the crude extracts, with values statistically comparable to mitragynine. These findings are particularly promising for anti-aging applications, as collagenase activity directly contributes to collagen degradation and the loss of skin structural integrity associated with visible signs of skin aging [19]. Previous studies have reported that the collagenase inhibitory effect of EGCG is primarily due to its polyphenolic structure. The hydroxyl groups of polyphenols play a crucial role by forming hydrogen bonds with the backbone amide and functional groups of the collagenase enzyme [20]. Additionally, hydrophobic interactions between the benzene rings of polyphenols and the enzyme may induce conformational changes, resulting in enzyme inactivation [21].

However, the collagenase inhibition observed with kratom leaf extracts may involve mechanisms distinct from the direct enzyme binding typically associated with polyphenols. The indole alkaloids from *M. speciosa*, particularly mitragynine, are known to exert their effects primarily through anti-inflammatory and antioxidant mechanisms [8,22]. Mitragynine has been shown to suppress the cyclooxygenase–prostaglandin E2 (COX-PGE2) signaling pathway, resulting in the down regulation of matrix metalloproteinase-1 (MMP-1), a collagenase responsible for collagen degradation [23,24]. This indirect mechanism suggests that mitragynine and kratom extracts can offer protection against collagen degradation and support anti-aging benefits, even in the absence of direct enzymatic inhibition. Moreover, kratom leaf extracts contain a diverse array of bioactive compounds, including flavonoids, polyphenols, and terpenoids, all known to contribute to various pharmacological activities [25]. The antioxidant properties demonstrated in this study suggest that these compounds may act synergistically, enhancing the overall inhibitory effect on collagenase activity. Such synergism may further support the potential utility of kratom leaf extracts in anti-aging skincare applications.

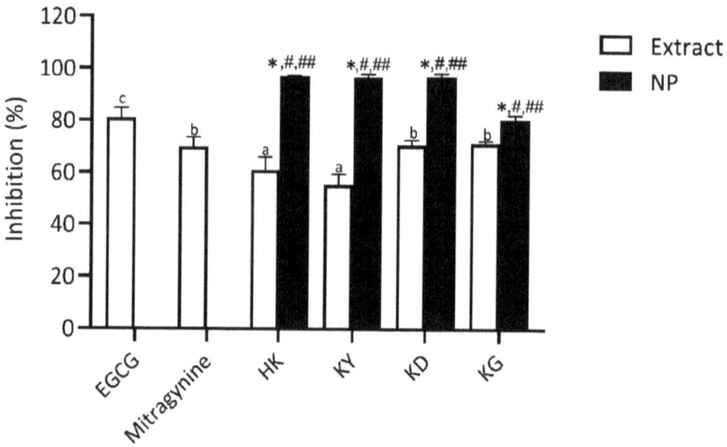

Figure 7. Collagenase inhibitory activity of kratom leaf extracts and their NP formulations compared to epigallocatechin gallate (EGCG) and mitragynine. All samples were tested at a concentration of 1 mg/mL. Bars represent mean ± SD (n = 3). Different letters indicate significant differences among the extracts ($p < 0.05$, one-way ANOVA with Tukey's test). Symbols from pairwise t-tests ($p < 0.05$): * vs. corresponding extract; # vs. mitragynine; ## vs. EGCG.

In the NP formulations, collagenase inhibitory activity significantly improved across all kratom leaf extracts. Notably, HK NP had the highest activity (97.2 ± 0.3%), which was not only significantly higher than its crude extract counterpart but also comparable to the inhibitory activity of EGCG and mitragynine. This suggests that the NP formulation of the HK extract achieves inhibition levels similar to those of well-established anti-aging agents. KY NP, KD NP, and KG NP also showed a strong inhibition in the range of 80.56 ± 1.60–97.23 ± 0.29%, all significantly higher than the activities of their respective crude extracts, EGCG, and mitragynine. The substantial enhancement in collagenase inhibition observed in the NP formulations may result from the improved dispersibility of particular active components upon encapsulation. These findings highlight the potential of kratom-derived nanoparticles—particularly HK NP—as promising candidates for anti-aging skincare applications targeting collagen breakdown and skin matrix preservation.

Elastase Inhibitory Activity

Elastin, a key component of the extracellular matrix (ECM), is essential for maintaining skin elasticity and structural integrity. Elastase, a serine protease, hydrolyzes elastin and other ECM proteins, and its dysregulation is closely associated with skin aging processes [26]. In this study, the elastase inhibitory activity of kratom leaf extracts and their NP formulations was evaluated and compared to EGCG and mitragynine. As shown in Figure 8, all tested samples exhibited varying degrees of elastase inhibition. EGCG demonstrated the highest elastase inhibitory activity (94.7 ± 0.1%), significantly greater than all other tested samples. Mitragynine exhibited moderate inhibition (54.5 ± 1.5%). Among the crude extracts, HK showed the highest elastase inhibition (44.3 ± 2.3%), followed by KY, KD, and KG, which demonstrated a progressively lower inhibition.

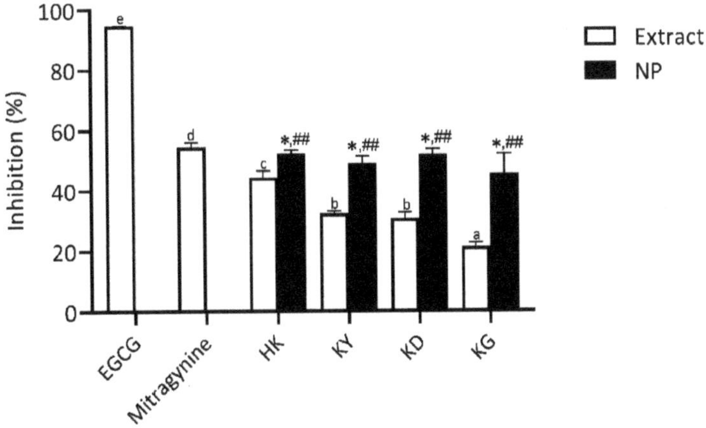

Figure 8. Elastase inhibitory activity of kratom leaf extracts and their NP formulations compared to epigallocatechin gallate (EGCG) and mitragynine. All samples were tested at a concentration of 1 mg/mL. Bars represent mean ± SD (n = 3). Different letters indicate statistically significant differences among extracts ($p < 0.05$, one-way ANOVA with Tukey's test). Symbols from pairwise t-tests ($p < 0.05$): * vs. corresponding extract; ## vs. EGCG.

Since elastase is a serine protease responsible for elastin degradation, its inhibition is critical for preserving skin firmness and elasticity [27]. Natural elastase inhibitors such as catechin and EGCG are well-documented, with polyphenols playing a significant role in this activity [28,29]. This study found that mitragynine, an indole alkaloid, and kratom leaf extracts exhibited a moderate but significant elastase inhibition; however, their efficacy was considerably lower than that of EGCG. The observed elastase inhibitory activity of kratom extracts may be influenced by their distinct composition of bioactive compounds, including indole alkaloids and polyphenols. While polyphenols are well known for their direct enzyme inhibition through specific molecular interactions, indole alkaloids may contribute to elastase regulation through alternative mechanisms, such as the suppression of inflammatory pathways and oxidative stress [30–32]. These complementary effects highlight the potential of kratom extracts in anti-aging formulations, particularly when combined with other compounds that provide direct enzyme-inhibitory activity.

All NP formulations exhibited significantly higher elastase inhibitory activity than their respective crude extracts, with inhibition levels ranging from approximately 45.46 ± 6.53% to 52.19 ± 1.20%. However, no significant differences were observed among the NP formulations, indicating that the improvement primarily arises from nanoencapsulation itself, which enhances the delivery and effectiveness of active compounds. Although the inhibition levels of the NP formulations remained lower than EGCG, their enhanced activity compared to crude extracts highlights the potential of nanotechnology in improving the efficacy of kratom extracts for dermal anti-aging applications targeting elastase inhibition.

Hyaluronidase Inhibitory Activity

Hyaluronic acid (HA) is a vital component of the extracellular matrix (ECM), playing a key role in maintaining skin hydration and moisture balance. Enhancing HA levels primarily involves inhibiting hyaluronidase activity, as this enzyme is responsible for HA degradation [4]. Therefore, the use of effective hyaluronidase inhibitors represents a targeted strategy to preserve HA content and support skin hydration. The hyaluronidase inhibitory activity of kratom leaf extracts and their NP formulations was evaluated and compared to EGCG, a well-established positive control, and mitragynine, the principal bioactive marker compound. As shown in Figure 9, all samples exhibited varying degrees of hyaluronidase inhibition. Mitragynine (83.33 ± 1.35%) and EGCG (77.75 ± 2.42%) exhibited high levels of inhibitory effects, both significantly greater than those of the kratom leaf extracts. The strong activity of mitragynine suggests a possible interaction with the enzyme, potentially preventing the degradation of hyaluronic acid, a crucial component for maintaining skin hydration and firmness The crude kratom extracts, including HK, KY, KD, and KG, demonstrated moderate hyaluronidase inhibition (61.56 ± 0.51–70.31 ± 1.66%), with no statistically significant difference among several samples, indicating a relatively similar bioactive compound profile. The notable hyaluronidase inhibitory effect of mitragynine observed in this study may be mechanistically associated with its well-documented anti-inflammatory properties, which may also contribute indirectly to the regulation of hyaluronidase activity.

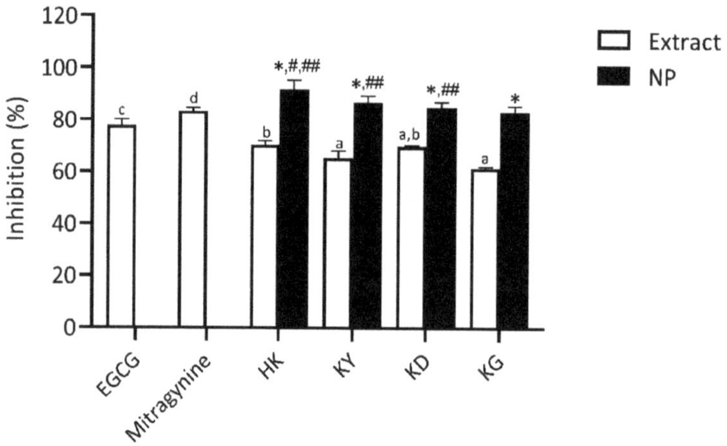

Figure 9. Hyaluronidase inhibitory activity of kratom leaf extracts and their NP formulations compared to epigallocatechin gallate (EGCG) and mitragynine. All samples were tested at a concentration of 1 mg/mL. Bars represent mean ± SD (n = 3). Different letters indicate significant differences among extracts ($p < 0.05$, one-way ANOVA with Tukey's test). Symbols from pairwise t-tests ($p < 0.05$): * vs. corresponding extract; # vs. mitragynine; ## vs. EGCG.

Following NP formulations, all kratom leaf extracts showed significantly enhanced hyaluronidase inhibitory activity. HK NP demonstrated the highest inhibitory activity (91.67 ± 3.56%), exceeding that of its corresponding extract and statistically comparable to mitragynine. Other NP samples—KY NP (86.83 ± 2.48%), KD NP (84.98 ± 2.06%), and KG NP (83.23 ± 2.34%) also showed marked improvements over their respective extracts. Other NP samples—KY NP (86.3 ± 2.5%), KD NP (84.9 ± 2.1%), and KG NP (83.2 ± 2.3%) also showed marked improvements over their respective extracts. These results suggest that nanoencapsulation effectively preserves or enhances the hyaluronidase-inhibiting activity of the bioactive constituents, highlighting their potential for skincare applications aimed at retaining skin moisture by protecting hyaluronic acid from enzymatic degradation.

The significant inhibition of hyaluronidase by mitragynine may also be attributed to its anti-inflammatory properties. Hyaluronidase not only degrades hyaluronic acid but also promotes inflammation by generating low-molecular-weight hyaluronic acid fragments that act as damage-associated molecular patterns (DAMPs). These fragments can trigger inflammation and tissue degradation [33]. Mitragynine, the main indole alkaloid in *Mitragyna speciosa*, has been shown to inhibit pro-inflammatory mediators such as TNF-α, IL-1β, and COX-2, while suppressing NF-κB signaling, all of which are closely involved in inflammation and aging processes [34,35]. Similarly, ginkgoside B dimethyl ester, an indole alkaloid N-glycoside derived from *Ginkgo biloba*, has been reported to protect dermal fibroblasts by reducing MMP-1 expression, restoring collagen levels, and inhibiting MAPK and Akt signaling pathways [36]. These findings suggest that mitragynine may exert protective effects on the skin through similar anti-inflammatory and anti-degradative mechanisms.

The anti-aging related enzyme activities indicate that while mitragynine contributes to enzyme inhibition, especially collagenase and hyaluronidase, its activity alone does not fully explain the observed effects. EGCG, used as a positive control, exhibited superior elastase and collagenase inhibition, reinforcing its role as a potent anti-aging standard. Therefore, the anti-enzyme effects in kratom extracts and NPs are likely involving not only mitragynine but also polyphenolic compounds.

Cytotoxicity Evaluation on HaCaT Cells

The cytotoxicity of kratom leaf extracts and their NP formulations was evaluated using the MTT assay in HaCaT human keratinocyte cells across a range of log concentrations. As shown in Figure 10A,B, both crude extracts and NP formulations demonstrated a dose-dependent reduction in cell viability, with varying IC_{50} values. Among the crude extracts, HK and KD exhibited the highest IC_{50} values (1265.20 $\pm$ 325.44 and 1252.57 $\pm$ 430.55 μg/mL, respectively), indicating a lower cytotoxicity. KG and KY extracts showed moderately lower IC_{50} values (896.70 $\pm$ 37.65 and 762.55 $\pm$ 15.34 μg/mL, respectively), suggesting a comparatively higher cytotoxicity. For the NP formulations, KD NP exhibited the highest IC_{50} value (1232.00 $\pm$ 140.50 μg/mL), followed by KG NP (718.33 $\pm$ 118.47 μg/mL), indicating an improved cytocompatibility relative to their respective crude extracts. In contrast, HK NP showed a reduced IC_{50} (447.67 $\pm$ 77.88 μg/mL), suggesting a slightly higher cytotoxicity than its crude extract. KY NP had an intermediate IC_{50} value of 550.87 $\pm$ 33.80 μg/mL. Overall, most extracts and NP formulations maintained over 80% cell viability at concentrations below 100 μg/mL. The findings suggest that kratom samples, particularly in nanoparticle form, are generally well tolerated by HaCaT cells at moderate concentrations. The improved IC_{50} values observed for KD NP and KG NP support their potential for safer application in skin-related formulations.

The IC_{50} values derived from the cytotoxicity assay against HaCaT keratinocyte cells were statistically analyzed using Tukey's multiple comparisons test. Comparisons among the crude extracts, HK, KD, KY, and KG, revealed no statistically significant differences ($p > 0.05$ for all pairwise comparisons), indicating comparable cytotoxicity profiles among these extracts. When comparing the nanoparticle formulations, significant differences were observed between KD NP and both KY NP and HK NP, suggesting that KY NP and HK NP exerted stronger cytotoxic effects than KD NP. However, no significant differences were found among the other nanoparticle groups, indicating a relatively similar cytotoxicity among KY NP, HK NP, and KG NP. When comparing each extract with its corresponding nanoparticle formulation, only HK NP demonstrated a significantly lower IC_{50} value compared to the HK extract ($p = 0.0031$), suggesting an enhanced cytotoxicity through nanoparticle delivery. In contrast, the differences between extract and nanoparticle forms for KD, KY, and KG were not statistically significant ($p > 0.05$). These findings suggest that nanoparticle encapsulation selectively enhanced the cytotoxic potential of the HK extract, while other formulations showed comparable activity to their crude counterparts.

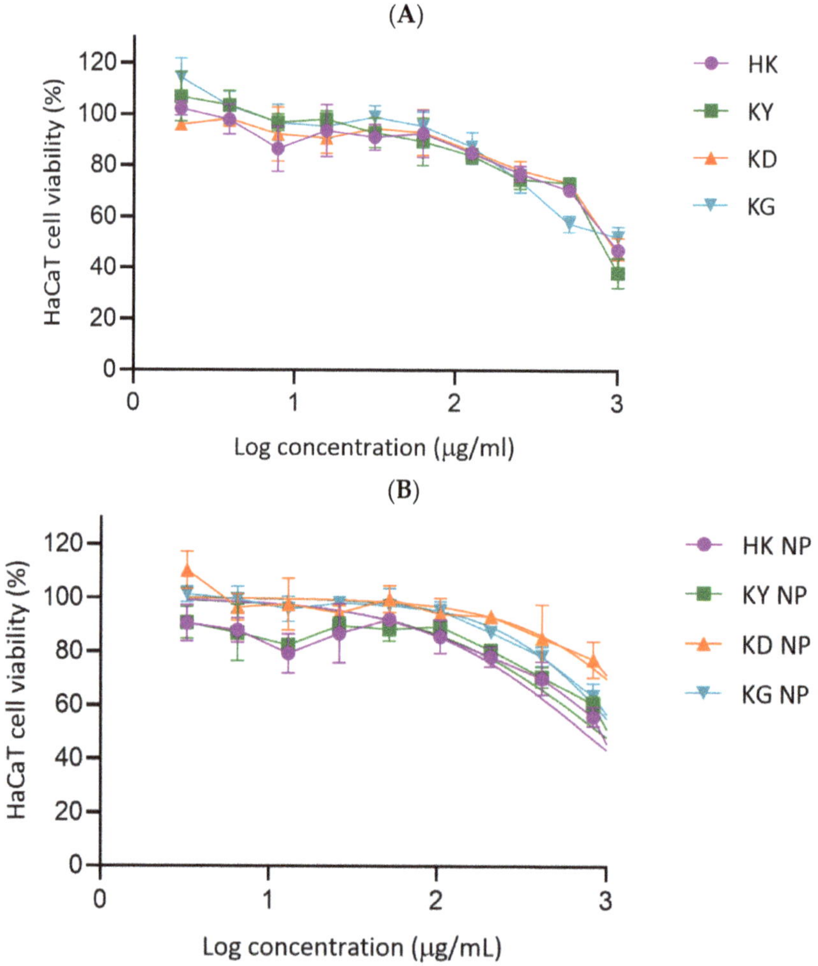

Figure 10. Cytotoxicity of kratom extracts and their NPs formulations in HaCaT cells. (**A**) Cell viability following exposure to increasing concentrations of crude extracts (HK, KY, KD, and KG). (**B**) Cell viability following exposure to nanoparticle formulations (HK NP, KY NP, KD NP, and KG NP). Cell viability was measured using the MTT assay and expressed as a percentage relative to untreated control cells. Data represent mean ± SD ($n = 3$).

Preliminary Physical and Sensory Evaluation of Kratom Nanoparticle-Loaded Gels

All four kratom NP-based gel formulations containing HK NP, KY NP, KD NP, and KG NP exhibited a clear to translucent, brownish appearance, indicating good clarity and the uniform dispersion of the NPs within the gel matrix (Figure 11). The assessment of physical and sensory attributes such as texture, spreadability, and tackiness was performed subjectively by the researchers through visual inspection and manual handling. While this preliminary evaluation suggested acceptable characteristics for topical application, no formal sensory panel or instrumental texture analysis was conducted, and no human volunteer testing was performed.

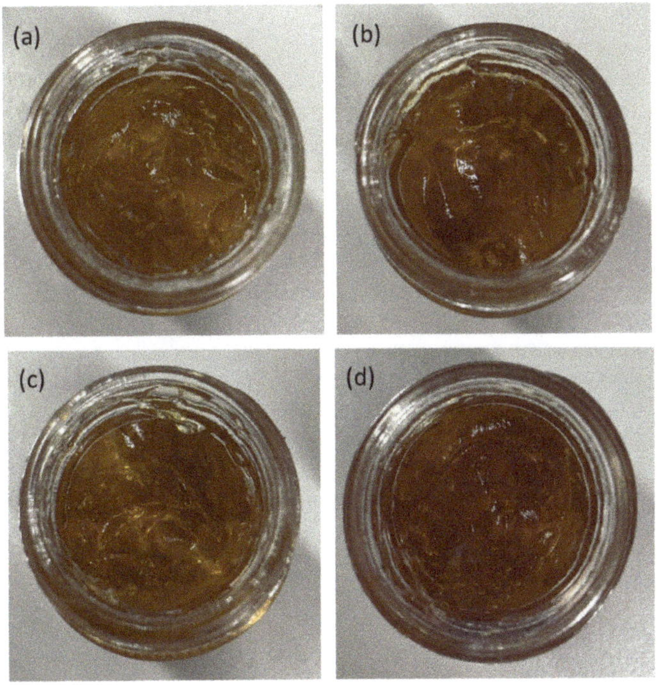

Figure 11. Macroscopic appearance of kratom NP-loaded gel formulations. Photographs of gels containing (**a**) HK NPs, (**b**) KY NPs, (**c**) KD NPs, and (**d**) KG NPs. All formulations exhibited a clear to translucent amber color with uniform consistency, indicating successful incorporation of kratom nanoparticles into the gel matrix. No phase separation, precipitation, or visual signs of instability were observed.

Characterization of Kratom Nanoparticle-Loaded Gels and Their Stability

Physical characteristics

After six heating/cooling cycles in the stability assay, the physical characteristics of all four kratom NP-loaded gel formulations had favorable physical properties. Each gel appeared clear and brownish, with a uniform texture and no visible phase separation, indicating good homogeneity and proper dispersion of active ingredients.

pH Measurement

All freshly prepared kratom nanoparticle-loaded gel formulations demonstrated near-neutral to slightly alkaline pH values. KD NP and KY NP gels both exhibited the initial highest pH, averaging 7.83 ± 0.05 and 7.83 ± 0.03, respectively, followed by KG NP gel at 7.56 ± 0.04 and HK NP gel at 7.27 ± 0.06. After undergoing six cycles of heating and cooling in a stability assay, a slight increase in pH was observed in most formulations. KD NP and KY NP gels again recorded the highest pH values at 7.88, while HK NP gel increased marginally to 7.33 ± 0.18. KG NP gel maintained a stable pH of 7.56 ± 0.09.

These findings suggest overall stability in pH across all formulations, with only minimal changes following thermal stress.

The pH values of the formulations ranged from 7.27 to 7.83, which are near-neutral and within the acceptable range for topical application (pH 5.3–7.6). These values suggest a low likelihood of skin irritation and support the potential for safe dermal use. Moreover, the low standard deviation among repeated measurements indicates good formulation consistency and batch reproducibility.

Rheological Evaluation

Upon fresh preparation, the KG NP-loaded gel displayed the highest viscosity (12.602 ± 0.294 Pa·s), followed closely by the KY NP-loaded gel (12.291 ± 0.038 Pa·s) and HK NP (11.898 ± 0.139 Pa·s). The KD NP-loaded gel had the lowest initial viscosity at 11.518 ± 0.072 Pa·s. After six cycles of heating and cooling in the stability assay, a slight reduction in viscosity was noted across all samples, though the relative order remained unchanged. The KG NP gel still showed the highest viscosity (11.674 ± 0.097 Pa·s), followed by HK NP (11.197 ± 0.163 Pa·s), KY NP (11.073 ± 0.233 Pa·s), and KD NP (10.719 ± 0.261 Pa·s). The blank gel formulation exhibited intermediate viscosity values at both time points (11.885 ± 0.225 Pa·s initially and 11.577 ± 0.071 Pa·s after thermal cycling), indicating that the presence of the kratom extract influenced the rheological behavior.

Collectively, these findings indicate that the kratom NP gel formulations maintained desirable physical characteristics under the studied conditions. The combination of stable appearance, acceptable pH, and consistent viscosity suggests that these formulations are suitable candidates for topical delivery. Nevertheless, accelerated and long-term stability testing under standard recommended storage conditions would be essential to confirm their robustness for commercial application.

3. Conclusions

Among the four *Mitragyna speciosa* extracts, KG exhibited the highest TPC and TFC. Antioxidant activities varied depending on the assay, with the KY extract showing the strongest DPPH and FRAP activities, while KG led in ABTS scavenging. KG showed the highest collagenase inhibition, whereas the HK extract was most effective against elastase and hyaluronidase. Upon nanoparticle formulation, all activities were enhanced. HK NP demonstrated the strongest DPPH and ABTS scavenging capacities as well as the greatest inhibitory effects against collagenase, elastase, and hyaluronidase. The kratom NP-loaded gels also exhibited desirable physicochemical properties, including suitable pH, consistent viscosity, and good visual stability, supporting their suitability for topical formulations.

4. Materials and Methods

4.1. Materials

Gallic acid, Griess reagent, DPPH (2,2-diphenyl-1-picrylhydrazyl), TPTZ (2,4,6-Tris(2-pyridyl)-s-triazine), and ABTS (2,2′-azino-bis(3-ethylbenzothiazoline-6-sulfonic acid)) were obtained from Sigma-Aldrich (St. Louis, MO, USA). Chemicals, including absolute ethanol, dimethyl sulfoxide (DMSO), sodium bicarbonate, sodium nitrate, acetic acid, and sodium hydroxide, were supplied by RCI Labscan (Bangkok, Thailand). Iron(III) chloride hexahydrate and 37% hydrochloric acid were purchased from Qrec Chemical Co., Ltd. (Auckland, New Zealand). The Folin–Ciocalteu phenol reagent, aluminum chloride, sodium acetate trihydrate, ferrous sulfate heptahydrate (purity ≥ 99%), and potassium persulfate were sourced from Loba Chemie (Mumbai, India). Ascorbic acid and poloxamer 407 were acquired from Chanjao Longevity Co., Ltd. (Bangkok, Thailand). Dulbecco's Modified Eagle's Medium (DMEM, high glucose), fetal bovine serum (FBS), penicillin-streptomycin, and trypsin-EDTA were provided by Gibco (Waltham, MA, USA). Clostridium histolyticum collagenase (EC.3.4.23.3), porcine pancreatic elastase (PE–E.C.3.4.21.36), bovine testis hyaluronidase (E.C.3.2.1.3.5), a synthetic peptide of N-[3-(2-furyl) acryloyl]-Leu-Gly-Pro–Ala (FALGPA), N-succinyl-Ala–Ala–Ala–p-nitroanilide (AAAPVN), epigallocatechin gallate (EGCG, 98% purity), and hyaluronic acid were purchased from Sigma-Aldrich (St. Louis, MO, USA).

4.2. Methods

4.2.1. Extraction of Kratom (*Mitragyna speciosa* (Korth.) Havil.)

Various kratom leaf samples were collected for this study, including Kan Daeng (KD), Yak Yai or Hang Kang (HK), Tai Bai-yao (KY), and Tang Kua or Kan Keaw (KG) (Figure 12). The kratom samples were collected from reputable plantations in Thailand, specifically from Surat Thani Province (9.00546° N, 98.94927° E for KD and 9.03296° N, 98.96132° E for KG) and Pathum Thani Province (14.10881° N, 100.42706° E for KY and HK). The plant materials were identified based on morphological characteristics, namely, leaf, flower, and fruit structures, by comparison with the taxonomic descriptions reported by Ngernsaengsaruay et al. [37]. To preserve thermolabile phytochemicals in kratom leaves, drying was conducted at 50 °C, a temperature shown to be effective in preventing the degradation of key alkaloids such as mitragynine. This is supported by Orio et al., who reported that extraction at 50 °C did not significantly compromise the alkaloid content in *Mitragyna speciosa* leaf extracts [38]. Dohárszky et al. demonstrated that ultrasound-assisted extraction at 50 °C yielded a high mitragynine content, comparable or superior to lower temperatures and substantially better than higher-temperature Soxhlet extraction, which caused alkaloid degradation [39]. Once dried, the leaves were ground into fine powder using an electric grinder. A total of 400 g of the powdered sample was subjected to extraction with 4000 mL of ethanol (10:1 solvent-to-sample ratio) using a Soxhlet apparatus (Haberg, Germany) for 20 h. The solvent was subsequently removed under reduced pressure using a rotary evaporator. The obtained crude extracts were stored in amber bottles and kept refrigerated for further experimental use.

Figure 12. The genetic variation of *Mitragyna speciosa* (kratom) analyzed in this study included the following accessions: (**A**) Kan Daeng (KD), (**B**) Hang Kang (HK), (**C**) Tai Bai-yao (KY), and (**D**) Kan Keaw (KG).

4.2.2. Formulation and Characterization of Kratom Nanoparticles

Kratom nanoparticles were prepared using the solvent replacement method. Briefly, kratom leaf extracts from KD, HK, KY, and KG (800 µL of 0.125 g/2 mL in 95% ethanol), centrifuged at 13,000 rpm for 3 min to precipitate the undissolved extract residue, were introduced dropwise into a mixture of 13.5 mL of 0.1% *w/v* poloxamer 407 and 1.5 mL of polyethylene glycol 400 (PEG 400) [12,16]. PEG 400 was used to enhance the solubilization of hydrophobic phytochemicals and improve nanoparticle stabilization. The addition was performed at a constant rate of 1 mL/h under continuous magnetic stirring at 800 rpm, and the system was maintained at room temperature (30 °C) to facilitate nanoparticle formation via rapid solvent diffusion and precipitation of the bioactive constituents. The resulting formulations were labeled as KD NP, HK NP, KY NP, and KG NP. To produce blank NPs, the same method was followed, omitting the kratom leaf extract. Kratom NPs were analyzed for particle size, polydispersity index (PDI), and zeta potential using a Zetasizer (Malvern Instruments, Worcestershire, UK). The dynamic light scattering (DLS) measurements were conducted using water as the dispersant, with a refractive index of 1.330 and a viscosity of 0.8872 cP at 25 °C. The material refractive index was set to 1.590 with an absorption value of 0.010, as selected from the instrument database. Measurements were performed at a backscatter angle of 173° using a DTS0012 cell. The samples were allowed to equilibrate for 120 s prior to measurement. To ensure an accurate particle size analysis and avoid multiple scattering, the formulations were ten-fold diluted with deionized water prior to measurement.

4.2.3. Total Phenolic Content Determination

The total phenolic content in kratom extracts (122.07–62,500 µg/mL) and kratom NPs (6.45–3300 µg/mL) was determined using a microplate-based Folin–Ciocalteu assay. Briefly, samples (50 µL) were added to 96-well plates, followed by 100 µL of 10% Folin–Ciocalteu reagent [40]. After 4 min incubation, 50 µL of 10% sodium carbonate was added, and the reaction was left in the dark at room temperature for 60 min. Absorbance was measured at 765 nm. Gallic acid (1.95–1000 µg/mL) served as the standard, and results were expressed as gallic acid equivalent (GAE).

4.2.4. Total Flavonoid Content Determination

Kratom extracts (122.07–62,500 µg/mL), kratom NPs (6.45–3300 µg/mL), and standards (quercetin: 1.95–1000 µg/mL; EGCG: 1.95–1000 µg/mL) were dispensed (50 µL/well) into 96-well plates. The concentration ranges for the extracts and nanoparticles differed due to differences in formulation constraints. For the nanoparticle formulations, lower concentrations were used because of the limited drug loading efficiency, which necessitated controlling the concentration to maintain an optimal particle size, polydispersity index, and physical stability. In contrast, higher concentrations of crude extracts were required to achieve measurable activity levels in the assay. To each well, 30 µL of 5% sodium nitrate was added, incubated for 5 min, followed by 50 µL of 2% aluminum chloride [40]. After a 6 min reaction time, 50 µL of 1 N sodium hydroxide was added, and the plate was incubated for 10 min. The absorbance was recorded at 510 nm. The flavonoid content was quantified using standard calibration curves and expressed as the quercetin equivalent (QE) for kratom extracts and EGCG equivalent (EGCGE) for kratom nanoparticle formulations.

4.2.5. HPLC Quantification of Bioactive Markers

The HPLC method for mitragynine analysis was adapted from a previously published report [41]. The HPLC analysis of the bioactive marker mitragynine was performed using a Waters Alliance e2695 system (Milford, MA, USA) equipped with a quaternary pump, autosampler, column thermostat, and photodiode array (PDA) detector. Data processing was conducted using the Empower 3 software (Waters, Milford, MA, USA). Chromatographic separation was achieved on a BDS Hypersil C18 analytical column (4.6 × 250 mm, 5 µm; Thermo Scientific, Waltham, MA, USA). The mobile phase consisted of an isocratic mixture of acetonitrile and 10 mM ammonium formate (65:35, *v*/*v*), delivered at a flow rate of 1.0 mL/min. The column was maintained at ambient temperature. Detection was carried out at 225 nm, which is the optimal wavelength for mitragynine. The injection volume for all standard and sample solutions was 5 µL per run. The mitragynine standard was used for calibration and quantification. Retention time and peak area were compared against the standard curve generated from known concentrations of mitragynine to determine its content in the samples.

4.2.6. DPPH Radical Scavenging Assay

The antioxidant capacity of kratom extracts, kratom NPs, mytragynine (1.95–1000 µg/mL), and the antioxidant standard (ascorbic acid: 1.95–1000 µg/mL) was assessed using the DPPH radical scavenging assay. Samples (100 µL) were pipetted into 96-well microplates, followed by an equal volume (100 µL) of 0.1 mM DPPH solution. After a 30 min incubation in the dark at room temperature, absorbance at 517 nm was recorded [42]. Radical scavenging activity was calculated using the following equation:

$$\text{DPPH Scavenging } (\%) = \left[1 - \left(\frac{Abs_{sample}}{Abs_{control}}\right)\right] \times 100$$

where $Abs_{control}$ represents the absorbance of the control reaction (reagent with DI water), and Abs_{sample} is the absorbance of the test sample (extracts or nanoparticles with reagent).

4.2.7. ABTS Radical Scavenging Assay

Briefly, 20 µL of each test sample (kratom extracts, kratom NPs, mytragynine, and ascorbic acid at 1.95–1000 µg/mL) were combined with 180 µL of 0.1 mM ABTS•+ solution in 96-well microplates. After 15 min incubation in the dark at room temperature, absorbance was measured at 734 nm [42]. Antioxidant activity was calculated as follows:

$$\text{ABTS Scavenging } (\%) = \left[1 - \left(\frac{Abs_{sample}}{Abs_{control}}\right)\right] \times 100$$

where $Abs_{control}$ represents the absorbance of the control reaction (reagent with DI water), and Abs_{sample} is the absorbance of the test sample (extracts or nanoparticles with the reagent).

4.2.8. Ferric Reducing Antioxidant Power Assay

The FRAP assay was used to assess the reducing power of kratom extracts, kratom NPs, mytragynine, and ascorbic acid (1.95–1000 µg/mL). A 20 aliquot of each sample was mixed with 180 µL of freshly prepared FRAP reagent (acetate buffer, 300 mM, pH 3.6; ferric chloride, 20 mM; and TPTZ, 10 mM in a 10:1:1 ratio). After 30 min of incubation at 37 °C, absorbance at 595 nm was measured [42]. A standard curve was generated using ferrous sulfate (9.8–5000 µM) for calculating FRAP values.

4.2.9. Collagenase Inhibitory Assay

The collagenase inhibitory activity of kratom extracts, kratom NPs, mytragynine, and EGCG was determined following the method described by Preedalikit et al. [29]. A collagenase enzyme solution was prepared using a 50 mM tricine buffer (pH 7.5) supplemented with 400 mM sodium chloride and 10 mM calcium chloride. The substrate used was the synthetic peptide FALGPA, which mimics the structural features of collagen. To prepare the substrate solution, FALGPA was dissolved in the same tricine buffer to achieve a final concentration of 2 mM. Test samples (10 μL at 1 mg/mL of kratom extracts, kratom NPs, mytragynine, and EGCG) were mixed with 40 μL of a collagenase enzyme solution and allowed to incubate at room temperature for 15 min. The enzymatic reaction was initiated by adding 50 μL of the substrate solution. The absorbance was then monitored immediately at 340 nm in kinetic mode using a microplate reader (Synergy H1, Agilent Technologies, Santa Clara, CA, USA). The percentage of collagenase inhibition was calculated using the following formula:

$$\text{Collagenase inhibition } (\%) = \frac{Abs_{control} - Abs_{sample}}{Abs_{control}} \times 100$$

where $Abs_{control}$ represents the absorbance of the control reaction (containing DI water, enzyme, and substrate), and Abs_{sample} corresponds to the absorbance of the test sample (containing extract or nanoparticles, enzyme, and substrate).

4.2.10. Elastase Inhibitory Assay

The elastase inhibitory activity was evaluated based on the method by Preedalikit et al. [29], using N-succinyl-Ala-Ala-Ala-p-nitroanilide (AAAPVN) as a specific substrate. Briefly, 50 μL of the kratom extracts, kratom NPs, mytragynine, or EGCG at a concentration of 1 mg/mL were pre-incubated with 25 μL of elastase enzyme solution (2 mg/mL, prepared in 100 mM Tris-HCl buffer, pH 8.0) at room temperature for 20 min. Following this, 25 μL of the AAAPVN substrate (4.4 mM in Tris-HCl buffer) was added to initiate the enzymatic reaction. The change in absorbance at 410 nm was monitored immediately in kinetic mode using a microplate reader (Synergy H1, Agilent Technologies, Santa Clara, CA, USA). The percentage inhibition of elastase activity was calculated using the following equation:

$$\text{Elastase inhibition } (\%) = \frac{Abs_{control} - Abs_{sample}}{Abs_{control}} \times 100$$

where $Abs_{control}$ is the absorbance of the control reaction (comprising DI water, elastase enzyme, and substrate), and Abs_{sample} is the absorbance of the test reaction containing the extract or nanoparticles, enzyme, and substrate.

4.2.11. Hyaluronidase Inhibitory Assay

The inhibitory effect of the samples on hyaluronidase activity was assessed using a turbidimetric method based on the procedure by Preedalikit et al. [29]. This assay evaluates the extent of hyaluronic acid degradation catalyzed by hyaluronidase enzyme. Residual, undegraded hyaluronic acid forms a turbid precipitate, and thus, the degree of turbidity correlates inversely with enzyme activity. Kratom extracts, kratom NPs, mitragynine, and EGCG (50 μL each) were pre-incubated with 100 μL of hyaluronidase solution (2 mg/mL) at 37 ± 5 °C for 10 min. After pre-incubation, 100 μL of 0.03% (*w*/*v*) hyaluronic acid in 300 mM phosphate buffer (pH 5.35) was added, and the mixture was incubated again at 37 ± 5 °C for 45 min. To precipitate the remaining hyaluronic acid, 1 mL of acetic albumin solution (prepared from sodium acetate, acetic acid, and bovine serum albumin, pH 3.75) was added. The mixture was left to stand at room temperature for 10 min to allow turbidity to develop. Absorbance was then measured at 600 nm using a microplate reader (Synergy H1, Agilent Technologies, Santa Clara, CA, USA). The percentage of hyaluronidase inhibition was calculated using the following equation:

$$\text{Hyaluronidase inhibition } (\%) = \left(\frac{Abs_{sample}}{Abs_{control}}\right) \times 100$$

where $Abs_{control}$ refers to the absorbance of the control reaction containing DI water, hyaluronic acid, and acetic albumin solution (without enzyme), and Abs_{sample} represents the absorbance of the reaction containing the extract or nanoparticles, enzyme, hyaluronic acid, and acetic albumin solution.

4.2.12. Cell Culture and Cytotoxicity Assay

The HaCaT cell line was cultured in high-glucose DMEM supplemented with 10% FBS and 1% penicillin-streptomycin. Cultures were maintained at 37 °C in a humidified 5% CO_2 incubator and subcultured every three days using 0.25% trypsin-EDTA. For the cytotoxicity assay, HaCaT keratinocytes were seeded in 96-well plates at a density of 2×10^3 cells per well and incubated for 24 h to allow cell attachment. Cells were then treated with kratom extracts and NPs for 24 h. After treatment, MTT solution (0.5 mg/mL, 100 μL) was added to each well and incubated at 37 °C for 2 h. Formazan crystals formed by viable cells were dissolved in 100 μL of dimethylsulfoxide (DMSO), and absorbance was measured at 550 nm using a microplate reader. The IC_{50} values were calculated using GraphPad Prism v8.0 (La Jolla, CA, USA), and cell viability was determined using the following formula:

$$\text{Cell viability } (\%) = \left(\frac{Abs_{sample}}{Abs_{control}}\right) \times 100$$

where $Abs_{control}$ refers to the absorbance of the untreated cells and Abs_{sample} represents the absorbance of the treated cells.

4.2.13. Preparation of the Kratom Nanoparticle-Loaded Gel

A 100 g batch of the nanoparticle-loaded gel was prepared using carbomer-based formulation. Purified water (83 g) was transferred into a beaker and stirred at 1200 rpm using a magnetic stirrer. Carbomer 980 (1.5 g) was gradually sprinkled into the water with continuous gentle stirring with a glass rod until a homogeneous dispersion was achieved. Glycerin (5 g) was then added and mixed thoroughly. Triethanolamine was added dropwise to adjust the pH until a clear and viscous gel base was formed. Subsequently, the previously prepared kratom NP suspension (10 g) was incorporated and mixed until uniformly dispersed. Finally, phenoxyethanol (0.5 g) was added as a preservative and the mixture was stirred until a homogeneous gel was obtained. The finished gel was filled into suitable containers for storage.

4.2.14. Characterization of Kratom Nanoparticle-Loaded Gel

pH Measurement

The pH of the kratom nanoparticle-loaded gel was measured in triplicate at room temperature using a calibrated pH meter (Thermo Fisher Scientific, Waltham, MA, USA). To evaluate the formulation's stability, pH measurements were also conducted after subjecting the samples to a heating/cooling stability test consisting of six complete cycles. Each cycle involved storing the gel for 24 h at 45 °C followed by 24 h at 4 °C.

Rheological Evaluation

The rheological properties of the kratom nanoparticle-loaded gel were assessed using a rotational rheometer (Anton Paar MCR 102e Rheometer, Graz, Austria) equipped with a plate–plate configuration. Measurements were performed at a constant shear rate of 100 s^{-1} while gradually increasing the temperature from 25 °C to 35 °C at a rate of 1 °C/min. Rheograms were generated using Rheocompass software version 1.33. To investigate thermal stability and viscoelastic consistency under stress conditions, the rheological behavior was also evaluated before and after the six-cycle heating/cooling stability test (as described above).

4.2.15. Statistical Analysis

All experimental data are expressed as the mean ± standard deviation (SD) from three independent experiments (n = 3). A statistical analysis was performed using a one-way analysis of variance (ANOVA), followed by Tukey's multiple comparisons post hoc test to determine significant differences between groups. A p-value of less than 0.05 was considered statistically significant. All analyses were conducted using GraphPad Prism version 8.02 (GraphPad Software, La Jolla, CA, USA).

The English language of this manuscript was reviewed and refined using AI-based language tools (ChatGPT (GPT-4), OpenAI, San Francisco, CA, USA) to improve clarity, grammar, and readability prior to submission.

Author Contributions: Conceptualization, M.N.T. and C.C.; Methodology, P.A. (Pimporn Anantaworasakul), W.P., P.A. (Phunsuk Anantaworasakul), S.S., A.I., W.A., M.N.T. and C.C.; Validation, M.N.T., S.Y. and C.C.; Formal analysis, P.A. (Pimporn Anantaworasakul), W.P., P.A. (Phunsuk Anantaworasakul), S.Y. and C.C.; Investigation, P.A. (Pimporn Anantaworasakul), W.P., P.A. (Phunsuk Anantaworasakul), S.S., A.I., W.A. and C.C.; Resources, P.A. (Phunsuk Anantaworasakul), A.I. and W.A.; Writing – original draft, P.A. (Pimporn Anantaworasakul), W.P., S.Y. and C.C.; Writing – review & editing, P.A. (Pimporn Anantaworasakul), W.P., P.A. (Phunsuk Anantaworasakul), S.S., A.I., W.A., M.N.T., S.Y. and C.C.; Project administration, C.C. All authors have read and agreed to the published version of the manuscript.

Funding: This research project was supported by the Fundamental Fund 2025, Chiang Mai University, and Thailand Science Research and Innovation (TSRI) (FRB680102/0162).

Institutional Review Board Statement: Not applicable.

Informed Consent Statement: Not applicable.

Data Availability Statement: The original contributions presented in this study are included in the article. Further inquiries can be directed to the corresponding author.

Acknowledgments: The authors acknowledge the use of ChatGPT (OpenAI) for assistance in improving the clarity and grammar of the manuscript's English language.

Conflicts of Interest: The authors declare no conflict of interest.

References

1. Hussein, R.S.; Bin Dayel, S.; Abahussein, O.; El-Sherbiny, A.A. Influences on Skin and Intrinsic Aging: Biological, Environmental, and Therapeutic Insights. *J. Cosmet. Dermatol.* **2025**, *24*, e16688. [CrossRef] [PubMed]
2. Papaccio, F.; D'Arino, A.; Caputo, S.; Bellei, B. Focus on the Contribution of Oxidative Stress in Skin Aging. *Antioxidants* **2022**, *11*, 1121. [CrossRef] [PubMed]
3. Jiratchayamaethasakul, C.; Ding, Y.; Hwang, O.; Im, S.-T.; Jang, Y.; Myung, S.-W.; Lee, J.M.; Kim, H.-S.; Ko, S.-C.; Lee, S.-H. In vitro screening of elastase, collagenase, hyaluronidase, and tyrosinase inhibitory and antioxidant activities of 22 halophyte plant extracts for novel cosmeceuticals. *Fish. Aquat. Sci.* **2020**, *23*, 6. [CrossRef]

4. Madan, K.; Nanda, S. In-vitro evaluation of antioxidant, anti-elastase, anti-collagenase, anti-hyaluronidase activities of safranal and determination of its sun protection factor in skin photoaging. *Bioorg. Chem.* **2018**, *77*, 159–167. [CrossRef]
5. Nawaka, N.; Lertcanawanichakul, M.; Porntadavity, S.; Pussadhamma, B.; Jeenduang, N. Kratom leaf extracts exert hypolipidaemic effects via the modulation of PCSK9 and LDLR pathways in HepG2 cells. *Sci. Rep.* **2025**, *15*, 15696. [CrossRef]
6. Singh, D.; Narayanan, S.; Müller, C.; Swogger, M.; Chear, N.; Dzulkapli, E.; Yusoff, N.; Ramachandram, D.; León, F.; McCurdy, C.; et al. Motives for Using Kratom (*Mitragyna speciosa* Korth.) among Regular Users in Malaysia. *J. Ethnopharmacol.* **2018**, *233*, 34–40. [CrossRef]
7. Hossain, R.; Sultana, A.; Nuinoon, M.; Noonong, K.; Tangpong, J.; Hossain, K.H.; Rahman, M.A. A Critical Review of the Neuropharmacological Effects of Kratom: An Insight from the Functional Array of Identified Natural Compounds. *Molecules* **2023**, *28*, 7372. [CrossRef]
8. Sornsenee, P.; Chimplee, S.; Romyasamit, C. Evaluation of Antibacterial, Antibiofilm, Antioxidant, and Anti-Inflammatory Activities of Kratom Leaves (*Mitragyna speciosa*) Fermentation Supernatant Containing Lactobacillus rhamnosus GG. *Probiotics Antimicrob. Proteins* **2025**, *17*, 328–340. [CrossRef]
9. Eastlack, S.C.; Cornett, E.M.; Kaye, A.D. Kratom-Pharmacology, Clinical Implications, and Outlook: A Comprehensive Review. *Pain Ther.* **2020**, *9*, 55–69. [CrossRef]
10. Ramanathan, S.; Parthasarathy, S.; Murugaiyah, V.; Magosso, E.; Tan, S.C.; Mansor, S.M. Understanding the physicochemical properties of mitragynine, a principal alkaloid of Mitragyna speciosa, for preclinical evaluation. *Molecules* **2015**, *20*, 4915–4927. [CrossRef]
11. Chittasupho, C.; Samee, W.; Na Takuathung, M.; Okonogi, S.; Nimkulrat, S.; Athikomkulchai, S. Clerodendrum chinense Stem Extract and Nanoparticles: Effects on Proliferation, Colony Formation, Apoptosis Induction, Cell Cycle Arrest, and Mitochondrial Membrane Potential in Human Breast Adenocarcinoma Breast Cancer Cells. *Int. J. Mol. Sci.* **2024**, *25*, 978. [CrossRef] [PubMed]
12. Chittasupho, C.; Umsumarng, S.; Srisawad, K.; Arjsri, P.; Phongpradist, R.; Samee, W.; Tingya, W.; Ampasavate, C.; Dejkriengkraikul, P. Inhibition of SARS-CoV-2-Induced NLRP3 Inflammasome-Mediated Lung Cell Inflammation by Triphala-Loaded Nanoparticle Targeting Spike Glycoprotein S1. *Pharmaceutics* **2024**, *16*, 978. [CrossRef] [PubMed]
13. White, J.M.; Calabrese, M.A. Impact of small molecule and reverse poloxamer addition on the micellization and gelation mechanisms of poloxamer hydrogels. *Colloids Surf. A Physicochem. Eng. Asp.* **2022**, *638*, 128246. [CrossRef] [PubMed]
14. Oh, K.; Bronich, T.; Kabanov, A. Micellar formulations for drug delivery based on mixtures of hydrophobic and hydrophilic Pluronic((R)) block copolymers. *J. Control. Release Off. J. Control. Release Soc.* **2004**, *94*, 411–422. [CrossRef]
15. Ghezzi, M.; Pescina, S.; Padula, C.; Santi, P.; Del Favero, E.; Cantù, L.; Nicoli, S. Polymeric micelles in drug delivery: An insight of the techniques for their characterization and assessment in biorelevant conditions. *J. Control. Release* **2021**, *332*, 312–336. [CrossRef]
16. Chittasupho, C.; Srisawad, K.; Arjsri, P.; Phongpradist, R.; Tingya, W.; Ampasavate, C.; Dejkriengkraikul, P. Targeting Spike Glycoprotein S1 Mediated by NLRP3 Inflammasome Machinery and the Cytokine Releases in A549 Lung Epithelial Cells by Nanocurcumin. *Pharmaceuticals* **2023**, *16*, 862. [CrossRef]
17. Tsarenko, E.; Schubert, U.S.; Nischang, I. Nanoparticle Formulation Composition Analysis by Liquid Chromatography on Reversed-Phase Monolithic Silica. *Anal. Chem.* **2023**, *95*, 565–569. [CrossRef]
18. Platzer, M.; Kiese, S.; Herfellner, T.; Schweiggert-Weisz, U.; Miesbauer, O.; Eisner, P. Common Trends and Differences in Antioxidant Activity Analysis of Phenolic Substances Using Single Electron Transfer Based Assays. *Molecules* **2021**, *26*, 1244. [CrossRef]
19. Bravo, K.; Alzate, F.; Osorio, E. Fruits of selected wild and cultivated Andean plants as sources of potential compounds with antioxidant and anti-aging activity. *Ind. Crops Prod.* **2016**, *85*, 341–352. [CrossRef]
20. Madhan, B.; Krishnamoorthy, G.; Rao, J.R.; Nair, B.U. Role of green tea polyphenols in the inhibition of collagenolytic activity by collagenase. *Int. J. Biol. Macromol.* **2007**, *41*, 16–22. [CrossRef]
21. Darwisa; Fachrialc, E. A Comparative Study of Peel and Seed Extract of Passion Fruit (*Passiflora edulis*) as. 2019. Available online: https://asrjetsjournal.org/index.php/American_Scientific_Journal/article/view/4687/1650 (accessed on 23 June 2025).
22. Annuar, N.A.K.; Azlan, U.K.; Mediani, A.; Tong, X.; Han, R.; Al-Olayan, E.; Baharum, S.N.; Bunawan, H.; Sarian, M.N.; Hamezah, H.S.; et al. An insight review on the neuropharmacological effects, mechanisms of action, pharmacokinetics and toxicity of mitragynine. *Biomed. Pharmacother.* **2024**, *171*, 116134. [CrossRef] [PubMed]
23. Utar, Z.; Majid, M.I.; Adenan, M.I.; Jamil, M.F.; Lan, T.M. Mitragynine inhibits the COX-2 mRNA expression and prostaglandin E_2 production induced by lipopolysaccharide in RAW264.7 macrophage cells. *J. Ethnopharmacol.* **2011**, *136*, 75–82. [CrossRef] [PubMed]
24. Rahmawati, S.I.; Indriani, D.W.; Ningsih, F.N.; Hardhiyuna, M.; Firdayani, F.; Ahmadi, P.; Rosyidah, A.I.; Septiana, E.; Dharmayanti, N.L.P.I.; Bayu, A.; et al. Dual anti-inflammatory activities of COX-2/5-LOX driven by kratom alkaloid extracts in lipopolysaccharide-induced RAW 264.7 cells. *Sci. Rep.* **2024**, *14*, 28993. [CrossRef] [PubMed]

25. Limcharoen, T.; Pouyfung, P.; Ngamdokmai, N.; Prasopthum, A.; Ahmad, A.R.; Wisdawati, W.; Prugsakij, W.; Warinhomhoun, S. Inhibition of α-Glucosidase and Pancreatic Lipase Properties of *Mitragyna speciosa* (Korth.) Havil. (Kratom) Leaves. *Nutrients* **2022**, *14*, 3909. [CrossRef]
26. Korkmaz, B.; Horwitz, M.S.; Jenne, D.E.; Gauthier, F. Neutrophil elastase, proteinase 3, and cathepsin G as therapeutic targets in human diseases. *Pharmacol. Rev.* **2010**, *62*, 726–759. [CrossRef]
27. Fulop, T.; Khalil, A.; Larbi, A. The role of elastin peptides in modulating the immune response in aging and age-related diseases. *Pathol. Biol.* **2012**, *60*, 28–33. [CrossRef]
28. Karim, A.A.; Azlan, A.; Ismail, A.; Hashim, P.; Abd Gani, S.S.; Zainudin, B.H.; Abdullah, N.A. Phenolic composition, antioxidant, anti-wrinkles and tyrosinase inhibitory activities of cocoa pod extract. *BMC Complement. Altern. Med.* **2014**, *14*, 381. [CrossRef]
29. Preedalikit, W.; Chittasupho, C.; Leelapornpisid, P.; Potprommanee, S.; Kiattisin, K. Comparison of Biological Activities and Protective Effects on PAH-Induced Oxidative Damage of Different Coffee Cherry Pulp Extracts. *Foods* **2023**, *12*, 4292. [CrossRef]
30. Jin, Q.; Zhao, Y.L.; Liu, Y.P.; Zhang, R.S.; Zhu, P.F.; Zhao, L.Q.; Qin, X.J.; Luo, X.D. Anti-inflammatory and analgesic monoterpenoid indole alkaloids of *Kopsia officinalis*. *J. Ethnopharmacol.* **2022**, *285*, 114848. [CrossRef]
31. Du, K.; Liang, Y.; Song, Z.; Zheng, C.; Lai, L.; Zong, K.; Wang, Y.; Meng, D. Monoterpenoid indole alkaloids from *Melodinus axillaris* W.T.Wang exhibit anti-inflammatory activities by inhibiting the NF-κB signaling pathways. *J. Ethnopharmacol.* **2024**, *324*, 117771. [CrossRef]
32. Jasiewicz, B.; Kozanecka-Okupnik, W.; Przygodzki, M.; Warżajtis, B.; Rychlewska, U.; Pospieszny, T.; Mrówczyńska, L. Synthesis, antioxidant and cytoprotective activity evaluation of C-3 substituted indole derivatives. *Sci. Rep.* **2021**, *11*, 15425. [CrossRef] [PubMed]
33. Ferreira, N.D.R.; Sanz, C.K.; Raybolt, A.; Pereira, C.M.; DosSantos, M.F. Action of Hyaluronic Acid as a Damage-Associated Molecular Pattern Molecule and Its Function on the Treatment of Temporomandibular Disorders. *Front. Pain Res.* **2022**, *3*, 852249. [CrossRef] [PubMed]
34. Pérez, R.; Burgos, V.; Cabrera-Pardo, J.R.; Ortiz, L.; Camins, A.; Ettcheto, M.; Schmidt, B.; Nchiozem-Ngnitedem, V.-A.; Paz, C. 9-Hydroxyaristoquinolone: A New Indole Alkaloid Isolated from *Aristotelia chilensis* with Inhibitory Activity of NF-κB in HMC-3 Microglia Cells. *Int. J. Mol. Sci.* **2025**, *26*, 2419. [CrossRef]
35. Cao, P.; Liang, Y.; Gao, X.; Li, X.M.; Song, Z.Q.; Liang, G. Monoterpenoid indole alkaloids from *Alstonia yunnanensis* and their cytotoxic and anti-inflammatory activities. *Molecules* **2012**, *17*, 13631–13641. [CrossRef]
36. Choi, Y.J.; Alishir, A.; Jang, T.; Kang, K.S.; Lee, S.; Kim, K.H. Antiskin Aging Effects of Indole Alkaloid *N*-Glycoside from Ginkgo Fruit (*Ginkgo biloba* fruit) on TNF-α-Exposed Human Dermal Fibroblasts. *J. Agric. Food Chem.* **2022**, *70*, 13651–13660. [CrossRef]
37. Ngernsaengsaruay, C.; Leksungnoen, N.; Boonthasak, W.; Utharatsamee, S.; Racharak, P.; Leetanasakskul, K.; Pongamorn, P.; Saengbuapuean, A. Additional knowledge on the genus Mitragyna (*Rubiaceae*) in Thailand. *Thai For. Bull.* **2022**, *50*, 20–39. [CrossRef]
38. Orio, L.; Alexandru, L.; Cravotto, G.; Mantegna, S.; Barge, A. UAE, MAE, SFE-CO2 and classical methods for the extraction of *Mitragyna speciosa* leaves. *Ultrason. Sonochem.* **2012**, *19*, 591–595. [CrossRef]
39. Dohárszky, A.; Vági, E.M.; Könczöl, Á.; Simon, A.; Várnagy, E.; Muratov, M.; Steiger, K.I.; Várnai, B.; Béni, S.; Riethmüller, E.; et al. Kratom Alkaloids: A Blood–Brain Barrier Specific Membrane Permeability Assay-Guided Isolation and Cyclodextrin Complexation Study. *Molecules* **2024**, *29*, 5302. [CrossRef]
40. Chittasupho, C.; Chaobankrang, K.; Sarawungkad, A.; Samee, W.; Singh, S.; Hemsuwimon, K.; Okonogi, S.; Kheawfu, K.; Kiattisin, K.; Chaiyana, W. Antioxidant, Anti-Inflammatory and Attenuating Intracellular Reactive Oxygen Species Activities of *Nicotiana tabacum* var. Virginia Leaf Extract Phytosomes and Shape Memory Gel Formulation. *Gels* **2023**, *9*, 78. [CrossRef]
41. Doloh, S.; Wungsintaweekul, B.; Wungsintaweekul, J. Phytochemical Analysis and Evaluation of Antioxidant and Antibacterial Activities of Kratom-Containing Formulations of Ya-Gae-Bid-Na-Ron. *J. Health Sci. Med. Res.* **2023**, *42*, 20231006. [CrossRef]
42. Tunit, P.; Thammarat, P.; Okonogi, S.; Chittasupho, C. Hydrogel Containing *Borassus flabellifer* L. Male Flower Extract for Antioxidant, Antimicrobial, and Anti-Inflammatory Activity. *Gels* **2022**, *8*, 126. [CrossRef] [PubMed]

gels

Article

In Situ-Prepared Nanocomposite for Water Management in High-Temperature Reservoirs

Hui Yang [1,*], Jian Zhang [2,3], Zhiwei Wang [2,3], Shichao Li [2,3], Qiang Wei [4], Yunteng He [5], Luyao Li [1], Jiachang Zhao [1], Caihong Xu [1] and Zongbo Zhang [1,*]

1 Key Laboratory of Science and Technology on High-Tech Polymer Materials, Institute of Chemistry, Chinese Academy of Sciences, Beijing 100190, China; 18840163221@163.com (L.L.); 18454320579@163.com (J.Z.); caihong@iccas.ac.cn (C.X.)
2 State Key Laboratory of Offshore Oil and Gas Exploitation, Beijing 102209, China; zhangjian@cnooc.com.cn (J.Z.); wangzhw28@cnooc.com.cn (Z.W.); lishch23@cnooc.com.cn (S.L.)
3 CNOOC Research Institute Co., Ltd., Beijing 100028, China
4 Institute of Geographic Sciences and Natural Resources Research, Chinese Academy of Sciences, Beijing 100101, China; weiq@igsnrr.ac.cn
5 Institute of Mechanics, Chinese Academy of Sciences, Beijing 100190, China; hyt@imech.ac.cn
* Correspondence: yanghui@iccas.ac.cn (H.Y.); zongbo@iccas.ac.cn (Z.Z.)

Abstract: In the field of enhanced oil recovery (EOR), particularly for water control in high-temperature reservoirs, there is a critical need for effective in-depth water shutoff and conformance control technologies. Polymer-based in situ-cross-linked gels are extensively employed for enhanced oil recovery (EOR), yet their short gelation time under high-temperature reservoir conditions (e.g., >120 °C) limits effective in-depth water shutoff and conformance control. To address this, we developed a hydrogel system via the in situ cross-linking of polyacrylamide (PAM) with phenolic resin (PR), reinforced by silica sol (SS) nanoparticles. We employed a variety of research methods, including bottle tests, viscosity and rheology measurements, scanning electron microscopy (SEM) scanning, density functional theory (DFT) calculations, differential scanning calorimetry (DSC) measurements, quartz crystal microbalance with dissipation (QCM-D) measurement, contact angle (CA) measurement, injectivity and temporary plugging performance evaluations, etc. The composite gel exhibits an exceptional gelation period of 72 h at 130 °C, surpassing conventional systems by more than 4.5 times in terms of duration. The gelation rate remains almost unchanged with the introduction of SS, due to the highly pre-dispersed silica nanoparticles that provide exceptional colloidal stability and the system's pH changing slightly throughout the gelation process. DFT and SEM results reveal that synergistic interactions between organic (PAM-PR networks) and inorganic (SS) components create a stacked hybrid network, enhancing both mechanical strength and thermal stability. A core flooding experiment demonstrates that the gel system achieves 92.4% plugging efficiency. The tailored nanocomposite allows for the precise management of gelation kinetics and microstructure formation, effectively addressing water control and enhancing the plugging effect in high-temperature reservoirs. These findings advance the mechanistic understanding of organic–inorganic hybrid gel systems and provide a framework for developing next-generation EOR technologies under extreme reservoir conditions.

Keywords: nanocomposite; high-temperature resistance; water management

1. Introduction

Waterflooding, a widely adopted technique for enhanced oil recovery (EOR), has been extensively applied for decades [1]. However, many mature oilfields now face critical challenges, including reservoir conformance issues and premature water breakthrough [2]. These problems lead to diminished sweep efficiency, elevated water cut, and increased residual oil saturation [3]. Consequently, they contribute to suboptimal economic returns from waterflooding [4], costly wastewater treatment [5], and significant environmental risks [6]. To address these inefficiencies, conformance control—a process utilizing chemical or mechanical interventions to mitigate excessive water production [7]—has emerged as a pivotal strategy for optimizing EOR performance in aging reservoirs worldwide [8].

Cross-linked polymer gels remain the most widely utilized technology for improving sweep efficiency, enhancing oil production, and controlling excessive water output in reservoir operations. Since the development of diverse cross-linking agents, the operational temperature limit of these gels has been extended to 121 °C for water shutoff treatments [9]. Additionally, hybrid gels incorporating solid particles—such as silica nanoparticles—have gained prominence due to their superior mechanical strength [10], enhanced thermal stability [11], and robust plugging performance compared to unmodified gel systems [12]. However, critical limitations persist: a short gelation time (GT) and field application constraints. For example, GT reaches 9–48 h at a temperature of 80 °C [13], decreases from 54.0 to 15.5 h with increasing temperature from 100 to 130 °C [14], or decreases from ~16 to ~13, ~11, and ~9 h when silica nanoparticles are added at concentrations of 0.10, 0.20, and 0.30 wt%, respectively [6]. Current GT profiles fail to meet the requirements of large-spacing well configurations, where extended gelation periods are essential for uniform propagation across heterogeneous reservoirs. The adjustability of composite gelation time at high temperatures is limited, primarily because the interactions between organic and inorganic components remain not fully revealed.

Considering the benefits of composite gels, we developed a silica-sol-reinforced PAM/phenolic aldehyde cross-linked polymer gel system (Figure 1), taking advantage of our silica sol's pH stability in a broad weakly alkaline range [15]. As the mixed system's pH remains mildly alkaline and changes little before and after the reaction, the inorganic component's addition is not expected to shorten the gelation time. This hybrid formulation leverages the synergistic effects of organic–inorganic components to address field application challenges. Its key performance advantages lie in the following: (1) Extended gelation time and thermal stability—the system exhibits prolonged gelation kinetics (adaptable to large well spacing) and maintains structural integrity at elevated temperatures. (2) High operational efficiency—liquid-phase delivery (polymer solution + nanoparticle suspension) simplifies field deployment compared to solid filler-based systems, enhancing safety and scalability [16]. (3) Gel degradability—the designed complex hydrogel, which can undergo reverse chemical reactions under specific conditions to contract or break down and restore fluidity, effectively mitigates risks in the plugging process.

The drive behind this work emanates from the existing knowledge gap in silica sol nanoparticles' role in enhancing polymer-based in situ-cross-linked gel functionality, especially in high-temperature settings. Presently, while silica sol has been sporadically utilized in gel systems, there is a paucity of in-depth studies comprehensively exploring its influence on gelation dynamics and internal water interactions, which are pivotal for high-temperature EOR applications. To address these needs, our study conducted a systematic investigation into how silica sol affects gelation time and strength. We explored how varying silica sol contents impact the gel's thermal stability, meticulously measured the fractions of free and bound water within the gels to understand water behavior within the matrix, and employed advanced techniques to observe and analyze the gel microstructures. Based on these findings, mechanisms for silica sol's enhancement of gel strength and thermal stability were proposed. Finally, as a typical application, the temporary plugging performance and gel-breaking response were assessed. This work establishes a tunable gel system with dual-phase reinforcement mechanisms, offering a practical solution for high-temperature reservoirs requiring delayed gelation and sustained mechanical performance.

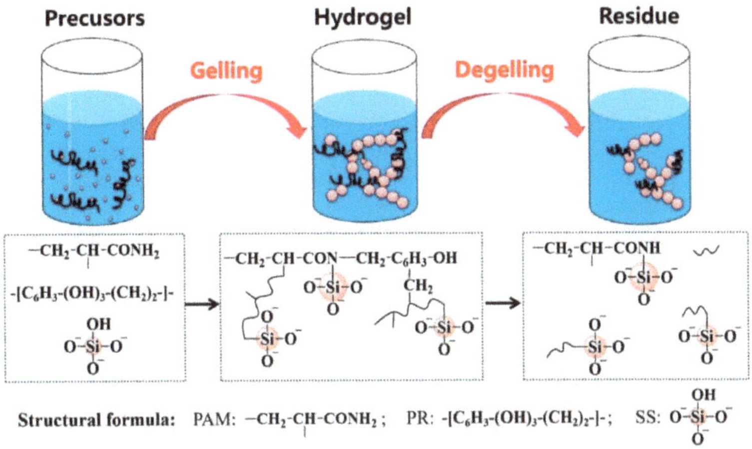

Figure 1. Diagrammatic representation of the gel formation and dissolution processes.

2. Results and Discussion

2.1. Gelation Performance

The precursor solutions containing varying concentrations of SS exhibited progressively darker coloration and were thermally cured at 130 °C. Mature composite gels formed after several days (Figure 2a), demonstrating a gradual increase in gel strength (GS) from Code C (0% SS) to Code G (6% SS). The gelation time (GT) was monitored via optical imaging (Table S1), revealing a minimum threshold exceeding 48 h.

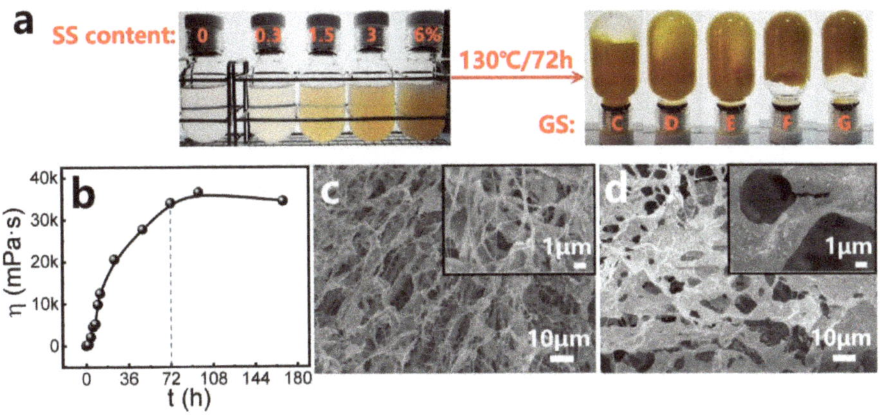

Figure 2. (**a**) Typical examples of composite gels, before and after gelation, at different SS contents of 0–6% and at 130 °C; (**b**) viscosity profile during different periods, at an SS content of 0.3% and at a reaction temperature of 130 °C. Micrographs of prepared gel samples: (**c**) without SS; (**d**) with 0.3% SS.

To precisely determine GT, twelve replicates at fixed SS concentrations were prepared and analyzed under controlled thermal conditions (130 °C). Time-lapsed images (45°-angled and inverted views, Table S2) showed reduced transparency and intensified coloration starting at 2 h, with negligible morphological changes beyond 72 h despite continued color darkening. Furthermore, the viscosity profiles of gel samples prepared at varying reaction times were measured as illustrated in Figure 2b. The viscosity values exhibit a linear increase with prolonged reaction time. Near 72 h, the viscosity growth rate plateaued, confirming a gelation time of approximately 72 h. To the best of our knowledge, the in situ cross-linking composite gel system presented here exhibits the longest GT at high temperature.

SEM analysis reveals critical insights into the micromorphological evolution of mature gels with varying SS contents (Figure 2c,d). From the micrographs, three-dimensional network structures are observed, in which polymer chain bunches with a thickness of several micrometers surround the pores with a coverage of tens of micrometers (Figure 2c). In the gelling solutions, amide groups ($-CONH_2$) from PAM cross-link with the hydroxyl groups ($-CH_2OH$) from PR, contributing to the porous network structure [17]. The distributed pores are the main storage spaces of water in gels. After adding SS, it can be observed that silica nanoparticles aggregate together and form varied arrangements. The aggregations and arrangements of silica nanoparticles are attached to the polymer chain bunches and meshes of the gel structure (Figure 2d), which strengthens the network structure of the gel.

2.2. Surface Modification

The in situ adsorption/desorption dynamics of mixed precursor solutions containing SS on silica surfaces was analyzed via quartz crystal microbalance with dissipation monitoring (QCM-D), tracking real-time shifts in frequency (Δf) and dissipation (ΔD) (Figure 3). The experiments were performed as follows (three main stages ((i)~(iii)) can be distinguished, with Figure 3a as an example): (i) Baseline stabilization: the quartz crystal sensor was equilibrated in formation brine to establish initial Δf and ΔD values. (ii) Precursor adsorption: the introduction of SS-containing solutions triggered abrupt Δf and ΔD decreases, attributed to the rapid surface adsorption of PAM, PR molecules, and SS nanoparticles. (iii) Desorption phase: the reintroduction of brine removed loosely bound molecules once adsorption saturation (Δf or ΔD plateau) was achieved.

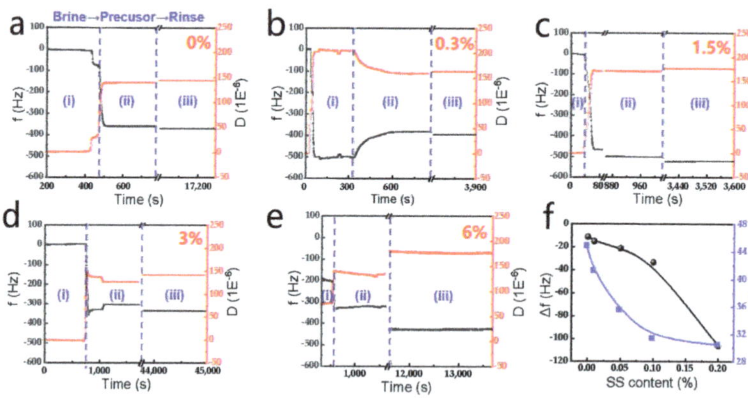

Figure 3. (**a–e**) Frequency and dissipation shift as a function of the adsorption time in the presence of different SS contents; (**f**) the wettability of the adsorbed surfaces in the presence of different SS contents.

It can be concluded that the adsorption process of the mixed precursor systems onto the solid surfaces (Figure 3b–e) is similar to that of the organic precursor system (Figure 3a). In addition, Δf decreases with the increase in SS content, which generally suggests an increase in the adsorbed mass. This indicates that SS prefers to adsorb on SiO_2's surface or facilitates the adsorption of the organic precursor. CA measurements reveal improved surface hydrophilicity, decreasing from 44.9° to 30.5° with rising SS content (Figure 3f), aligning with the goal of reservoir wettability modification. This SS-driven hydrophilicity enhancement suggests potential for tailoring gel–surface interactions in reservoirs.

2.3. Enhancement Properties and Mechanisms

To elucidate the correlation between viscoelastic properties and SS content, frequency-dependent rheological analyses were conducted on the mature gels (Figure 4a,b). Both storage modulus (G') and loss modulus (G'') increase monotonically across the tested frequency range (0.1–100 rad/s) as the SS content rises from 0% to 6%. The average $\overline{G'}$ and $\overline{G''}$ values exhibit a linear dependence on SS loading (Figure 4c), confirming enhanced elastic and viscous contributions from SS incorporation. At equivalent shear rates, G'' consistently exceeds G', indicating a viscosity-dominated response characteristic of weakly cross-linked or liquid-like gel systems [18]. The amplified viscoelasticity aligns with macroscopic gel strength trends (Figure 4a), where SS nanoparticles act as multifunctional cross-linkers to improve network rigidity while maintaining dynamic fluidity.

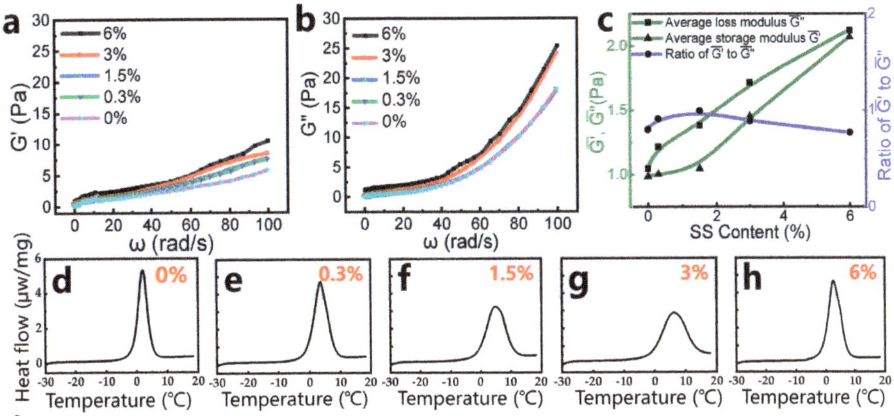

i

SS content (%)	0	0.3	1.5	3	6
ΔH (J/g)	261.1	257.7	245.6	234.6	215.0
w_f (wt%)	78	77	74	70	64
w_b (wt%)	22	23	26	30	36

Figure 4. (**a**) The storage modulus (G') and (**b**) loss modulus (G'') of gel samples prepared with different SS contents, within the frequency range from 0.1 rad/s to 100 rad/s. (**c**) Average storage modulus ($\overline{G'}$) and average loss modulus ($\overline{G''}$), within the frequency range from 0.1 rad/s to 100 rad/s. (**d–h**) DSC curves of the gel samples prepared with different SS contents of 0, 0.3, 1.5, 3, and 6%; the scanning temperatures range from −30 to 20 °C. (**i**) Mass fractions of free water and bound water in the gel samples.

In cross-linked polymer gels, water exists in two distinct states: free water (mobile, unbound) and bound water (immobilized via polymer–water interactions) [6]. Differential scanning calorimetry (DSC) was employed to quantify these water states in gels containing varying SS contents, with the following observations. The solid–liquid phase transition of free water occurs at about 0 °C. So, when the samples are heated gradually near 0 °C, the heat absorption is caused by the melting of frozen free water. Meanwhile, the solid–liquid transition of bound water occurs at temperatures away from 0 °C [19], because of the strong interactions between bound water and polymer molecules [20]. DSC measurements were applied to investigate the content of different types of water within the gels prepared with different contents of SS. In this part, the scanning temperature ranged from −30 to 20 °C and all of the gel samples fully gelled. As shown in Figure 4d–h, single endothermic peaks appear at about 0–10 °C, attributable to the melting of frozen free water in the gel samples.

By the integration of heat flow with time, the endothermic enthalpies required to melt the free water during the DSC scanning process were calculated. On the basis of the melting enthalpy of free water, the mass fractions of free water and bound water in the gel samples can be obtained from the following formulas [6]:

$$w_f = \Delta H / \Delta H_0 \quad (1)$$

$$w_b = 1 - w_f \quad (2)$$

where w_f and w_b are the mass fractions of free water and bound water, respectively (%); ΔH is the enthalpy (J/g); and ΔH_0, with a constant value of 333.5 J/g, is the melting enthalpy of free water (J/g).

The thermal characteristics and water distribution in the gel samples are quantified in Figure 4i, revealing two key trends: (1) Water state modulation by SS content—A progressive decrease in free water's endothermic enthalpy and a concomitant increase in the bound water mass fraction (23–36% vs. 22% in SS-free gel) are observed with elevated SS content (0.3–6%). (2) The hydrophilic mechanism of silica nanoparticles—The hydroxyl-rich surfaces of SS nanoparticles enhance water retention via hydrogen bonding, directly increasing bound-water proportions [21]. This aligns with the improved hydrophilic behavior demonstrated in Figure 3f. (3) Thermal stability implications—Elevated bound water content reduces dehydration susceptibility, suggesting enhanced thermal resilience in SS-modified gels [6]. Additionally, we conducted thermogravimetric analysis (TGA) to evaluate the thermal stability of gels with varying silica nanoparticle (SS) contents (0–6 wt%). As demonstrated in Figure S1a, the aging temperature of the pure gel (0% SS) is 120 °C, with a residual mass of 0.4% corresponding to hydrate water loss. Intriguingly, increasing the SS content to 0.3–6% elevates the aging temperature to 135–148 °C, while simultaneously reducing the residual mass to 2.4–10.9% (Figure S1b–e). Furthermore, the final mass loss at 1000 °C exhibits a clear SS-dependent trend, increasing from 0% (pure gel) to 1.2–9.0% for composite gels. These results indicate that SS incorporation enhances thermal resistance during the aging phase but accelerates decomposition at high temperatures, due to organic–inorganic interactions.

In the PAM/water-soluble PR system, hydroxyl groups (from PR) undergo co-condensation with amide groups (from PAM), forming C−N bonds (Figure 5a ①) that facilitate the development of a three-dimensional organic gel network [22]. The incorporation of silica nanoparticles into the gelling solution further enhances two aspects of this structural framework: gel strength and thermal stability. As illustrated in Figure 5a ②–④, silica nanoparticles aggregate into arrangements during the cross-linking process. By setting the initial energy of isolated reactants PAM and PR to zero, computational analysis reveals distinct reaction pathways with and without SS (Figure 5b). In the absence of SS, weak hydrogen bonding occurs between the –OH group of PAM and the carbonyl oxygen of PR (O–H bond length: 1.76 Å), resulting in a system energy reduction to −85.91 kJ·mol^{-1}.

The subsequent release of one H_2O molecule generates a complex, accompanied by an energy increase of 21.66 kJ·mol^{-1}. In contrast, the presence of SS enables simultaneous dual hydrogen bonding between the Si–OH cluster and both PAM and PR, significantly lowering the total system energy to −154.58 kJ·mol^{-1}—a thermodynamically favorable pathway. The strengthened interfacial interactions between Si–OH, PAM, and PR reduce the energy barrier for complex formation, yielding a smaller energy increase of 14.48 kJ·mol^{-1}. Furthermore, stabilizing interactions between the complex and the Si–OH cluster maintain the system energy at −140.10 kJ·mol^{-1}, demonstrating the catalytic role of SS in facilitating the organic–inorganic reaction and promoting long-chain molecular assembly.

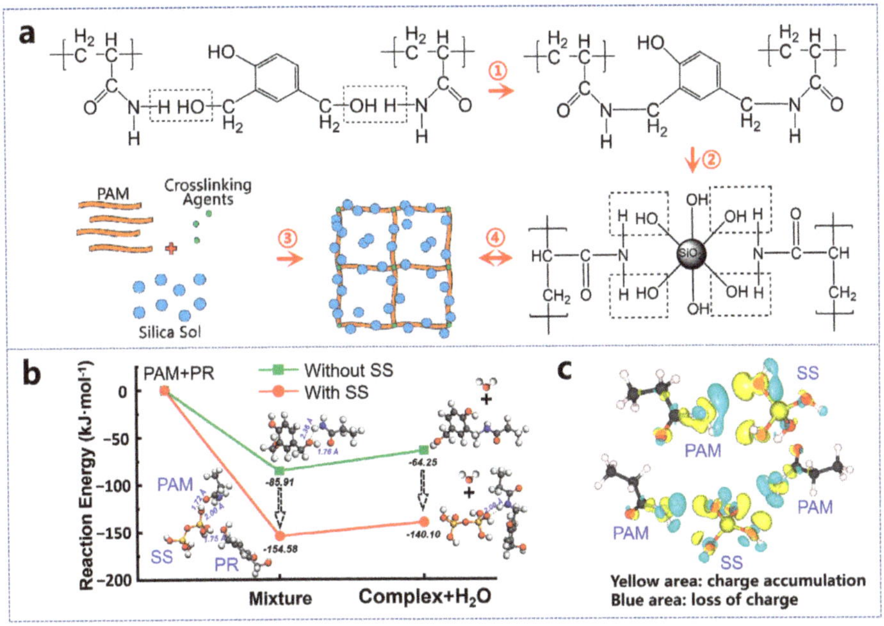

Figure 5. (**a**) Strengthening mechanism of silica nanoparticles in the gel network; (**b**) calculation of binding energy between PAM and SS molecules; (**c**) charge distribution analysis of the intermolecular interactions.

Additionally, the binding energy between -NH_2 groups and the Si_2-OH cluster is stronger than that with Si-OH (Figure S2), indicating that hydrogen bonding facilitates silica particle growth. The oxygen atom acquires a partial negative charge due to its stronger electronegativity in attracting shared electrons, while the hydrogen atom bears a partial positive charge. This charge asymmetry generates Coulomb attraction between O and H, forming a weak hydrogen bond (Figure 5c). These structured assemblies integrate into the polymer chain clusters and meshes, significantly improving the gel's mechanical integrity. The resultant structural reinforcement leads to a marked increase in gel strength. This mechanistic interpretation is corroborated by FTIR spectroscopy (Figure S3). The broad absorption at 3300 cm^{-1} integrates N–H (PAM), O–H (PR), and Si–OH (SS) stretching vibrations, with peak broadening indicative of extensive hydrogen bonding networks [23]. The amide I band at 1637 cm^{-1} (C=O stretching) and aromatic C=C vibrations (PR) confirm organic phase integrity [24], while the blue-shifted Si–O stretching peak at 1128 cm^{-1} ($\Delta\nu$ = +28 cm^{-1} from 1100 cm^{-1}) directly evidences interfacial hydrogen bonding between organic components and the silica network [25]. Critically, the progressive intensity enhancement of the 1128 cm^{-1} peak with increasing SS content quantifies the concentration-dependent strengthening of organic–inorganic interactions, aligning with computational predictions. Collectively, these results establish hydrogen bonding as the dominant interfacial linkage mechanism, synergistically reinforced by SS to optimize composite gel performance.

The thermal stability enhancement is driven by two synergistic interactions: hydrogen bonding and electrostatic interactions [26]. Surface hydroxyl (-OH) groups on silica nanoparticles form hydrogen bonds with water molecules, converting free water into bound water [27]. Negatively charged silica nanoparticles attract hydronium ions (H_3O^+) through electrostatic forces, further stabilizing the bound water phase [28]. The combined effects yield a higher bound water ratio compared to silica-free gels, correlating with stronger hydrophilicity, superior water retention, and enhanced thermal resistance [29].

In the study of gelation kinetics for organic–inorganic hybrid hydrogels, silica sol systems have a longer gelation time than those with directly added silica nanoparticles. This is due to the following reasons: the nanoparticles in silica sols are highly pre-dispersed, giving the system excellent colloidal stability, and the system's pH stays stable at 8–9 throughout the gelation process. This allows it to stay in a metastable, non-equilibrium liquid-phase state for a long time. The homogeneous dispersion of nanoparticles affects reaction kinetics in two main ways: (1) The suppression of local concentration gradients: Pre-dispersed particles eliminate microscopic concentration fluctuations caused by agglomeration, reducing the driving force for heterogeneous nucleation [30]. (2) Dynamic interfacial equilibrium: Uniformly exposed hydroxyl groups establish adsorption–desorption equilibrium at reaction interfaces, slowing down condensation reactions [31].

In contrast, exogenous silica nanoparticles, limited by intrinsic factors such as surface hydroxyl passivation and insufficient dispersion, are prone to forming discrete "reactive hotspots" within organic matrices. These heterogeneous microdomains facilitate localized condensation via autocatalytic mechanisms, resulting in premature gel network locking and macroscopic phase separation.

The delayed gelation behavior of silica sol systems fundamentally originates from their distinctive multi-scale synergistic effects: Colloidal stability ensures sustained particle dispersion through electrostatic repulsion and steric hindrance, while controlled condensation kinetics enables self-regulated reaction rates via the gradual consumption of surface hydroxyl groups. This spatiotemporally ordered cross-linking process establishes an optimal thermodynamic window for the hierarchical assembly of three-dimensional networks, and it is thereby expected to achieve the synergistic optimization of structural homogeneity, mechanical integrity, and functional tunability.

2.4. *Application Performance*

2.4.1. Aging Resistance

The thermal resistance of hydrogels with varying silica sol contents (0% to 6%) was systematically evaluated using DSC measurements. During testing, the temperature was ramped from 50 °C to 200 °C, with the peak endothermic rate (indicating the maximum water loss rate) identified as the sharpest endothermic peak in the DSC thermograms (Figure 6a). Below the temperature inflection point, heat absorption increases steadily with rising temperature due to the gradual evaporation of free and bound water within the hydrogel matrix. Above the inflection temperature, the heat flow curve declines sharply, signifying near-complete dehydration of the hydrogel. This structural dehydration correlates with irreversible damage to the hydrogel network. The incorporation of nanosilica sol significantly enhances the thermal resistance. The maximum tolerable temperature rises from 114 °C (0% SS) to 156 °C (6% SS).

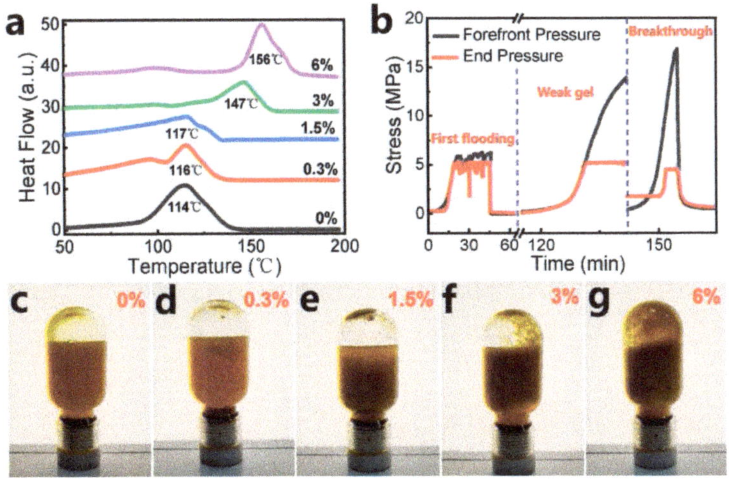

Figure 6. (**a**) DSC curves of hydrogels with different SS contents of 0~6%; (**b**) conductivity of PAM the hydrogel system containing 1% silica sol; (**c**–**g**) mature gels with different SS contents of 0, 0.3, 1.5, 3, and 6% aging for 60 days at 130 °C.

Further evaluation demonstrated the robust stability of the hydrogel under prolonged thermal stress, which was performed as shown in Figure 6c–g. After 60 days of high-temperature exposure, the hydrogels exhibited negligible alterations in macroscopic appearance, indicating structural integrity. Less significant syneresis (water expulsion) was observed, confirming effective water retention within the gel matrix. The viscosity retention rate of the hydrogel systems exceeded 90%, validating the material's resistance to thermal degradation and sustained rheological performance. As the SS content increased, the hydrogels exhibited a more cohesive texture and a higher viscosity retention rate. The improvement stems from the reinforcement effect of SiO_2 nanoparticles: mechanical strengthening and water retention, i.e., silica particles fortify the hydrogel skeleton—reducing thermal degradation—and nanoparticles act as anchoring sites for water molecules, improving hydration stability.

2.4.2. Practical Application

The simulated core conductivity experiments provide dual insights: (1) Injection feasibility—assessing the injectability of hydrogel systems in real reservoir formations. (2) Plugging efficiency—evaluating the hydrogel's capacity to block fluid pathways within the core. The hydrogel system containing 0.3% SS demonstrates effective advancement within the core. Due to the weak gelation capacity of the low-strength weak gel system, it exhibits favorable deep migration and transport capabilities under the subsequent flushing force of water flooding. During the initial water flooding stage, the pressure difference between the front and tail ends of the system measured 0.6 MPa. After 5 days of curing, this pressure difference significantly increased to 12.31 MPa (Figure 6b). As quantified by Equation (4) (Section 4.3.9), the core permeability exhibited a reduction of approximately 92.4% between pre- and post-plugging conditions. The system achieves superior sealing performance compared with in situ-cross-linked polymer gels [32].

2.4.3. Gel Degradation

The 10% GB solution was introduced into the gelled composite system maintained at 130 °C. Within 30 min of thermal treatment, the gel system exhibited initial degradation with visible structural heterogeneity (Figure 7). Complete degradation was achieved after 150 min, demonstrating a weight loss ratio of 93.8% (calculated as follows: [Initial weight − Residual weight]/Initial weight [33]). This degradation behavior confirms that the gelled composite system retains the organic component's degradability, enabling conventional methods to effectively mitigate plugging risks while maintaining temperature-responsive functionality [34]. Notably, the synergistic organic–inorganic network structure ensures controlled decomposition without blockage risk.

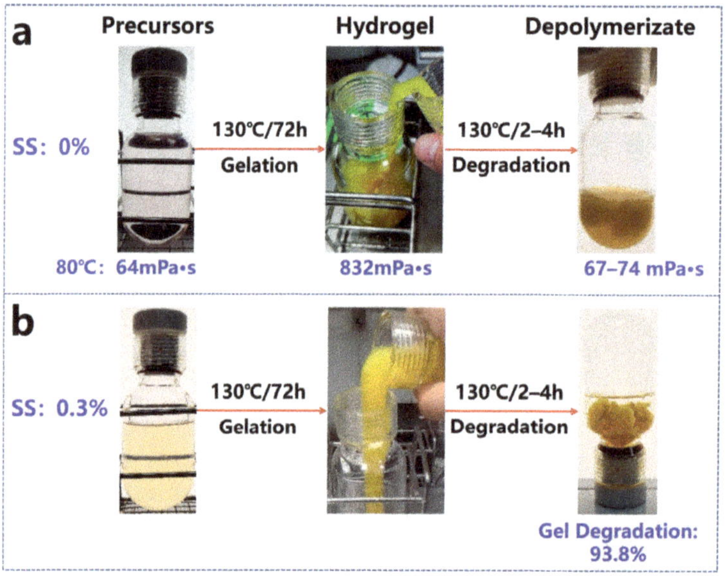

Figure 7. Gel breaking process of the composite gel with (**a**) 0% SS and (**b**) 0.3% SS.

3. Conclusions

This study systematically investigates a silica-reinforced cross-linked polymer gel, focusing on its gelation behavior, thermal resilience, structural reinforcement mechanism, and related applications. The key findings are as follows:

(1) Gelation Performance Enhancement: Silica sol extended the gelation time to 72 h while significantly improving gel strength, attributed to the structured assemblies integrate into the polymer chain clusters and meshes. Both storage and loss modulus increased compared to silica-free gels, confirming mechanical reinforcement via hybrid organic–inorganic networks.

(2) Thermal Stability and Water Retention: Bound water content rose from 22% (SS-free) to 36% in SS-modified gels, correlating with hydroxyl-mediated hydration layers on nanoparticle surfaces. The maximum tolerable temperature rose from 114 °C (0% SS) to 156 °C (6% SS), owing to stabilized water–polymer interactions and restricted chain mobility.

(3) Strengthening Mechanism: A three-dimensional network architecture was observed, where silica aggregates form an interfacial interaction with polymer bundles, mechanically reinforcing the gel matrix. The synergistic interplay of hydrogen-

bonded hydration layers and electrostatic interactions elevates the bound-water ratio, correlating with stronger hydrophilicity, superior water retention, and enhanced thermal resistance.

(4) Field-Relevant Performance: The core permeability reduction of 92.4% post-gelation demonstrates effective pore throat sealing, critical for water management in fractured reservoirs. The synergistic network design enables controlled decomposition without secondary contamination, addressing environmental concerns in oilfield applications.

This work establishes a framework for designing nanoparticle-enhanced gels with tunable rheological and thermal properties for harsh reservoir conditions.

In terms of future research directions, the following proposals can be made:

(1) Further exploration of the hydrogel's properties under different dehydration conditions to better understand the correlation between structural dehydration and network damage.
(2) Investigation of ways to enhance the hydrogel's stability and reversibility under varying environmental conditions.
(3) Exploration of hydrogels' potential applications in specific fields to validate their practical utility.

4. Materials and Methods

4.1. Materials

The polyacrylamide (PAM) with an average molecular weight of 12,000,000 g/mol, phenolic resin (PR) with a weight average molecular weight of 11,411 g/mol, and the gel breaker (GB) were provided from the laboratory of CNOOC. The inorganic precursor of silica sol (SS, 30%), with a narrow particle size distribution of about 50.0 nm, was prepared in our laboratory [15]. Sodium chloride, potassium chloride, calcium chloride, magnesium chloride hexahydrate, sodium sulfate, and sodium bicarbonate were AR-grade materials. The synthetic formation brine used contained 2323.52 mg/L Na^+, 26.57 mg/L K^+, 115.53 mg/L Ca^{2+}, 52.29 mg/L Mg^{2+}, 3689.88 mg/L Cl^-, 11.16 mg/L SO_4^{2-}, and 619.22 mg/L HCO_3^-.

4.2. Method for Gel Preparation and Gel Degradation

4.2.1. Gel Preparation

Using the synthetic formation brine, the silica sol was diluted to the concentration of 0%, 0.3%, 1.5%, 3%, and 6%, respectively. We set the stirring speed to 200RPM in the diluted silica sol, and kept the temperature at 45 °C. The high-purity nitrogen was aerated to the system at a flow rate of 10mL/min. The PAM powder was added into the system little by little. Two hours later, the mixed system was cooled down to room temperature. A certain amount of PR was added into the mixed system at a stirring speed of 200 RPM for 15min. The concentrations of PAM and PR were both set to 2500 mg/L. The precursor solution was degassed by purging with high-purity nitrogen (99.999%) at 1 L/min for 30 s, followed by static settlement in a sealed flask for 12 h to allow bubble removal. Prior to sealing, a final nitrogen purge was applied to establish an oxygen-free atmosphere, ensuring stability for subsequent reactions. The sealing precursor solution was placed in an oven at a temperature of 130 °C as an example, for a period of time until a mature gel was formed.

4.2.2. Gel Degradation

After the formation of a mature gel, GB at a concentration of 10% was added into the gel system at 130 °C, for a period of time until the gel was broken.

4.3. Characterization

The experimental workflow is depicted in Figure S4.

4.3.1. Measurement of the Gelation Time (GT) and Strength (GS)

The gelation time and strength of the gel were measured through bottle tests and evaluated by the gel strength code. As introduced in a previous report [35], Codes A through I are described as follows: Code A—no detectable continuous gel formed; Code B—highly flowing gel; Code C—flowing gel; Code D—moderately flowing gel; Code E—barely flowing gel; Code F—highly deformable non-flowing gel; Code G—moderately deformable non-flowing gel; Code H—slightly deformable non-flowing gel; and Code I—rigid gel. For water shutoff gel, the initial gelation time is usually considered the period of time when the gelation solution in Code A turns to flowing gel in Code C. The frequency of gel code determination is every 2–24 h.

4.3.2. Viscosity and Rheology Measurements

Viscosity measurements of the nanocomposite during the growth process were performed using a modular compact rheometer (MCR-302, Anton Paar, Physica, Graz, Austria). The test temperature was 80 °C and shear rates ranged from 0.1 to 1000 s^{-1}. The rheological properties of the gelling solution and mature gel were measured at a shear rate of 200–0.1 s^{-1}.

4.3.3. Scanning Electron Microscopy (SEM) Scanning

The morphology features of different samples were analyzed by SEM (S-4800, HITACHI, Tokyo, Japan), operating at an accelerating voltage of 10 kV and at an electric current of 10 μA. The prepared gels were frozen to −110 °C using liquid nitrogen and metal coated for the later observation.

4.3.4. Calculation Methods

To investigate the interaction between SiO_2 nanoparticles and -NH_2 units of PAM, we conducted density functional theoretical (DFT) calculations using VASP (Vienna Ab initio Simulation Package) [36]. The projector-augmented wave pseudopotential (PAW) method was employed to describe the interaction between ionic cores and valences [37]. The exchange–correlation effects were calculated within the generalized gradient approximation (GGA) using the Perdew−Burke−Ernzerhof (PBE) functionals, with a cutoff energy of 400 eV for the plane-wave basis set [38]. The convergence criteria for ionic and electronic calculations were set to 0.02 eV $Å^{-1}$ and 10^{-5} eV, respectively [39]. The optimization of Si_n-OH clusters and -NH_2 units was performed in a cubic box of 30 Å in each direction. The Brillouin zone integration was carried out using a single k-point at the Γ point.

The binding energy (ΔE_b) was calculated by the following equation:

$$\Delta E_b = \Delta E - \Delta E_{Si\text{-}OH} - \Delta E_{\text{-}NH2} \quad (3)$$

where ΔE is the total electronic energy of the complex. $\Delta E_{Si\text{-}OH}$ and $\Delta E_{\text{-}NH2}$ refer to the energies of Si_n-OH cluster and -NH_2 unit, respectively. Bader charge analysis was conducted to investigate the charge transfer between these two species [40].

4.3.5. Differential Scanning Calorimetry (DSC) Measurement

DSC measurements of the gel samples were conducted by a differential scanning calorimeter (STA 449 F3, Netzsch, Selb, Germany). In each measurement process, the gel sample, measuring about 30~45 mg, was placed in a crucible with a nitrogen purge rate of

50 cm^3/min. The temperature change rates were 5 °C/min for all measurements. In the high-temperature test, the scanning temperature range was 50–200 °C, and the scanning temperature range was −30–20 °C when the proportion of bound water was measured.

4.3.6. Quartz Crystal Microbalance with Dissipation (QCM-D) Measurement

QCM-D measurements were conducted to measure the interaction between precursors and rock surfaces. The resulting frequency difference Δf is inversely proportional to the adsorbed mass [41]. The injection sequence of each sample was as follows: formation water ⇒ precursor solution ⇒ formation water ⇒ distilled water rinsing, in which the SS contents were 0%, 0.3%, 1.5%, 3%, and 6%. Commercial SiO_2 sensors (QSX 303, Biolin Scientific, Solna, Sweden) were selected to simulate reservoir rock surfaces for the QCM-D experiments.

4.3.7. Contact Angle (CA) Measurement

The water contact angles of the adsorbed surfaces were obtained using a DSA 100 (Kruss DSA CA goniometer, Hamburg, Germany) drop shape analysis system, and 3 mL of distilled water was used.

4.3.8. Injectivity and Temporary Plugging Performance Evaluation

The injectivity of the gel precursor was evaluated using the experimental setup shown in Figure S5. An artificial core with a permeability of ~100 mD was saturated with formation brine and connected to the apparatus. The injection process involved three sequential phases:

(1) Initial water flooding: Brine solution was injected at 0.5 mL/min and 90 °C, with intermediate container 3 open and container 4 closed. Injection rates and nodal pressures (front/middle sections) were recorded until pressure stabilization.
(2) Gel precursor injection: Container 3 was closed, and container 4 was opened to inject the gel precursor solution. Nodal pressures were monitored to assess flow resistance during precursor delivery.
(3) Post-flooding evaluation: Brine solution was reinjected to measure residual injectivity, with pressure data collected until equilibrium.

4.3.9. Plugging Performance Evaluation

Plugging efficiency was quantified through a three-stage protocol:

(1) Baseline brine flooding: The brine-saturated core underwent brine injection (container 3 open, container 4 closed) until stable nodal pressures were achieved.
(2) Gel formation: The gel precursor was injected, followed by a 3-day aging period to enable in situ cross-linking.
(3) Post-treatment evaluation: Water flooding resumed until pressure stabilization, and the plugging rate (%) was calculated using the following equation [42]:

$$K = \frac{QL\mu}{A\Delta P} \tag{4}$$

Q is the medium flow rate (cm^3/s) passing through the core under pressure difference ΔP; A is the cross-sectional area of the core (cm^2); L is the length of the core (cm); μ is the viscosity of the injected medium fluid (mPa·s); ΔP is the pressure difference between the front and back in the direction of core length (10^{-4} MPa); and K is the proportional coefficient, i.e., the absolute permeability (D) of the porous medium.

Supplementary Materials: The following supporting information can be downloaded at: https://www.mdpi.com/article/10.3390/gels11060405/s1, Figure S1: TGA and DTG curves of mature gel with different SS content of (a) 0%, (b) 0.3%, (c) 1.5%, (d) 3%, (e) 6%; Figure S2: Calculated binding energies between $-NH_2$ with Si-OH (with one Si species) and Si_2-OH (with two Si species) clusters; Figure S3: (a) FTIR spectra and (b) magnified spectral region (500–1500 cm^{-1}) of composite gels with varying SS contents (0–6%); Figure S4: The flow chart of the experiment; Figure S5: Setup diagram for plugging evaluation experiment; Table S1: Optical images (placing upright, at a fixed angle of 45°, and upside down) during gelation process at different SS content and 130 °C; Table S2: Optical images (at a fixed angle of 45° and upside down) during gelation process, at the SS content of 0.3% and at 130 °C.

Author Contributions: Methodology, Q.W.; investigation, H.Y., Z.W., Y.H., L.L. and J.Z. (Jiachang Zhao); data curation, Y.H.; writing—original draft preparation, H.Y.; writing—review and editing, H.Y. and Z.Z.; supervision, J.Z. (Jian Zhang), S.L., C.X. and Z.Z.; funding acquisition, J.Z. (Jian Zhang). All authors have read and agreed to the published version of the manuscript.

Funding: This research was funded by the National Natural Science Foundation of China [U22B6005].

Institutional Review Board Statement: Not applicable.

Informed Consent Statement: Not applicable.

Data Availability Statement: The data presented in this study are available on request from the corresponding authors.

Conflicts of Interest: Authors Jian Zhang, Zhiwei Wang, Shichao Li were employed by the company CNOOC Research Institute Co., Ltd. The remaining authors declare that the research was conducted in the absence of any commercial or financial relationships that could be construed as a potential conflict of interest.

References

1. Sydansk, R.D.; Southwell, G.P. More Than 12 Years of Experience with a Successful Conformance-Control Polymer Gel Technology. *SPE Prod. Facil.* **2000**, *15*, 270–278. [CrossRef]
2. Zhu, D.Y.; Hou, J.R.; Chen, Y.G.; Wei, Q.; Zhao, S.D.; Bai, B.O.J. Evaluation of Terpolymer-Gel Systems Crosslinked by Polyethylenimine for Conformance Improvement in High-Temperature Reservoirs. *SPE J.* **2019**, *24*, 1726–1740. [CrossRef]
3. Wang, W.; Liu, Y.Z.; Gu, Y.G. Application of a Novel Polymer System in Chemical Enhanced Oil Recovery (EOR). *Colloid Polym. Sci.* **2003**, *281*, 1046–1054. [CrossRef]
4. Liu, Y.F.; Dai, C.L.; Wang, K.; Zhao, M.W.; Gao, M.W.; Yang, Z.; Fang, J.C.; Wu, Y.N. Investigation on Preparation and Profile Control Mechanisms of the Dispersed Particle Gels (DPG) Formed from Phenol Formaldehyde Cross-linked Polymer Gel. *Ind. Eng. Chem. Res.* **2016**, *55*, 6284–6292. [CrossRef]
5. Sang, Q.; Li, Y.J.; Yu, L.; Li, Z.Q.; Dong, M.Z. Enhanced Oil Recovery by Branched-Preformed Particle Gel Injection in Parallel-Sandpack Models. *Fuel* **2014**, *136*, 295–306. [CrossRef]
6. Liu, Y.F.; Dai, C.L.; Wang, K.; Zou, C.W.; Gao, M.W.; Fang, Y.C.; Zhao, M.W.; Wu, Y.N.; You, Q. Study on a Novel Cross-Linked Polymer Gel Strengthened with Silica Nanoparticles. *Energy Fuels* **2017**, *31*, 9152–9161. [CrossRef]
7. Liu, J.; Seright, R.S. Rheology of Gels Used for Conformance Control in Fractures. *SPE J.* **2001**, *6*, 120–125. [CrossRef]
8. Bai, B.; Zhou, J.; Yin, M. A Comprehensive Review of Polyacrylamide Polymer Gels for Conformance Control. *Pet. Explor. Dev.* **2015**, *42*, 525–532. [CrossRef]
9. Hutchins, R.D.; Dovan, H.T.; Sandiford, B.B. Field Applications of High Temperature Organic Gels for Water Control. In Proceedings of the SPE/DOE Improved Oil Recovery Symposium, Tulsa, Oklahoma, 21–24 April 1996.
10. Hu, Z.; Lu, X.; Gao, J.; Wang, C. Polymer Gel Nanoparticle Networks. *Adv. Mater.* **2000**, *12*, 1173–1176. [CrossRef]
11. Liu, Y.; Zhu, M.F.; Liu, X.L.; Zhang, W.; Sun, B.; Chen, Y.M.; Adler, H.J.P. High Clay Content Nanocomposite Hydrogels with Surprising Mechanical Strength and Interesting Deswelling Kinetics. *Polymer* **2006**, *47*, 1–5. [CrossRef]
12. Aalaie, J.; Vasheghani-Farahani, E.; Rahmatpour, A.; Semsarzadeh, M.A. Effect of Montmorillonite on Gelation and Swelling Behavior of Sulfonated Polyacrylamide Nanocomposite Hydrogels in Electrolyte Solutions. *Eur. Polym. J.* **2008**, *44*, 2024–2031. [CrossRef]

13. Bai, Y.; Xiong, C.; Wei, F.; Li, J.; Shu, Y.; Liu, D. Gelation Study on a Hydrophobically Associating Polymer/Polyethylenimine Gel System for Water Shut-off Treatment. *Energy Fuels* **2015**, *29*, 447–458. [CrossRef]
14. He, H.; Wang, Y.; Sun, X.; Zhang, P.; Li, D. Development and Evaluation of Organic/Inorganic Combined Gel for Conformance Control in High Temperature and High Salinity Reservoirs. *J. Pet. Explor. Prod. Technol.* **2015**, *5*, 211–217. [CrossRef]
15. Yang, H.; Li, L.; Dong, G.; Wei, Q.; Zhang, J.; Wang, X.; Wang, Z.; Xu, C.; Zhang, Z. High-Temperature-Resistant and Gelation-Controllable Silica-Based Gel for Water Management. *Energy Fuels* **2024**, *38*, 23398–23406. [CrossRef]
16. Lakatos, I.J.; Lakatos-Szabo, J.; Szentes, G.; Vago, A.; Bodi, T. New Alternatives in Conformance Control: Nanosilica and Liquid Polymer Aided Silicate Technology. In Proceedings of the SPE European Formation Damage Conference and Exhibition, Budapest, Hungary, 3–5 June 2015; pp. 1–24.
17. Zhao, G.; Dai, C.; Zhao, M.; You, Q. The Use of Environmental Scanning Electron Microscopy for Imaging the Microstructure of Gels for Profile Control and Water Shutoff Treatments. *J. Appl. Polym. Sci.* **2013**, *131*, 1001–1007. [CrossRef]
18. Xie, K.; Cao, W.; Lu, X.; Song, K.; Liu, Y.; Zhang, Y.; Liu, J.; Lv, J.; Wang, W.; Na, R. Influence of Water Dilution on Performance of Chromium Polymer Weak Gel in Porous Medium. *J. Dispers. Sci. Technol.* **2020**, *41*, 1549–1558. [CrossRef]
19. Capitani, D.; Mensitieri, G.; Porro, F.; Proietti, N.; Segre, A.L. NMR and Calorimetric Investigation of Water in a Superabsorbing Crosslinked Network Based on Cellulose Derivatives. *Polymer* **2003**, *44*, 6589–6598. [CrossRef]
20. Mccrystal, C.B.; Ford, J.L.; Rajabi-Siahboomi, A.R. Water Distribution Studies within Cellulose Ethers using Differential Scanning Calorimetry. 1. Effect of Polymer Molecular Weight and Drug Addition. *J. Pharm. Sci.* **1999**, *88*, 792–796. [CrossRef]
21. Dai, C.; Chen, W.; You, Q.; Wang, H.; Yang, Z.; He, L.; Jiao, B.; Wu, Y. A Novel Strengthened Dispersed Particle Gel for Enhanced Oil Recovery Application. *J. Ind. Eng. Chem.* **2016**, *41*, 175–182. [CrossRef]
22. Liu, Y.; Dai, C.; Wang, K.; Zhao, M.; Zhao, G.; Yang, S.; Yan, Z.; You, Q. New Insights into the Hydroquinone (HQ)–Hexamethylenetetramine (HMTA) Gel System for Water Shut-off Treatment in High Temperature Reservoirs. *J. Ind. Eng. Chem.* **2016**, *35*, 20–28. [CrossRef]
23. Xie, Z.; Luo, R.; Ying, T.; Gao, Y.; Song, B.; Yu, T.; Chen, X.; Hao, M.; Chai, C.; Yan, J.; et al. Dynamic-to-Static Switch of Hydrogen Bonds Induces a Metal–Insulator Transition in an Organic–Inorganic Superlattice. *Nat. Chem.* **2024**, *16*, 1803–1810. [CrossRef]
24. Yang, N.; Yang, S. Mechanistic Insights into Sludge Inorganics Suppressing Methane Yield Through Organic–Inorganic Interactions During Anaerobic Digestion. *Bioresour. Technol.* **2025**, *432*, 132659. [CrossRef] [PubMed]
25. Li, W.; Sun, Z. Second Hydration Shell of Mg^{2+}: Competition between Ion–Water Interaction and Hydrogen Bonding Interaction. *J. Phys. Chem. Lett.* **2025**, *16*, 503–509. [CrossRef] [PubMed]
26. Son, H.A.; Yoon, K.Y.; Lee, G.J.; Cho, J.W.; Choi, S.K.; Kim, J.W.; Im, K.C.; Kim, H.T.; Lee, K.S.; Sung, W.M. The Potential Applications in Oil Recovery with Silica Nanoparticle and Polyvinyl Alcohol Stabilized Emulsion. *J. Pet. Sci. Eng.* **2015**, *126*, 152–161. [CrossRef]
27. Khoder, H.; Siboulet, B.; Ollivier, J.; Baus-Lagarde, B.; Rébiscoul, D. How Cation–Silica Surface Interactions Affect Water Dynamics in Nanoconfined Electrolyte Solutions. *J. Phys. Chem. C* **2024**, *128*, 12558–12565. [CrossRef]
28. Gomes, O.P.; Rheinheimer, J.P.C.; Dias, L.F.G.; Batagin-Neto, A.; Lisboa-Filho, P.N. Revisiting the Hydroxylation Phenomenon of SiO_2: A Study Through "Hard-Hard" and "Soft-Soft" Interactions. *J. Mol. Model.* **2022**, *28*, 115. [CrossRef]
29. Liu, J.; Cao, H. Sub-Ambient Water Wettability of Hydrophilic And Hydrophobic SiO_2 Surfaces. *J. Chem. Phys.* **2024**, *161*, 184701. [CrossRef]
30. Dehne, H.; Reitenbach, A.; Bausch, A.R. Reversible and Spatiotemporal Control of Colloidal Structure Formation. *Nat. Commun.* **2021**, *12*, 6811. [CrossRef]
31. Kegel, W.K. Direct Observation of Dynamical Heterogeneities in Colloidal Hard-Sphere Suspensions. *Science* **2000**, *287*, 290–293. [CrossRef]
32. Yin, H.; Yin, X.; Cao, R.; Zeng, P.; Wang, J.; Wu, D.; Luo, X.; Zhu, Y.; Zheng, Z.; Feng, Y. In Situ Crosslinked Weak Gels with Ultralong and Tunable Gelation Times for Improving Oil Recovery. *Chem. Eng. J.* **2022**, *432*, 134350. [CrossRef]
33. Xie, K.; Su, C.; Liu, C.; Cao, W.; He, X.; Ding, H.; Mei, J.; Yan, K.; Cheng, Q.; Lu, X. Synthesis and Performance Evaluation of an Organic/Inorganic Composite Gel Plugging System for Offshore Oilfields. *ACS Omega* **2022**, *7*, 12870–12878. [CrossRef] [PubMed]
34. Takeuchi, M.; Nakajima, M.; Kojima, M.; Fukuda, T. Evaluation and Application of Thermoresponsive Gel handling towards Manipulation of Single Cells. In Proceedings of the 2011 IEEE/RSJ International Conference on Intelligent Robots and Systems, San Francisco, CA, USA, 25–30 September 2011; pp. 457–462.
35. Jia, H.; Chen, H.; Zhao, J.Z. Development of a Highly Elastic Composite Gel through Novel Intercalated Crosslinking Method for Wellbore Temporary Plugging in High-Temperature Reservoirs. *SPE J.* **2020**, *25*, 2853–2866. [CrossRef]
36. Kresse, G.; Furthmüller, J. Efficiency of ab-Initio Total Energy Calculations for Metals and Semiconductors Using a Plane-Wave Basis Set. *Comput. Mater. Sci.* **1996**, *6*, 15–50. [CrossRef]

37. Kresse, G.; Joubert, D. From Ultrasoft Pseudopotentials to the Projector Augmented-Wave Method. *Phys. Rev. B* **1999**, *59*, 1758–1775. [CrossRef]
38. Kohn, W.; Sham, L.J. Self-Consistent Equations Including Exchange and Correlation Effects. *Phys. Rev.* **1965**, *140*, 1133–1138. [CrossRef]
39. Liu, X.; Cui, C.; Wei, S.; Han, J.Y.; Zhu, X.; Ge, Q.; Wang, H. The Synergy of In Situ-generated Ni^0 and Ni_2P to Enhance CO Adsorption and Protonation for Selective CH_4 Production from Photocatalytic CO_2 Reduction. *Green Chem.* **2024**, *26*, 531–541. [CrossRef]
40. Cui, C.; Zhang, H.; Cheng, R.; Huang, B.; Luo, Z. On the Nature of Three-Atom Metal Cluster Catalysis for N_2 Reduction to Ammonia. *ACS Catal.* **2022**, *12*, 14964–14975. [CrossRef]
41. Liu, F.H.; Yang, H.; Yang, M.; Wu, J.Z.; Yang, S.Y.; Yu, D.F.; Wu, X.; Wang, J.Y.; Gates, I.; Wang, J.B. Effects of Molecular Polarity on the Adsorption and Desorption Behavior of Asphaltene Model Compounds on Silica Surfaces. *Fuel* **2021**, *284*, 118990. [CrossRef]
42. Yang, Z.; Li, X.; Li, D.; Yin, T.; Zhang, P.; Dong, Z.; Lin, M.; Zhang, J. New Method Based on CO_2-Switchable Wormlike Micelles for Controlling CO_2 Breakthrough in a Tight Fractured Oil Reservoir. *Energy Fuels* **2019**, *33*, 4806–4815. [CrossRef]

gels

MDPI

Review

Food-Grade Bigel Systems: Formulation, Characterization, and Applications for Novel Food Product Development

Konstantina Zampouni *, Dafni Dimakopoulou-Papazoglou and Eugenios Katsanidis

Department of Food Science and Technology, School of Agriculture, Faculty of Agriculture, Forestry and Natural Environment, Aristotle University of Thessaloniki, 54124 Thessaloniki, Greece; ddimakop@agro.auth.gr (D.D.-P.); ekatsani@agro.auth.gr (E.K.)

* Correspondence: zampounikona@gmail.com

Abstract: Bigels are characterized as biphasic systems consisting of two structured phases of different polarity, namely the oleogel and hydrogel phases. These systems have been widely used in pharmaceuticals and cosmetics, owing to their enhanced physicochemical stability compared to other structured systems and their ability to simultaneously deliver both hydrophilic and lipophilic compounds. Considering the above advantages, bigels could have considerable potential for the food industry. This review aims to provide a detailed description of the edible components used for structuring the oleogel and hydrogel phases and the preparation methods applied for the formation of food-grade bigels with application-specific tailored properties. The impact of the processing parameters, such as the oleogel-to-hydrogel ratio, methodology used for component mixing, and cooling rates, is presented. Moreover, the most applicable bigel characterization techniques, such as rheology, DSC, texture analysis, etc., are critically discussed. Finally, different bigel applications in foods as animal fat substitutes or as complex delivery systems for both polar and non-polar bioactive compounds are examined.

Keywords: bigels; biphasic systems; oleogels; hydrogels; food applications

1. Introduction

The term "bigels" or "hybrid gels" was first suggested in 2008 by Almeida and coworkers [1] to explicate the combination of two discrete structured systems, an oleogel and a hydrogel, that are differentiated by their opposing polarity [2]. In most instances, the addition of emulsifiers is not required, as the mixing of the two phases results in the formation of a stable gel. Given that oleogels and hydrogels are recognized as effective systems for the delivery and release of bioactive compounds, it is evident that bigels possess the potential to transport both lipophilic and hydrophilic components individually and simultaneously [3]. The greater permeability of lipophilic bioactive substances through the skin when using bigel systems as carriers, compared to simple emulsions, oleogels, and hydrogels, is a reason for the extensive recent application of bigels in the field of pharmaceuticals and cosmetics [4,5].

The main advantage of bigels is their improved stability compared to emulsions, emulsion gels, hydrogels, and oleogels. This enhanced physicochemical stability is explained by the colloidal dispersion of one phase within the other, which renders these systems highly suitable and ideal for utilization in various food products [1,6]. Additionally, bigel systems typically do not exhibit any phase separation throughout storage at ambient temperatures, providing improved stability when incorporated into shelf-stable foods [7]. However, bigel stability can be influenced by several factors, such as the composition of the bigel, storage conditions, and interactions with food ingredients. Conversely, emulsion systems suffer from destabilization and phase separation over time. Thus, it is considered necessary to add an emulsifier to achieve the stabilization of an emulsion, in contrast to bigels [8]. However, it should be noted that, despite these advantages, bigels tend to destabilize when exposed to high temperatures; thereby, generally, they are not classified as thermoreversible systems [9].

Recently, the application of bigels in the field of food science has gained significant attention due to their potential use as fat-replacement systems. The benefit of the formation of a bigel-based fat-mimetic system is the ability to tailor its mechanical and functional properties. The utilization of bigels, which consist partly of a low-calorie hydrogel, can result in a reduced overall fat content compared to traditional solid fats. Moreover, bigels present an alternative solution for the development of nutritionally enriched foods. The biphasic nature of bigels allows the transport of various bioactive substances (e.g., lipids, pigments, vitamins), while their distinct physicochemical properties facilitate the controlled release of these components, providing them with additional protection [10–12]. Modulating the physicochemical and structural characteristics of bigels is very important to their successful incorporation into diverse food matrices [13].

This review focuses on recent advances in the development of the two structured phases (oleogels and hydrogels) of bigels and on the techniques for the successful preparation of a functional bigel system. Also, considering the importance of bigels for the food industry, an overview of the most used characterization methods for the assessment of the functionality of these systems is presented. Finally, the utilization of bigels in several food applications is also summarized.

2. Bigel Composition and Preparation Methods

2.1. Oleogels Used for Bigel Preparation

Over the past few years, oleogelation has increasingly attracted the interest of the food industry as an alternative to partial hydrogenation for the preparation of gels that successfully mimic hard fats. Oleogelation is a natural process which converts liquid edible oils into semi-solid gels, without modifying the chemical characteristics of the used oils [14]. The supramolecular organization of low- or high-molecular-weight oleogelators results in the formation of a thermoreversible three-dimensional (3D) network which restricts the mobility of hydrophobic liquid edible oils [15].

Various molecular interactions are responsible for the formation of oleogels, such as hydrogen bonds, π-π stacking, electrostatic interactions, and van der Waals interactions [16]. Proteins, polysaccharides, fatty acids, phytosterols, etc., can be used as oleogelators, depending on the desired physical properties of the food systems into which the obtained oleogels are to be incorporated. Process variables, such as the temperature, time, and mixing rate, play a crucial role in customizing the mechanical properties of the final oleogel. Also, the type and concentration of the oleogelator, the type of oil, and the proportions of the different structurants can modify the microstructure of oleogels, thereby impacting their textural parameters. Different methodologies can be applied to structure oil and form oleogels. The most frequently used method is direct dispersion, where lipid-soluble oleogelators are directly dissolved in the oil at elevated temperatures. Upon cooling, low-molecular-weight oleogelators self-assemble to form a 3D network which efficiently entraps the oil, resulting in an oleogel structure [17,18]. Nevertheless, to broaden the scope of applicable oleogelators like hydrophobic proteins and other polymers (except ethylcellulose), alternative, indirect methods have been investigated. In the indirect approach, the oleogelation process involves two separate stages and includes techniques such as the emulsion template, foam template, and solvent exchange method [19]. The emulsion template technique involves, first, the formation of an oil-in-water emulsion, after which the emulsion undergoes a drying process to remove the water phase and prepare the oleogel network [20]. Similarly, in the foam template technique, the removal of the aqueous phase leads to a porous polymer structure that entraps the liquid oil and forms an oleogel [21]. On the other hand, in the solvent exchange method, a secondary solvent is used to introduce gelators. Once the gelators are adequately dispersed, the solvent is removed by evaporation, resulting in the stabilization of the oil phase [22].

2.1.1. Oleogelators

An ideal oleogelator must be edible, economical, adaptable, effective (able to structure lipids at relatively low concentrations), and capable of forming structures with comparable physical properties to the lipid material it is intended to replace. Such properties may include hardness/plasticity at a specific temperature, spreadability, a specific melting point, and a specific melting profile [23].

Various oleogelators have been used to structure different oils through the direct approach, such as monoglycerides (MGs), diglycerides (DGs), stearic acid (SAC), and different waxes. For example, MGs have been used in various concentrations to structure canola oil [6,24–26], corn oil [27–30], soybean oil [31], sunflower oil [13,32,33], and olive oil [34–36]. Binary gelator systems have also been investigated, such as combinations of glycerol monostearate (GMS) with rice bran wax (RBW) in corn oil [37] and soybean oil [38,39] or GMS with beeswax (BW) in soybean oil [11,40]. Also, a mixture of MGs and DGs, along with RBW, has been used for the preparation of the oleogel phase in gelatin (GEL) bigels [41]. Moreover, SAC has been used as an oleogelator in soybean oil in different bigel systems with [12,42,43] or without the addition of soy lecithin (SL) [44,45]. Different waxes have been selected as oleogelators of the oleogel phase of bigels, such as candelilla wax (CW) [46,47] and BW in canola oil [48], corn oil [29,49], medium-chain triglycerides (MCT) oil [50], sunflower oil [51,52], sesame oil [53], and grape seed oil [54]. Moreover, RBW was used to structure soybean oil [41,55] and high oleic soybean oil oleogels [56], while carnauba wax (CBW) has been used to form sunflower oil and olive pomace oil oleogels [57,58].

The use of polymeric oleogelators for the formation of the oleogel phase in bigels has also been evaluated. For example, ethylcellulose (EC) [59] and whey protein isolate (WPI) [60] oleogels were produced for the preparation of food-grade bigels.

2.1.2. Oils

The type of oil (acting as the solvent in the oleogel system) also significantly affects the thermal, rheological, and mechanical properties, as well as the nutritional characteristics of the produced oleogels and, in turn, the final properties of the bigels. The chemical characteristics of the oil phase (such as the dielectric constant, polarity, and viscosity of the oil) are mainly determined by the saturated/unsaturated fatty acids (SFA/UFA) ratio, the length of the fatty acid chains, the conformation of the triglycerides (TGs) chains, and the presence of polar/non-polar functional groups [61]. Martins et al. [62] reported that the composition of TGs in the used oils is an important parameter for tailoring the properties of oleogels. A variety of vegetable oils has been used in the oleogel phase of bigels, including canola oil [6,24,25,46,47], corn oil [27–30,37], soybean oil [31,42,63,64], sunflower oil [32,51,59,65], sesame oil [53] and olive oil [34–36]. Also, oils, such as coconut oil, avocado oil, and pomegranate oil, that contain bioactive compounds, have been evaluated as the lipid phase of bigels [66,67].

2.2. Hydrogels Used for Bigel Preparation

Polymeric hydrogels are aqueous networks of polymer chains cross-linked via chemical or physical bonds. Hydrogels are formed using macromolecules with hydrophilic groups like -OH, -COOH, -SO_3H, -CONH-, and -$CONH_2$- that either grafted to or embedded in their polymeric backbones [68]. Generally, hydrogels are described as intermediate entities which exhibit both elastic (solid-like) and viscous (liquid-like) features. Due to the presence of hydrophilic regions and groups, hydrogels have the ability to absorb and retain amounts of water that are several hundred times their dry weight within the 3D network they form [69]. Hydrogels play significant roles in food formulations, primarily as thickeners and stabilizers [70]. Also, hydrogels are used in the development of edible food packaging materials, delivering nutraceuticals while ensuring their bioavailability, regulating the intake of calories from foods, and structuring foods to attain desired sensory textures. Parameters that affect the hydrogelation of formulations include the temperature, pH, ionic strength, and gelling agent concentration [71].

Hydrogelators

For the preparation of food-grade bigels, hydrogels have been formulated from natural polymers, such as polysaccharides and proteins. Vershkov and Davidovich-Pinchas [50] and Zheng et al. [72] prepared xanthan gum (XG) hydrogels, while other researchers [11,13,40,73] have utilized gellan gum (GG) hydrogels as a part of bigels. Additionally, bigels with guar gum (GGM) [26], wheat starch (WS) [59], sweet potato (SPS) and chayote tuber starch (CTS) [31,48], sodium alginate (SA) [50], chitosan [74], and tapioca (TAP) [52] have also been used. Composite hydrogels with κ-carrageenan (KC) and locust bean gum (LBG) [33], KC and GEL [35], KC and XG [49], KC and SA [38], or konjac glucomannan (KG) and GEL [45] have been used for bigel production. Moreover, polymers such as carboxymethyl cellulose (CMC) [66,67] and hydroxypropyl methylcellulose (HPMC) [63] have also been utilized in bigel preparation. WPI [12,24,42,43,53,58], whey protein concentrate (WPC) [24], collagen (COL) [57,58], and GEL [6,27,32,75–77] have also been exploited to structure the hydrogel phase of edible bigels.

2.3. Bigel Preparation Techniques

A deep understanding of bigel processing parameters allows for the customization of bigel properties to provide desirable characteristics for use as solid fat replacers or as carriers of bioactive compounds. First, as discussed above, the oleogels and hydrogels are prepared separately using various oleogelators and hydrogelators, respectively. Several current studies have focused on the conditions of the mixing of the oleogel with the hydrogel phase to form a bigel under mechanical stirring at a pre-determined mixing rate (Table 1). Recent work done by Vershkov and Davidovich-Pinhas [47] focused on the effect of the homogenization temperature on bigels' properties and characteristics.

Table 1. Examples of edible bigels (composition, oleogel-to-hydrogel ratio, and bigel mixing parameters) that can be used for food applications.

Oleogel (O)		Hydrogel (H)	O:H Ratio	Bigel Mixing Parameters	Reference
Edible Oil	**Oleogelator**	**Hydrogelator**			
Canola oil	MGs * (10%)	KC (4%) WPC 80 (15%)	80:20	4000 rpm at 90 °C for 2 min Sonication at 90 °C for 2 min 10,000 rpm for 2 min + Sonication at 90 °C	[24]
	MGs (0–35%) SL (2–20%)	GEL (0.5–3%) Glycerol (5–15%)	20:80, 30:70, 40:60, 50:50, 60:40, 70:30, 80:20	1600 rpm at 70 °C for 3 min	[6]
	MGs (25%) SL (1.6%) SEs (2%)	GEL (1%) Glycerol (5%)	50:50	O:70 °C, H:90 °C 16,000 rpm for 1 min	[25]
	CW (8%) SEs	XG (0.5%)	55:45, 65:35, 75:25, 85:15	20,000 rpm at 30–60 °C for 2.5 min	[47]
	MGs (10%) PGPR (1.5%)	GGM (1%) Tween 60 (1.7%)	67:33, 83:17, 58:15	600/800 rpm for 20 min	[26]
	BW (5%)	SPS (10%) CTS (10%)	30:70, 40:60, 50:50	Sonication at 90 °C for 2 min	[48]

Table 1. *Cont.*

Oleogel (O)		Hydrogel (H)	O:H Ratio	Bigel Mixing Parameters	Reference
Edible Oil	**Oleogelator**	**Hydrogelator**			
Corn oil	GMS (4–8%) BW (4–8%)	KC (0.75%) Tween 20 (0.5%)	50:50	300 rpm at 45 °C	[29]
MCT oil	BW (3%, 6%)	SA (2%)	50:50, 80:20, 90:10, 95:5, 99:1	600 rpm at 22 °C for 45 min	[50]
Soybean oil	SL, SAC (7:3, 20%)	WPC 80 (15–25%)	30:70, 50:50, 70:30	23,000 rpm for 3 min	[42]
	RBW (10%)	GEL (5, 7, 10%)	20:80, 30:70, 40:60, 50:50	24,000 rpm at 90 °C for 2 min	[41]
	RBW (7.5%)	GEL (7%)	40:60, 50:50, 60:40, 70:30	24,000 rpm at 85 °C for 20 sec 13,000 rpm at 85 °C for 30 sec	[55]
	RBW (10%) MGs/DGs (0.5–3%)	GEL (7%)	60:40, 70:30, 80:20	24,000 rpm for 1 min	[64]
	RBW (9%) MGs (0–2%)	SA (1%) KC (0.5%)	70:30, 80:20	15,000 rpm at 85 °C for 1 min	[38]
	GMS (8%)	WPS (8%)	20:80, 30:70, 40:60, 60:40, 70:30, 80:20	10,000 rpm at 25 °C for 3 min	[31]
	SAC (3%)	KG/GEL (2.25%)	20:80, 30:70 40:60, 50:50	9000 rpm at 75 °C for 3 min	[45]
Sunflower oil	GMS (5–15%)	GG (1–1.5%)	20:80, 35:65, 50:50, 65:35, 80:20	500–1500 rpm at 25 °C for 10 min	[13]
	GMS (5–20%)	KC/LBG (1:1, 0.5–2.5%)	10:90, 20:80, 30:70, 40:60, 50:50	2000 rpm for 10 min	[33]
	Fatty acids mixture (10%)	XG (1%)	40:60, 50:50, 60:40, 70:30, 80:20	14,000 rpm at 85 °C for 3 min	[72]
	BW (6%)	TAP (5–10%)	25:75, 40:60, 50:50, 60:40, 75:25	14,000 rpm at 85 °C for 1 min	[52]
	WPI (40 mg/mL)	WPI (60, 80 mg/mL)	30:70, 50:50, 70:30	800 rpm at 24 ± 2 °C for 10 min	[60]
Sunflower oil Olive pomace oil	CBW (8%) SL (0.5%)	COL (40%, 60%)	40:60, 50:50, 60:40	11,000 rpm at 85 °C for 3 min	[58]
Olive oil	MGs (15%)	GEL (8–12%) KC (1%)	20:80, 40:60	300 rpm at 50 °C or 80 °C for 15 min	[35]
Olive oil Coconut oil	MGs Ps	GEL	40:60, 60:40	600 rpm at 50 °C for 15 min	[77]

* BW: beeswax, CBW: carnauba wax, COL: collagen, CTS: chayote tuber starch, CW: candelilla wax, DGs: diglycerides, GEL: gelatin, GG: gellan gum, GGM: guar gum, GMS: glycerol monostearate, KC: κ-carrageenan, MCT: medium-chain triglycerides, MGs: monoglycerides, PGPR: polyglycerol ricin alkyd ester, Ps: phytosterols, RBW: rice bran wax, SA: sodium alginate, SAC: stearic acid, SEs: sucrose esters, SL: soy lecithin, SPS: sweet potato starch, TAP: tapioca starch, XG: xanthan gum, WPC: whey protein concentrate, WPI: whey protein isolate.

2.3.1. Cold Homogenization

The process of cold homogenization necessitates an extended period of mixing or a high mixing rate to attain a homogeneous system [13]. The mixing of cold, pre-set GMS oleogels and SPS hydrogels at 25 °C at 10,000 rpm for 3 min has been suggested by Barroso and coworkers [31]. The rapid homogenization process resulted in the efficient dispersion of the oleogel phase in the hydrogel phase, forming the bigel structure. In contrast, lower mixing rates require longer processing times to ensure the homogeneity of the system. For example, BW oleogels were mixed with SA hydrogels at 600 rpm at 22 °C for 45 min [50]. By adjusting the mixing speed and time of the two phases, the final texture and stability of the bigel can be significantly affected.

2.3.2. Hot Homogenization

Bigels can also be formed by applying hot emulsification to blend the oleogel phase with the hydrogel phase [35,38,39,45,52,64]. This methodology typically requires a shorter mixing time but a longer setting period for the bigels to solidify, in contrast to cold homogenization. For instance, Samui et al. [6] used the hot emulsification method to prepare bigels based on a GEL hydrogel and a GMS oleogel, with the incorporation of SL as a surfactant and glycerol as a co-surfactant. Also, in the study by Cho et al. [63], a hot RBW oleogel was homogenized with a hot GEL hydrogel at 85 °C either for 20 s at 24,000 rpm or 30 s at 13,000 rpm. This approach facilitates the dispersion and integration of the oleogel and hydrogel phases, leading to the formation of stable bigel structures which can be applied in food products.

3. Bigel Characterization

Figure 1 presents the most commonly used techniques for the characterization of a bigel system, including the evaluation of its physical properties, examination of the microstructure to determine its morphology, structural organization, and chemical composition, textural and rheological analysis to assess its mechanical and viscoelastic properties, differential scanning calorimetry (DSC) to assess its thermal behavior, etc. Each technique, along with findings from studies on food-grade bigel systems, is discussed in the sections below.

3.1. Physical Properties

Biphasic formulations that exhibit structural matrix continuity and a self-standing ability are considered successfully formed bigels. The tube inversion method has been mainly used to evaluate the self-standing ability of bigels [31,33,35,72]. The formation, appearance, and color of a bigel system are related to the composition and mixing ratio of both phases. The color of the bigel is a crucial factor for various food applications, as it may significantly impact the appearance of the final product, affecting the consumer perception and acceptance. In the study conducted by Bollom et al. [42], it was observed that the studied bigels displayed an opaque and off-white visual appearance, irrespective of the composition of their protein or water content. An increase of the oleogel proportion resulted in bigels with a more yellowish hue and that were greasier to the touch, which can be attributed to the presence of olive oil [35,78]. On the contrary, higher hydrogel ratios resulted in bigels with a whiter color [26]. According to data obtained from the evaluation of color parameters using a colorimeter, the lightness (L^* value) and redness (a^* value) decreased, while the yellowness (b^* value) increased, for bigels structured with higher CBW oleogel fractions [58]. The incorporation of KC or WPC80 hydrogels in MGs oleogel resulted in a light color and smooth texture [24]. Also, Samui et al. [6] reported that, depending on the selected composition, bigels may exhibit a white opaque appearance and a variety of textures, from soft "mayonnaise" to hard "butter". According to Saffold and Acevedo [64], the addition of MGs and DGs to RBW/GEL bigels contributed to softer and more malleable bigels without visual differences compared to the control. Bigels that were produced with GG as the hydrogelator and GMS/BW as the oleogelators remained stable without alterations in color and appearance for 3 months [11], while those produced with EC with XG and GGM were visually stable for 2 months [78].

Figure 1. The most used characterization techniques for bigel systems.

3.2. Microstructure

3.2.1. Microscopy Methods

The characteristics of a bigel system are contingent upon the structural distribution of the oleogel and hydrogel phases within the bigel matrix, as well as the average size of the dispersed phase droplets. The types of formed bigels have been investigated in many studies by examining their microstructures with techniques such as optical and polarized light microscopy (PLM) [36,77], confocal laser scanning microscopy (CLSM) [37,39], and scanning electron microscopy (SEM) [28,49]. In general, bigels can be categorized as oleogel-in-hydrogel, hydrogel-in-oleogel, or bicontinuous systems [79] (Figure 2).

Oleogel-in-hydrogel systems, where the hydrogel is the continuous phase and the oleogel is the dispersed phase, are possibly the most extensively researched bigel systems in the literature. For example, the formation of an oleogel-in-hydrogel RBW/GEL bigel for all the studied oleogel-to-hydrogel ratios was documented through CLSM imaging [41]. The second type of bigel, known as hydrogel-in-oleogel, comprises a system in which the hydrogel phase is dispersed throughout the continuous oleogel matrix. Microscopy studies confirmed that, as the oleogel ratio increased over 50% in EC/XG/GGM bigels, an inversion in the continuous phase was detected and the type of bigel changed from oleogel-in-hydrogel to hydrogel-in-oleogel [77]. Additionally, other authors have determined the formation of hydrogel-in-oleogel-type bigels using GMS in high oleic sunflower oil with GG, MGs in canola oil with KC and WPC80, and MGs with RBW in corn oil and SA [13,24,37]. The bicontinuous-type bigels are considered as more complex metastable structures, where there is no discrete continuous or dispersed phase due to the co-existence of the two structured phases in the bigel matrix [6,77]. In a study by Zheng et al. [72], very few oil droplets were visible in KC/MGs bigels with a 60% oleogel fraction, with the micrographs showing that the oleogel phase had a dispersal that could be considered as continuous.

In addition, a significant percentage of the hydrogel phase was also combined with the oleogel phase and did not form discrete droplets. Thus, such a bigel was considered to form a bicontinuous structure [30,77]. Variation in the oleogel and hydrogel ratios, specifically in the proportions of 30:70, 50:50, and 70:30, in the presence of various emulsifiers, such as MGs and SL, resulted in a significant modification of the structural compositions of the bigels [75]. This transformation involved a shift from the conventional oleogel-in-hydrogel configuration to either a hydrogel-in-oleogel arrangement or a bicontinuous state. Guo et al. [40] reported that, as the oleogel ratio increased from 62% to 70%, the oleogel oil droplets started to aggregate and form a continuous oil phase with a small number of oleogel droplets. Simultaneously, the hydrogel phase underwent a transition from a continuous state to exhibiting block-like regions, and some spherical droplets of hydrogel were also formed, confirming that it was a bicontinuous-type of bigel. When the hydrogel fraction decreased below 30%, a hydrogel-in-oleogel structure prevailed [40]. Several researchers have studied the impact of the oleogel and hydrogel ratio on the microstructures of bigels. Micrographs of bigels revealed that the oil droplets became greater in size, more asymmetrical in form, and more tightly packed inside the hydrogel matrix as the oleogel percentage increased [35,41,77]. Similarly, smaller-sized GMS oleogel particles were obtained by increasing the KC/LBG hydrogel content in bigels. Likewise, an increase in polymer concentration led to the formation of smaller, more uniformly distributed oleogel oil droplets, thereby enhancing the structural stability of bigels [33]. A comparable observation has been reported for MGs/BW and high-acyl gellan gum (HAGG) bigels with up to a 60% oleogel ratio and using CLSM. Zhu et al. [11] attributed the differences observed in the micrographs of their bigel to the enhanced viscoelastic properties of the system due to the increased quantity of oleogel, which hindered homogenization and resulted in the formation of larger droplets. For hydrogel-in-oleogel bigels (oleogel:hydrogel ratios of 80:20 and 60:40), Chen et al. [63] reported that BW soybean oil oleogel acted as the continuous phase in HPMC bigels, allowing sufficient BW crystal growth to form thin needle-like crystals, similar to those in plain BW–soybean oil oleogels. On the other hand, in bigels with oleogel:hydrogel ratios of 50:50, 40:60, and 20:80, the oleogel phase was identified as the dispersed phase, where the limited size of the oil droplets led to smaller crystalline structures organized in clusters. On the other hand, in the study by de Los Santos-Trinidad et al. [48], a higher oleogel ratio in SPS/CTS and BW bigels resulted in a greater oleogel distribution within the continuous hydrogel phase, resulting in the formation of smaller and more spherical droplets. Also, Habibi et al. [24] documented that the properties of hydrogelators (e.g., KC and WPC80) affect the crystallization of MGs in hydrogel-in-oleogel bigels, given that the same oleogel-to-hydrogel ratio and homogenization settings were used for the bigel preparation.

SEM has been used mostly to examine the surface morphology of freeze-dried bigels and bigel beads [28,51]. The addition of the oleogel phase in hydrogel resulted in the formation of continuous networks characterized by dispersed particles [51]. The bigel beads containing 10% BW oleogel exhibited a rough surface and an irregular form according to SEM images. Also, the degree of protuberance decreased, and the surface became smoother as the oleogel concentration increased [28].

3.2.2. Fourier Transform Infrared Spectroscopy

Fourier transform infrared spectroscopy (FTIR) has emerged as a powerful technique, capable of furnishing valuable insights regarding the functional groups, molecular bonds, and general chemical characteristics of bigels. Through the analysis of their FTIR spectra (typically in the range of 4000–400 cm^{-1}), researchers can gain insights into the molecular interactions and structural properties of bigel systems. These observations hold great importance, as they facilitate comprehension of the properties and, thus, the potential applications of the formed bigels. The FTIR spectrum of a bigel system is sensitive to changes in bigel composition, mixing speed, preparation methods, and the concentration of the components. Specifically, the preparation technique can affect the molecular interactions, phase distribution, and chemical environment within the bigel matrix. For example, hot emulsification can cause phase transitions or crystallization in certain components, and these can be observed as peak shifts or changes in the FTIR spectrum. In many studies, the impact of the various oleogel-to-hydrogel ratios on the FTIR spectra of bigels was observed through reductions in hydrogen bonding and alterations in the intensities of specific peaks [75,77]. In one study, as the oleogel ratio increased, the intensity of the signals at ~2920, 2850, and 1465 cm^{-1}, which represent the stretching vibrations of C-H or C=H from the fatty acyl chains in the vegetable oils phase, also increased [33]. Furthermore, the absorption signals representing C=O stretching vibrations in the oil phase exhibited an increase as the oleogel ratio increased. Zhu et al. [11] reported that some of these signals were entirely lost in bigels due to a low oleogel content (<20%). In contrast, a higher hydrogel ratio in the bigel system was related to an increase in the intensity of the absorption peak in the broad band (3690–3000 cm^{-1}), which is associated with O-H stretching vibrations [11,30]. Similar changes were observed by Saffold and Acevedo [41] for RBW/GEL bigels. Chen et al. [63] determined the type of HPMC/BW bigels as oleogel-in-hydrogel, identifying an increase in the absorption intensity of two peaks of HPMC, around 3480 and 1056 cm^{-1}, in the bigels' spectra. Overall, the researchers did not detect major shifts or new peaks resulting from the interaction between the two structured phases in the studied bigels. This confirms the absence of a chemical affinity between the structured phases and underlays the dominance of the physical interactions in the formation of bigels [13,48].

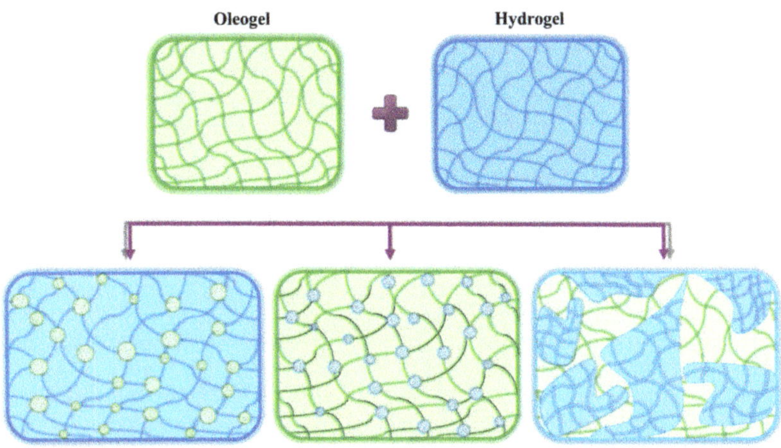

Figure 2. Schematic illustration of the three types of bigels that were formed by mixing an oleogel and a hydrogel phase in different ratios.

3.2.3. X-Ray Diffraction

X-ray diffraction (XRD) analysis can be conducted on bigel systems to describe the nature of the formed structures by identifying amorphous or crystalline regions. Various studies have documented the crystallinity of different bigel systems using an X-ray diffractometer equipped with a Cu tube. XRD analysis has suggested that bigels exhibit different crystalline structures depending on the type of the oleogelator [29]. Bigels formed with 50% MG oleogels had the highest crystallinity [30]. Ghorghi et al. [54] noted negligible differences in the diffraction patterns among bigels with various BW oleogel-to-SA hydrogel ratios and confirmed that the oleogel phase constituted the crystalline structure of the studied bigel. According to Quilaqueo et al. [46], the formed bigels in their study were semi-crystalline solids, characterized both by crystalline and amorphous components. The incorporation of GG hydrogel in bigels decreased the crystallinity of the system due to its amorphous state [13]. In contrast, XRD analysis in KC with LBG and GMS bigels revealed that they had a mainly amorphous structure, demonstrating a loss of crystallinity compared to the gelator compounds and oleogels used to form the bigels [33]. Similar findings were also reported earlier by Martins et al. [50] for SA with BW bigels, as no polymorphism was observed for all examined oleogel-to-hydrogel ratios.

3.3. *Textural and Rheological Properties of Bigels*

Texture profile analysis (TPA) is conducted to evaluate the textural attributes of bigel systems. The mechanical properties of bigels can be customized and optimized depending on the desired attributes of the final product and the specific processing conditions applied. The mechanical characteristics of bigels, such as hardness, spreadability, and adhesiveness, were found to depend on the bigel composition [6,13], as well as the processing parameters [26]. Additionally, rheology has been used to provide information about the structure and viscoelasticity of bigels and their resulting suitability for applications in the food industry.

3.3.1. Oleogel-to-Hydrogel Ratio

The majority of research on biphasic systems has identified a clear correlation between the mechanical characteristics of bigels and the oleogel-to-hydrogel ratio used. For this reason, a more extensive discussion regarding the impact of modulating the ratio of the two phases is provided in this section. Typically, mechanical properties such as the storage modulus, stiffness, and fracture stress of the bigels have been improved with higher contents of oleogel. Also, better mechanical characteristics were reported when the type of bigel was modified from oleogel-in-hydrogel to bi-continuous or hydrogel-in-oleogel [30]. The effect of the ratio of the two structured phases on the textural properties of GEL and GMS bigels was investigated by Samui et al. [6]. The findings of the study demonstrated a noticeable increase in the hardness of the bigel when the oleogel ratio was increased up to 60% [6]. Also, higher firmness values were observed for WPI bigels with 70% oleogels, demonstrating a clear positive correlation between the firmness of the bigels and their oleogel proportion [60]. A significant increase in the stiffness and fracture stress of MGs/KC bigels has been noted as the MGs oleogel percentage increased, while a strong linear correlation between the two examined parameters was also identified [30]. An increase in the EC oleogel concentration led to a subsequent increase in the hardness and gumminess, while it decreased the adhesiveness, springiness, stringiness, and chewiness of the obtained bigels [78]. Similar findings have been reported by Zheng et al. [30] and Zhu et al. [11]. Habibi et al. [24] noted that the addition of 20% KC hydrogel in bigel weakened the structural integrity of the MGs crystalline network in the oleogel, thus reducing the final hardness of the bigel. According to Han et al. [37], increased GMS/RBW oleogel fractions in bigels led to an increasing trend in their cohesiveness, consistency, and viscosity index. These findings were consistent with previous research on MGs/BW with HAGG bigels [11]. Also, in the study conducted by Mata-Mota et al. [26], it was observed that the firmness of the obtained bigels significantly decreased when the GGM hydrogel proportion was increased from 63% to 87%. The results reported by other researchers further support the impact of the oleogel-to-hydrogel ratio on the textural profile attributes of bigels [33]. However, no discernible changes were detected in the springiness parameter of the bigels, indicating a high recovery capability of the system [33]. On the other hand, Martins et al. [50] found an opposite trend for SA/BW bigels with oleogel-to-hydrogel ratios varying from 50:50 to 99:1, as a reduction in the hardness and spreadability values was observed when increasing the oleogel proportion. Also, bigels with a 65:35 CW oleogel-to-XG hydrogel ratio were, on average, more cohesive compared to the other examined ratios. Interestingly, the findings of Vershkov and Davidovich-Pinchas [47] pointed out a weak correlation between bigel mechanical properties and the oleogel-to-hydrogel ratio, in contrast with the aforementioned studies.

Similarly, the most important variable influencing the rheological characteristics of bigel systems is the oleogel-to-hydrogel ratio. In one study, bigel formulations exhibited a non-Newtonian shear behavior with pseudoplastic characteristics, as demonstrated by a reduction in viscosity (η) and an increase in shear stress (τ) when the shear rate (γ) was increased [46]. Several studies have documented that, in bigel systems, a larger oleogel fraction leads to a higher η [49,51,63]. Additionally, rheological analysis has indicated a solid-like rather than a viscous behavior for bigels, with the storage modulus (G′) having higher values than the loss modulus (G″) in the linear viscoelastic region (LVR), without a cross-over point between G′ and G″. These results were confirmed for MG-BW and HAGG bigels [11], EC and XG/GGM bigels [77], KC and MG bigels [30], BW with agar/XG or GEL/XG bigels [51], and RBW and SA/KC bigels [34], identifying the synergistic effect of the two structured phases on the viscoelastic properties of final bigels and the formation of a network structure with a higher density. At 20 °C, bigel formulations with RBW and GEL showed frequency independence and a more solid than liquid (G′ > G″) character, regardless of the oleogel-to-hydrogel ratio. Additionally, a crossover point between G′ and G″ was not found in the angular frequency range of 0–100 rad/s, suggesting that all bigel formulations may be categorized as gels in the examined frequency range, and that they have enhanced mechanical properties compared to their individual oleogel and hydrogel phases [41]. Moreover, the G′ values of bigels was shown to increase with increasing ratios of GMS/BW-based oleogels; this may be due to the rigid structures of the studied oleogels. Fasolin et al. [13] used GG and high-oleic sunflower oil with GMS to form bigel. They concluded that the obtained bigels exhibited superior strength compared to the original oleogels and hydrogels, displaying higher values of complex moduli (G*) and lower phase angle δ values (defined by $\tan\delta = G''/G'$), particularly when the oleogel content was increased, indicating a synergistic effect between the two phases. Also, the results of the mechanical parameters were in agreement with rheological results, as the bigels exhibited high levels of hardness, shear work, and adhesion work [13]. A reduction in the LVR of bigels was observed with an increase in the oleogel content corresponding to the transition of the type of bigel from oleogel-in-hydrogel to hydrogel-in-oleogel [78]. Similar results have been reported in another study where the linear viscoelastic behavior of the bigels was mainly affected by the continuous oleogel phase in hydrogel-in-oleogel bigel [24]. Bigels containing a greater proportion of hydrogel phase exhibited enhanced rheological characteristics due to having a higher critical strain. Such gels have the ability to withstand greater shear forces, providing a more desired semi-solid behavior compared to plain oleogels [42]. Additionally, Zheng et al. [25] reported that the G′ slightly increased as the frequency increased, suggesting that bigels with greater hydrogel proportions had higher η values and less elastic characteristics than bigels with higher oleogel content.

3.3.2. Type and Concentration of the Gelators

Several authors have reported the effects of different concentrations and types of structurants on the rheological and mechanical properties of bigels. Overall, increasing the gelator concentration results in enhanced mechanical properties of the oleogel or hydrogel, and, consequently, of the bigel. For example, the gel strength of the BW- and GMS-based bigels increased progressively with higher oleogelator levels. The increased hardness of the bigels was attributed to the interaction forces between oleogelator molecules, causing the formation of oleogel networks [29]. Bollom et al. [42] also documented that a higher concentration of SL/SAC in the oleogel phase contributed to higher G′ values of the obtained bigels. This observation agreed with the results of Cho et al. [55] and Nutter et al. [38]. Additionally, Lu et al. [27] confirmed that the concentration of the oleogelator affects the hardness and the η of bigel systems. The higher content of GMS in the oleogel phase formed a stronger oleogel and, in turn, resulted in greater mechanical properties of the resultant bigels. Also, in bigels fabricated using oleogels structured with WPI, the firmness increased with increasing WPI levels. The firmness value significantly increased from 40.80 to 110.90 as the WPI level increased from 6% to 8%. This finding suggested that an increase in the WPI level reinforces the gel network of the continuous matrix in oleogel-in-hydrogel bigels, enhancing their mechanical strength [60]. Golodnizky and Davidovich-Pinhas [25] documented that the hydrophilic–lipophilic balance (HLB) value of sucrose esters (SEs) had a notable influence on the rheological properties of bigels. For example, bigels fabricated using SEs with lower HLB values showed higher elasticity and exhibited solid-like behavior in comparison with bigels prepared with SEs that had higher HLB values. This difference in rheological behavior was attributed to the low average droplet size and to the narrow particle size distribution of the bigels, as documented by CLSM [25]. The mechanical properties, such as adhesiveness, spreadability, and hardness, and rheological characteristics of bigels are influenced by the type of hydrogel used [13,51]. Martins et al. [33] documented a significant impact of the polymeric structurants content (KC/LBG) of the hydrogel on the consistency of bigels. Similarly, another study emphasized the significance of the composition of the hydrogel phase on the cohesiveness and hardness of bigels. When KC was incorporated into the GEL hydrogel phase of a bigel system, it resulted in increased hardness and reduced cohesiveness due to the cross-linking of the structurants [35]. Regarding rheological parameters, bigels with agar and XG showed higher G' and G'' moduli than bigels with GEL and XG. This difference was attributed to the distinct chemical-physical architectures of the used polymers or the more substantial intermolecular interactions between the elements of the bigel matrix [51].

3.3.3. Process Parameters and Storage

The mixing rate and mixing time of the oleogel with the hydrogel also affect the textural characteristics of the resulting bigel. For instance, bigels with MGs and GGM exhibited improved mechanical properties when prepared with a lower mixing time and higher mixing rate [26]. According to Fasolin et al. [13], the mechanical properties of guar gum/GMS bigels were significantly affected by increasing the mixing speed from 500 to 1500 min^{-1} for 10 min. Moreover, the mechanical properties of bigels are also dependent on the storage temperature. Bigels with CW and XG stored at 4 °C showed higher firmness, cohesiveness, and plasticity compared to bigels stored at 25 °C, due to the more organized crystal network of the oleogel phase at this temperature [47]. Also, GEL/GMS bigels that were set at a temperature of 4 °C exhibited greater hardness values compared to bigels kept at 25 °C [6]. Zampouni et al. [36] documented that KC/GEL/MGs bigels exhibited textural stability for up to 14 days at 4 °C. Similarly, other studies reported that storage for up to 90 days did not significantly affect the hardness of GEL/MGs bigels [6].

3.4. *Thermal Properties of Bigels*

Thermal characterization aims to interpret the physical characteristics of bigels as a function of temperature and time under certain conditions through the use of appropriate experimental techniques. An important property of polymeric gelled systems is the temperature-dependent gel–sol (solid-like to liquid-like) transition, which is crucial for food applications. Several studies have examined this transformation using various techniques, such as differential scanning calorimetry (DSC) and thermogravimetric analysis (TGA), to evaluate the thermal transition that occurs during heating and the contribution of bound and unbounded water to this process, respectively.

DSC was used for the thermal analysis of bigels with RBW and GEL in the study by Cho et al. [55]. The thermographs revealed two characteristic endothermic peaks during the heating cycle; one corresponded to the melting of GEL around 35 °C, and the other, a sharp endothermic peak at 66 °C, was attributed to the melting of RBW of the oleogel phase. In RBW/SA/KC bigels, the heating and cooling thermograms, regardless of the oleogel-to-hydrogel ratio, revealed prominent peaks at around 70.0 °C and ~65.6 °C, confirming rice bran wax phase transitions. A similar melting profile has been reported for bigels prepared by combining an SA/KC hydrogel with a RBW oleogel [38]. However, other researchers have suggested that the thermodynamic stability of bigels was enhanced by the presence of HAGG hydrogel, as the studied bigels exhibited higher melting temperatures than plain GMS/BW-based oleogels [11]. On the contrary, Ghorghi et al. [54] documented that the thermal properties of SA/BW bigels were improved by increasing the hydrogel proportions in the biphasic system. The DSC analysis of bigels provided further confirmation regarding the co-existence of two structurally organized phases within the bigel system [35]. In the study by Zheng et al. [72], the melting temperatures of bigels with varying oleogel/hydrogel ratios did not show any significant differences, indicating that the thermal behavior of oleogels with fatty acid mixtures remained stable even after they were mixed with XG hydrogels. This demonstrated that bigels can preserve their oleogel-like thermal properties. Habibi et al. [24] reported that the incorporation of KC/WPC80 hydrogel in MG oleogel in an 80:20 ratio decreased the melting point of the MGs. In contrast, the melting behavior of bigel systems remained unchanged when the size of structured aqueous droplets was reduced by the application of different mixing methods [24]. Additionally, several studies have reported that the enthalpies of the endothermic peaks were higher as the proportion of oleogel in the bigels was increased, due to the extended crystallization [35,72].

3.5. *Swelling Capacity*

Assessing the swelling capacity of bigels can provide valuable information on the way that a bigel performs in the digestion process and whether it can subsequently release incorporated bioactive components. The swelling capacity is the measure of the ability of a bigel to absorb water and provides insights into the bigel's structure, functionality, and potential applications. According to Zheng et al. [30], a 50:50 oleogel-to-hydrogel ratio exhibited a high swelling capacity in bigels with KC and MGs. A higher ratio of oleogel increased the hydrophobic nature of bigels and hindered water absorption in the biphasic system [30,35,78]. Thus, the most important factor that affected the swelling ratio was the bigel type. When the continuous phase was comprised of hydrogel, the absorption of water was promoted, while, when oleogel was the continuous phase, the penetration of the water was hindered [11]. Also, Kaimal and Singhal [78] reported that a higher hydrophilicity of the hydrogelator facilitated the absorption of water and allowed bigel to swell. Interestingly, a low swelling capacity of bigels was observed when KC with GEL was used as a hydrogelator, due to the formation of a stiff network that did not allow the cross-linked polymeric chains to expand and absorb water [35].

3.6. *Stability*

3.6.1. Centrifugation Stability

The assessment of the solvent holding capacity (SHC) of bigels was employed to evaluate the ability of the bigel matrix to retain a significant amount of solvent when subjected to centrifugation forces (e.g., $10\times g$ for 15 min or $9200\times g$ for 15 min) at room temperature [6,33]. The results indicated that treatments with GEL hydrogel and GMS oleogels mixed in ratios of 60:40 and 50:50 exhibited full stability under high centrifugal forces [6]. According to the results of a similar study, the extent of solvent retention within the bigel structure demonstrated an enhanced degree of stability and the dominance of the polymeric network of locust bean gum with KC hydrogel, even in bigel treatments with low GMS concentrations [33].

3.6.2. Oxidative Stability

The determination of the oxidative state and stability of bigel systems is crucial for their use as fat replacers, because they could affect the flavor, the aroma, and the nutritional value of food products during manufacturing and storage. The development of oxidation is influenced by several factors, such as the composition and origin of the oil used in the oleogel phase, the ratio of the oleogel to hydrogel, the oleogel and bigel preparation temperatures, the type of the formed bigel, the duration and the temperature of storage, and the addition of pro- and anti-oxidant agents. Several methods have been used for the evaluation of the oxidative stability of bigel systems. Quilaqueo et al. [46] measured the peroxide values (PVs) and p-anisidine values (AVs) to examine the extent of primary and secondary oxidation in bigels, respectively. The authors concluded that the use of CMC hydrogel in bigel did not offer protection against the development of primary oxidation products during the bigel preparation process. However, lower PVs were measured in bigel systems compared to plain BW-canola oil oleogels due to the lower lipid content in bigels. Similar results were also reported by Baltuonytė et al. [57], as bigels structured with GEL and agar showed lower oxidation compared to a mixture of sunflower oil and olive pomace oil after 14 days of storage at 37 °C. Also, other studies have obtained lower PVs during 30 days of storage at 4 °C for SA/BW bigels by elevating the hydrogel content in the biphasic system [54]. Cho et al. [41] reported that the PVs of bigels remained low during storage time (180 days) at 22 ± 1 °C, even when 1 mg Cu^{2+}/kg oil was added as a pro-oxidant. Also, the primary and secondary oxidation products were determined in GEL/MGs bigels by measuring the PVs and thiobarbituric acid (TBA) values [6]. Regarding PVs, the bicontinuous GEL/MGs bigels showed a behavior that was dependent on time, temperature, and the hydrogel content of the bigels. According to Samui et al. [6], the hydrogel phase enabled oxygen diffusion to the oleogel phase, promoting the oxidation process. Also, it is interesting to note that the TBA values of bigels exhibited a gradual rise during storage at different temperatures. On the contrary, the study of Quilaqueo et al. [46] reported that the AVs of CMC/BW bigels were lower than those of BW oleogels, indicating that the presence of the hydrogel phase at a 50% ratio provided protection against the development of secondary oxidation products.

3.6.3. Freeze–Thaw Stability

The freeze–thaw stability of bigels is crucial for frozen foods and is directly dependent on the degree of the collapse of the hydrogel network and the formation of ice crystals during freezing. The freeze–thaw stability of bigels has been evaluated by measuring the oil holding capacity (OHC) and water holding capacity (WHC) of bigels after freeze–thaw cycles [29,63]. According to Martins et al. [33], the degree of the oil holding capacity was indicative of the bigel stability. Different oleogelators formed distinct crystalline networks, which, in turn, affected the bigel oil binding capacity [45]. Additionally, the WHC was strongly influenced by both the architecture and the stability of the formed gel network [60]. Generally, the OHC and the WHC have been evaluated using centrifugation for a certain time [29,45]. The OHC of bigels was improved by the incorporation of WPC80 hydrogel into the MGs oleogel matrix [24]. Also, an increase in the homogenization time from 1 to 3 and 5 min also increased the OHC in GEL/MG-based bigels [6]. Yang et al. [29] documented that the type of the formed bigels, the strength of the oleogel phase, and the type of oleogelator affected the freeze–thaw stability. In GMS/KC bigels, both the WHC and OHC increased with an increase in GMS content, due to the formation of stronger oleogel networks and more closely packed oil droplets. On the contrary, in BW/KC bigels, only the OHC significantly increased with an increase in the BW content. However, phase separation of the hydrogel and oleogel was observed after freeze–thawing, indicating the disruption of the interfacial layer due to the growth of ice crystals [29].

4. Bigel Applications

Table 2 provides an overview of the current food applications of bigels, highlighting their potential in various sectors of the food industry. For instance, according to the literature, bigels can be used as fat substitutes, edible coatings, or effective delivery systems for bioactive compounds. Also, bigels have been investigated as innovative food matrices and edible food packaging materials, using 3D food-printing technology.

Table 2. Bigel systems (composition, oleogel-to-hydrogel ratio, and bigel mixing parameters) and their current application in food products.

Application		Oleogel (O) Edible Oil	Oleogelator	Hydrogel (H) Hydrogelator	O:H ratio	Bigel Mixing Parameters	Reference
Delivery system	Bioactive fatty acids	Coconut oil Avocado oil Pomegranate oil	MGs * Tween 80	CMC (2%)	Non reported	18.000 rpm for 1 min + sonication for 1 min	[67]
	Bioactive fatty acids in yogurt	Avocado oil	MGs+DGs Tween 80	CMC (2%)	Non reported	18.000 rpm for 1 min + sonication for 1 min	[66]
	β-Carotene	Corn oil	MGs (20%)	KC (1.5%)	25:75, 40:60, 50:50, 60:40, 75:25	300 rpm at 80 °C for 20 min + 500 rpm at 80 °C for 2 min	[31]
	Curcumin EGCG	Corn oil	GMS (4–8%)	GEL (2.5%) Tween 20 (0.5%)	50:50	300 rpm at 60 °C for 20 min	[27]
	Curcumin EGCG in bigel beads	Corn oil	GMS (10%)	SA (1%) Tween 20 (0.5% in bigel)	10:90, 15:85, 20:80, 25:75, 30:70	500 rpm	[28]
	Lutein	Sunflower oil	EC (15%)	XG GGM (1:1, 1.5%)	25:75, 50:50, 75:25	10,000 rpm at 75 °C for 5 min	[78]
	Lycopene	Soybean oil	GMS (1%) BW (1%)	GG (0.3%)	10:90, 20:80, 30:70, 40:60, 50:50, 60:40	10,000 for 1 min	[11]
	Probiotics	Soybean oil	SL, SAC (1:1, 20%)	WPC 80 (20%)	80:20	13,500 rpm at 85 °C for 2 min	[26]
	Probiotics in yogurt	Soybean oil	SL, SAC (1:1, 16%)	WPC 80 (20%)	75:25	13,500 rpm at 85 °C for 2 min	[12]
	Quercetin	Soybean oil	SAC (2.5%)	KG, GEL (2.25%)	Non reported	9000 rpm at 75 °C for 3 min	[44]
Fat substitution	Beef burger	Sunflower oil	EC (10%)	WS (10%)	25:75, 50:50, 75:25	2000 rpm for 2 min + 2500 rpm for 2 min	[59]
	Coarse-ground sausages	High oleic soybean oil	RBW (7.5%)	GEL (7- 8%)	70:30, 80:20	24,000 rpm for 1.5 min	[56]
	Fermented sausages	Olive oil	MGs (15%)	GEL (10%) KC (1%)	20:80, 40:60	600 rpm at 80 °C for 15 min	[34]
	Semi-dry sausages	Olive oil	MGs (15%)	GEL (4%) KC (2%)	20:80	300 rpm at 80 °C for 15 min 12,000 rpm for at 80 °C 15 min	[36]
	Bread	Corn oil	GMS (1%) RBW (2%)	SA (2%)	20:80, 30:70, 40:60, 50:50, 60:40	10,000 rpm at 85 °C for 3 min	[37]
	Cookies	Canola oil	BW (10%)	CMC SA (3%)	50:50	2500 rpm at 30 °C	[46]
	Shortbread	Soybean oil	RBW (9%) MGs (0–2%)	KC (0.5%) SA (1%)	80:20	15,000 rpm for 1 min at 85 °C	[38]
	Butter spread	Sesame oil	BW (10%)	SA (3%) WPC 80 (25%)	Non reported	900 rpm for 2 min	[53]
	Chocolate	Grape seed oil	BW (20%)	SA (2%)	99:1, 95:5, 90:10	600 rpm for 45 min	[54]

Table 2. *Cont.*

Application		Oleogel (O) Edible Oil	Oleogelator	Hydrogel (H) Hydrogelator	O:H ratio	Bigel Mixing Parameters	Reference
3D printing		Corn oil	BW (15%)	KC XG (1:1, 1.5%)	20:80, 30:70, 50:50, 70:30, 80:20	3000 rpm for 2 min + 6000 rpm for 3 min	[49]
		Soybean oil	BW (10%) PGPR (1%)	HPMC (3%)	20:80, 40:60, 50:50, 60:40, 80:20	8000 rpm at 85 °C	[63]
		Soybean oil	BW (10%) GMS (2%)	GG (3%)	30:70, 40:60, 50:50, 60:40, 65:35, 70:30, 80:20	8000 rpm at 80 °C for 3 min + 100 rpm for 10 min	[40]
		High oleic acid sunflower seed oil	CW (3%), MGs (1%), PGPR (1%), SL (1%)	Fish GEL (5%)	30:70, 50:50, 70:30	10,000 rpm for 1 min	[76]
		Sunflower oil	BW (5%)	GEL (10%) XG (1%) Agar (15%)	5:95, 10:90, 20:80	2000 rpm at 50 °C for 10 min	[51]
3D printing	Delivery system of quercetin	High oleic acid sunflower seed oil	CW (3%) MGs (1%) SL (1%)	Fish GEL	30:70, 50:50, 70:30	10,000 rpm for 1 min	[75]
	Food packaging	Sunflower oil	BW (6%) GMS (2%)	Agar (0.5–2%)	90:10, 80:20, 70:30, 60:40	12,000 rpm for 1 min	[65]
Other applications	Edible coating with rosemary extract	Sunflower oil	MGs (15%)	GEL (10%)	20:80	300 rpm min at 70 °C for 15	[32]
	Spread with lingonberry pomace	Sunflower oil Olive pomace oil	CBW (10%) SL (0.5–1.5%)	COL (15.6%) GEL (5%) Agar (2.5%)	60:40	15,000 rpm for 2 min at 85 °C	[57]
	Cream analogue	MCT oil	GMS Cinnamaldelhyde Span 80 (0–0.9%)	Chitosan	20:80, 33:67, 40:60	200 rpm for 20 min	[74]
	Foam-based bigels	Soybean oil	BW (10%) GMS (2%)	GG (1.2%)	20:80, 30:70, 40:60, 50:50, 60:40, 70:30, 80:10, 90:10	8000 rpm for 2 min	[73]

* BW: beeswax, CBW: carnauba wax, CMC: carboxymethyl cellulose, COL: collagen, CW: candelilla wax, DGs: diglycerides, EC: ethylcellulose, EGCG: epigallocatechin gallate, GEL: gelatin, GG: gellan gum, GMS: glycerol monostearate, HPMC: hydroxypropyl methylcellulose, KC: κ-carrageenan, MCT: medium-chain triglycerides, MGs: monoglycerides, PGPR: polyglycerol ricin alkyd ester, RBW: rice bran wax, SA: sodium alginate, SAC: stearic acid, SL: soy lecithin, XG: xanthan gum, WPC: whey protein concentrate, WS: wheat starch.

4.1. Bigels as Delivery Systems for Bioactive Compounds

Several studies have utilized in vitro digestion protocols to investigate the potential of bigel systems to act as vehicles for the efficient delivery of bioactive substances. For example, GMS/BW and HAGG bigels in various oleogel-to-hydrogel ratios were evaluated for the delivery of lycopene [11]. After gastric and intestinal phase digestion, a higher percentage of lycopene was released from bigels containing a lower oleogel ratio, indicating a slower and more controlled release of the compound. Zhu et al. [11] suggested that the oleogel enhanced the bigel structure, resulting in slower penetration of the gastric fluids into the bigel network, and practically inhibiting the release of lycopene during gastric digestion. On the other hand, the collapse of the hydrogel network during intestinal digestion enabled pancreatic enzymes and bile salts to successfully break down the oleogel, causing the controlled release of lycopene. Kaimal and Singhal [78] developed EC and XG/GGM bigels to study the effectiveness of biphasic systems in the delivery and protection of another carotenoid substance, lutein, through a simulated gastrointestinal tract. The results of this study were in agreement with the results of Zhu et al. [11], as the release of lutein from different bigels exhibited statistically significant variations, showing a greater release of lutein as the hydrogel fraction in the bigel increased to 75%. On the contrary, Zheng et al. [30], who studied the utilization of MG/KC bigels as a delivery system of β-carotene, found that both the release rate and the total (%) β-carotene release were increased for bigels with a higher oleogel ratio. Interestingly, the bigel containing 75% oleogel showed the highest release of β-carotene. This was supported by the fact that, in this type of bigel, the continuous phase was the oleogel, which was readily exposed to the enzymes for lipid hydrolysis. Therefore, based on the above, it could be concluded that the types of gelator used for structuring the oleogel and hydrogel phase could modulate the release profile of each substance more effectively than the ratio of the two phases and the formed type of the final bigel. Also, according to Lu et al. [27], the concentration of the oleogel structurant modified the release of the lipophilic curcumin from GMS/GEL bigel, as the increased concentration of GMS resulted in a significantly reduced rate. On the contrary, when epigallocatechin gallate (EGCG) was encapsulated within the hydrogel phase, the impact of this GMS hydrolysis on the release of EGCG was minimal. The release of EGCG occurred rapidly within the gastric juice, as the hydrogel phase was susceptible to pepsin degradation, and subsequently, the release rate remained relatively constant [27]. These findings support the idea that bigel systems have the ability to simultaneously transport bioactive components with varying polarities. For example, Yang et al. [28] suggested that GMS/SA bigel beads with different oleogel-to-hydrogel ratios can succeed in the co-delivery and protection of curcumin and EGCG. An increase in the oleogel content resulted in a gradual decline in the encapsulation efficacy of EGCG, attributed to the relatively diminished water phase content, in contrast with the encapsulation efficacy of curcumin, which showed an opposite trend. Also, a less dense alginate network with a less rigid configuration facilitated the effective diffusion of EGCG through SA gel microstructures and also improved the release of the bioactive substance. After a 40-day storage test, the retention ratio of loaded curcumin was sufficiently high, indicating that the bigel structure prevented the compound from being degraded [28]. Liu et al. [44] investigated the gastrointestinal digestive release of the hydrophobic substance quercetin from KG/GEL bigels. According to the study, an increase in the KG-to-GEL ratio resulted in a decrease in the bioaccessibility of the quercetin due to the formation of a more compact and cohesive bigel. In the same study, the release of volatile compounds (i.e., limonene) in the oral cavity from the bigel structure was evaluated using an electronic nose. The results demonstrated that the modification of the gel strength of the hydrogel phase led to different volatile release behaviors of bigels, with stronger gels preventing the release of the substances [44].

Moreover, MG/CMC bigel systems have been evaluated concerning the ability to offer protection for bioactive fatty acids of vegetable oils (coconut oil, avocado oil, pomegranate oil) during transit through the in vitro gastrointestinal environment. Throughout all stages of digestion, the concentration of the main bioactive fatty acids in each oil remained higher in bigel formulations in comparison to the corresponding oils [67]. Similarly, bigels with avocado oil were incorporated into yogurt to partially modify the profile of saturated fatty acids in milk fat and achieve a more balanced ratio of saturated and unsaturated lipids. Yogurt containing avocado oil bigel exhibited increased stability throughout the simulated gastrointestinal tract, indicating a greater recovery percentage of oleic acid in comparison with yogurt with avocado oil in a free state. The outcomes derived from the in vitro study demonstrated that these functional yogurts possessed the ability to increase the lipolytic rate in adipocytes, as evidenced by an approximate 20% increase in the glycerol release in yogurts with avocado oil bigel [66].

Bollom et al. [38] studied the capability of a bigel system to protect probiotic microorganisms (i.e., *Bifidobacterium lactis* and *Lactobacillus acidophilus*) during in vitro digestion, and also ascertained the impact of phospholipids on the survival of probiotics. The results showed that the bigel structure that results from the addition of the gelators (SL/SAC and WPC80) is crucial for the survival of the probiotics through the gastric phase. In contrast, the same study showed that the addition of phospholipids in the bigel did not provide significant protection to the viability of probiotics during digestion because of their enzymatic breakdown by digestive enzymes. Similarly, other studies have evaluated bigels using the aforementioned gelators for the protection of probiotics in two yogurt products, indicating that the bigel structure enhanced the probiotic viability [12].

4.2. Bigels as Fat Substitutes

4.2.1. Meat Products

Bigel systems were incorporated as fat substitutes in coarse-ground fully cooked sausages [56]. The bigels were composed of a RBW oleogel and GEL hydrogel, and were mixed in 70:30 and 60:40 ratios. In this study, the authors reported that the bigel remained stable after all the preparation stages of the sausages and the bigel with a 70:30 oleogel-to-hydrogel ratio exhibited a higher resemblance to the characteristics of pork fat. In an attempt to improve the nutritional properties of semi-dry sausages, Zampouni et al. [36] used two different hydrogel-in-oleogel bigels composed of olive oil oleogels (20%) and KC or KC plus GEL hydrogels (80%) for the partial replacement of pork backfat. The weight loss, moisture content, and water activity were shown to differ significantly between the sausages with fat replacement and the control. These variations were ascribed to the bigels' composition, the applied dehydration conditions, and the bigels' greater initial moisture content. Analysis of the textural characteristics of the semi-dry sausages revealed that the fat replacement did not affect the textural attributes of the sausages. Ghiasi and Golmakani [59] reported that EC and starch bigel with a 75% oleogel fraction exhibited excellent performance as a fat substitute in low-fat burgers. According to the results, all the low-fat burgers had superior cooking characteristics compared to the animal fat control burgers, especially when the replacement percentage was higher. The replacement of up to 50% animal fat with bigel resulted to low-fat burgers with acceptable sensory properties, even though the addition of bigel increased lipid oxidation in the burgers. Also, bigels with KC, GEL, and MGs were used to replace animal fat in fermented sausages. The addition of the bigel did not affect the water activity and microbial populations of the produced sausages. Interestingly, the color, texture, juiciness, flavor, taste, and overall acceptability of the sausage treatments did not significantly differ, according to consumer sensory evaluation studies [34].

4.2.2. Bakery Products

Recently, Nutter et al. [38] assessed the complete substitution of solid fats in a high-fat (30% *w*/*w*) short dough product using plant-based bigels. Bigels were formulated using 20% SA/KC hydrogel and 80% RBW soybean oil oleogel, with or without the addition of MGs. The findings of the study revealed that the obtained bigels effectively imitated the functionality of butter and shortenings by providing the desired shortening effect, leading to the formation of inelastic doughs. The desired fat functionality was preserved even after baking, as the bigel cookies exhibited a similar structure, moisture content, and water activity compared to traditional shortbread. Furthermore, the bigel cookies were softer than those made with commercial solid fats such as butter and shortening. Similarly, Quilaqueo et al. [46] evaluated the application of bigel systems comprised of 50% BW oleogel and 50% SA or CMC hydrogel as fat substitutes in cookies. They found that cookies with bigels exhibited similar hardness but significantly higher fracturability compared to cookies made with the original shortening. Regarding changes in the cookie shape after baking, while the width of the bigel cookies remained unchanged, the thickness was significantly increased. In another study, butter was replaced with hydrogel-in-oleogel bigel with RBW, GMS, and SA to enhance the quality characteristics of the bread [37]. The bread produced with bigels exhibited a greater specific volume and softer texture than the bread with butter, regardless of the oleogel ratios in the bigels. Additionally, the researchers concluded that, when the proportion of the oleogel reached 80%, it resulted in bread with an ideal specific volume and the most favorable textural attributes [37]. The above findings indicate the great potential of bigels as a fat replacer in bakery products.

4.3. 3D Printing

Additive manufacturing, also known as 3D printing, is an innovative technology that has garnered significant interest across various scientific fields. This manufacturing process involves the layer-by-layer construction of complex solid or semi-solid shapes. The potential of 3D food printing is supported by the ability to create food products that meet consumers' requirements with both economic and environmental advantages. However, to achieve the formulation of novel printed structures, it is necessary to first understand the rheological properties of the printing materials (food inks) to ensure the feasibility of the printing process [80]. Recently, research has been conducted regarding the development and evaluation of food-grade bigel systems for 3D printing applications. The study by Xie et al. [75] included the fabrication of 3D-printed bigels with a complicated structure which would have the potential to encapsulate catechin and quercetin and enhance the nutritional and functional properties of foods. The main components utilized in the formation of the 3D-printed bigels were GEL hydrogels, CW-based oleogel, and emulsifiers, such as MGs and SL [75]. According to another study by Xie et al. [76], the ratio of oleogel to hydrogel in bigels significantly affected the quality of printing, with better results obtained for higher oleogel ratios. For example, a bigel consisting of MG and SL with an oleogel/hydrogel ratio of 70:30 exhibited a printing performance comparable to that of pure hydrogel, with a smooth surface and good shape retention after the printing process. However, despite its desired shape, there was some oil leakage in the printed object. These 3D printing results indicated that the hydrogel mostly provided the necessary mechanical strength to support the intended shape during the printing process. However, the increased ratio of hydrogel in the bigel could potentially obstruct the printing nozzle, resulting in an unacceptably low printing accuracy of the bigel. Similarly, Chen et al. [63] evaluated the capability of printing by testing the ability of different types of bigels to form structures for replacing conventional solid fats and for the fabrication of food products with specific appearance attributes. Analysis of the printing results revealed that hydrogel-in-oleogels bigels were more suitable for flat stacked models, whereas oleogel-in-hydrogel bigels performed better in models with small-area monolayer structures. Furthermore, TPA tests indicated that the extrusion process had a significant effect on hydrogel-in-oleogel and bicontinuous bigels, but not on oleogel-in-hydrogel bigels.

Packaging

The work of Zhai et al. [65] presented an innovative approach to developing colorimetric gas sensors for intelligent food packaging that were based on anthocyanins, using 3D-printable hydrogel-in-oleogel bigel. The bigel used was composed of agar, purple sweet potato anthocyanins, BW, glyceride monooleate, and sunflower oil. Then, the composite film was fabricated by extruding the bigel onto polyvinylidene fluoride (PVDF) film using 3D printing technology. The final PVDF–bigel film was successfully utilized as a volatile amines sensor to monitor meat and fish freshness. While bigel systems offer a lot of advantages such as improved stability and the ability to deliver bioactive compounds, there are several challenges to consider before their use as packaging materials. For example, bigels may exhibit lower mechanical properties than traditional packaging materials or exhibit lower moisture or oxygen barrier properties, which can lead to stability problems and degradation over time.

4.4. Other Applications

Edible Coating

An earlier study by Kanelaki et al. [26] evaluated the effectiveness of GEL/MGs bigel as an edible coating for sardine fillets. The results suggested that, when bigels loaded with rosemary extract were used as coatings, they delayed lipid oxidation and total volatile basic nitrogen (TVB-N) formation in sardine fillets, but they did not affect microbial growth during storage.

5. Conclusions and Future Perspectives

Bigel systems offer numerous benefits over other structured systems, such as oleogels, hydrogels, or even emulsion gels. The characteristics of these biphasic systems are predominantly contingent upon parameters that include the oleogel-to-hydrogel ratio, the type and the concentration of the structural components of each phase, and the methods of fabrication and storage. Therefore, the modification and optimization of the above variables is necessary for the formation of a bigel system that exhibits the necessary and desirable characteristics for use in a food matrix. Nevertheless, it is necessary to determine the resulting microstructure, the mechanical and thermal properties, and the stability of bigels using specific techniques before their utilization for food applications. To date, the versatility of bigel systems has led to the development of applications across a wide range of food products, although still on a lab scale. For instance, bigels have demonstrated potential efficacy as delivery systems of bioactive compounds, fat substitutes in meat, bakery and confectionery products, and edible and biodegradable packaging materials. The encapsulation of various substances (e.g., vitamins, flavorings, antioxidants) in bigels and the assessment of their controlled release and diffusion rate could offer the opportunity for the development of functional and sustainable new food products, meeting current and future consumer demands. Also, the utilization of bigels in 3D printing applications appears to be an innovative approach not only to customize and tailor bigels' properties but also for the formulation of new products with enhanced functionality. Thus, the future of bigels in the food field is very promising, suggesting potential for a wide range of innovations regarding food formulation, processing, and packaging.

Author Contributions: Conceptualization, K.Z. and E.K.; methodology, K.Z.; data curation, K.Z.; writing—original draft preparation, K.Z.; writing—review and editing, D.D.-P. and E.K.; visualization, K.Z.; supervision, E.K.; All authors have read and agreed to the published version of the manuscript.

Funding: This research received no external funding.

Institutional Review Board Statement: Not applicable.

Informed Consent Statement: Not applicable.

Data Availability Statement: Not applicable.

Conflicts of Interest: The authors declare no conflicts of interest.

References

1. Almeida, I.F.; Fernandes, A.R.; Fernandes, L.; Pena Ferreira, M.R.; Costa, P.C.; Bahia, M.F. Moisturizing effect of oleogel/hydrogel mixtures. *Pharm. Dev. Technol.* **2008**, *13*, 487–494. [CrossRef] [PubMed]
2. Deng, G.; Ma, Q.; Yu, H.; Zhang, Y.; Yan, Z.; Liu, F.; Jiang, H.; Chen, Y. Macroscopic organohydrogel hybrid from rapid adhesion between dynamic covalent hydrogel and organogel. *ACS Macro Lett.* **2015**, *4*, 467–471. [CrossRef] [PubMed]
3. Shakeel, A.; Lupi, F.R.; Gabriele, D.; Baldino, N.; De Cindio, B. Bigels: A unique class of materials for drug delivery applications. *Soft Mater.* **2018**, *16*, 77–93. [CrossRef]
4. Singh, V.K.; Qureshi, D.; Nayak, S.K.; Pal, K. Bigels. In *Polymeric Gels*; Elsevier: Amsterdam, The Netherlands, 2018; pp. 265–282. [CrossRef]
5. Zulfakar, M.H.; Chan, L.M.; Rehman, K.; Wai, L.K.; Heard, C.M. Coenzyme Q10-loaded fish oil-based bigel system: Probing the delivery across porcine skin and possible interaction with fish oil fatty acids. *AAPS PharmSciTech.* **2018**, *19*, 1116–1123. [CrossRef]
6. Samui, T.; Goldenisky, D.; Rosen-Kligvasser, J.; Davidovich-Pinhas, M. The development and characterization of novel in-situ bigel formulation. *Food Hydrocoll.* **2021**, *113*, 106416. [CrossRef]
7. Di Michele, L.; Fiocco, D.; Varrato, F.; Sastry, S.; Eiser, E.; Foffi, G. Aggregation dynamics, structure, and mechanical properties of bigels. *Soft Matter* **2014**, *10*, 3633–3648. [CrossRef]
8. Hu, Z.; Patten, T.; Pelton, R.; Cranston, E.D. Synergistic stabilization of emulsions and emulsion gels with water-soluble polymers and cellulose nanocrystals. *ACS Sustain. Chem. Eng.* **2015**, *35*, 1023–1031. [CrossRef]
9. Singh, V.K.; Banerjee, I.; Agarwal, T.; Pramanik, K.; Bhattacharya, M.K.; Pal, K. Guar gum and sesame oil based novel bigels for controlled drug delivery. *Colloids Surf. B Biointerfaces* **2014**, *123*, 582–592. [CrossRef]
10. Martinez, R.M.; Magalhaes, W.V.; da Silva Sufi, B.; Padovani, G.; Nazato, L.I.S.; Velasco, M.V.R.; da Silva Lannes, S.C.; Baby, A.R. Vitamin E-loaded bigels and emulsions: Physicochemical characterization and potential biological application. *Colloids Surf. B Biointerfaces* **2021**, *201*, 111651. [CrossRef]

11. Zhu, Q.; Gao, J.; Han, L.; Han, K.; Wei, W.; Wu, T.; Jinling, L.; Zhang, M. Development and characterization of novel bigels based on monoglyceride-beeswax oleogel and high acyl gellan gum hydrogel for lycopene delivery. *Food Chem.* **2021**, *365*, 130419. [CrossRef]
12. Zhuang, X.; Clark, S.; Acevedo, N. Bigels—Oleocolloid matrices—As probiotic protective systems in yogurt. *J. Food Sci.* **2021**, *86*, 4892–4900. [CrossRef] [PubMed]
13. Fasolin, L.H.; Martins, A.J.; Cerqueira, M.A.; Vicente, A.A. Modulating process parameters to change physical properties of bigels for food applications. *Food Struct.* **2021**, *28*, 100173. [CrossRef]
14. Luo, S.Z.; Hu, X.F.; Jia, Y.J.; Pan, L.H.; Zheng, Z.; Zhao, Y.Y.; Mu, D.D.; Zhong, X.Y.; Jiang, S.T. Camellia oil-based oleogels structuring with tea polyphenol-palmitate particles and citrus pectin by emulsion-templated method: Preparation, characterization and potential application. *Food Hydrocoll.* **2019**, *95*, 76–87. [CrossRef]
15. Davidovich-Pinhas, M.; Barbut, S.; Marangoni, A.G. Development, characterization, and utilization of food-grade polymer oleogels. *Annu. Rev. Food Sci. Technol.* **2016**, *7*, 65–91. [CrossRef]
16. Okesola, B.O.; Vieira, V.M.; Cornwell, D.J.; Whitelaw, N.K.; Smith, D.K. 1,3:2,4-Dibenzylidene-D-sorbitol (DBS) and its derivatives–efficient, versatile and industrially-relevant low-molecular-weight gelators with over 100 years of history and a bright future. *Soft Matt.* **2015**, *11*, 4768–4787. [CrossRef]
17. Onacik-Gür, S.; Żbikowska, A.; Przybysz, M.; Kowalska, M. Assessment of physical properties of structured oils and palm fat. *Mater. Plast.* **2017**, *54*, 800–805. [CrossRef]
18. Dimakopoulou-Papazoglou, D.; Zampouni, K.; Prodromidis, P.; Moschakis, T.; Katsanidis, E. Microstructure, physical properties, and oxidative stability of olive oil oleogels composed of sunflower wax and monoglycerides. *Gels* **2024**, *10*, 195. [CrossRef]
19. Demirkesen, I.; Mert, B. Recent developments of oleogel utilizations in bakery products. *Crit. Rev. Food Sci. Nutr.* **2020**, *60*, 2460–2479. [CrossRef]
20. Meng, Z.; Qi, K.; Guo, Y.; Wang, Y.; Liu, Y. Macro-micro structure characterization and molecular properties of emulsion-templated polysaccharide oleogels. *Food Hydrocoll.* **2018**, *77*, 17–29. [CrossRef]
21. Abdollahi, M.; Goli, S.A.H.; Soltanizadeh, N. Physicochemical properties of foam-templated oleogel based on gelatin and xanthan gum. *Eur. J. Lipid Sci. Technol.* **2020**, *122*, 1900196. [CrossRef]
22. De Vries, A.; Jansen, D.; van der Linden, E.; Scholten, E. Tuning the rheological properties of protein-based oleogels by water addition and heat treatment. *Food Hydrocoll.* **2018**, *79*, 100–109. [CrossRef]
23. Scharfe, M.; Flöter, E. Oleogelation: From scientific feasibility to applicability in food products. *Eur. J. Lipid Sci. Technol.* **2020**, *122*, 2000213. [CrossRef]
24. Habibi, A.; Kasapis, S.; Truong, T. Effect of hydrogel particle size embedded into oleogels on the physico-functional properties of hydrogel-in-oleogel (bigels). *LWT* **2022**, *163*, 113501. [CrossRef]
25. Golodnizky, D.; Davidovich-Pinhas, M. The effect of the HLB value of sucrose ester on physiochemical properties of bigel systems. *Foods* **2020**, *9*, 1857. [CrossRef] [PubMed]
26. Mata-Mota, J.D.; Gallegos-Infante, J.A.; Pérez-Martínez, J.D.; Rocha-Guzmán, N.E.; González-Laredo, R.F. Effect of hydrogel/oleogel ratio, speed and time of mixing, on the mechanical properties of bigel materials and the application of Cox-Merz rule. *Food Mater. Res.* **2023**, *3*, 24. [CrossRef]
27. Lu, Y.; Zhong, Y.; Guo, X.; Zhang, J.; Gao, Y.; Mao, L. Structural modification of O/W bigels by glycerol monostearate for improved co-delivery of curcumin and epigallocatechin gallate. *ACS Food Sci. Technol.* **2022**, *2*, 975–983. [CrossRef]
28. Yang, J.; Song, J.; Miao, S.; Gao, Y.; Mao, L. Alginate-based gel beads with bigel structures: Preparation, characterization and bioactive encapsulation. *Food Hydrocoll.* **2024**, *146*, 109294. [CrossRef]
29. Yang, J.; Zheng, H.; Mo, Y.; Gao, Y.; Mao, L. Structural characterization of hydrogel-oleogel biphasic systems as affected by oleogelators. *Food Res. Int.* **2022**, *158*, 111536. [CrossRef]
30. Zheng, H.; Mao, L.; Cui, M.; Liu, J.; Gao, Y. Development of food-grade bigels based on κ-carrageenan hydrogel and monoglyceride oleogels as carriers for β-carotene: Roles of oleogel fraction. *Food Hydrocoll.* **2020**, *105*, 105855. [CrossRef]
31. Barroso, L.A.; Karatay, G.G.B.; Hubinger, M.D. Effect of potato starch hydrogel: Glycerol monostearate oleogel ratio on the physico-rheological properties of bigels. *Gels* **2022**, *8*, 694. [CrossRef]
32. Kanelaki, A.; Zampouni, K.; Mourtzinos, I.; Katsanidis, E. Hydrogels, oleogels and bigels as edible coatings of sardine fillets and delivery systems of rosemary extract. *Gels* **2022**, *8*, 660. [CrossRef] [PubMed]
33. Martins, A.J.; Guimarães, A.; Fuciños, P.; Sousa, P.; Venâncio, A.; Pastrana, L.M.; Cerqueira, M.A. Food-grade bigels: Evaluation of hydrogel: Oleogel ratio and gelator concentration on their physicochemical properties. *Food Hydrocoll.* **2023**, *143*, 108893. [CrossRef]
34. Siachou, C.; Zampouni, K.; Katsanidis, E. Bigels as fat replacers in fermented sausages: Physicochemical, microbiological, sensory, and nutritional characteristics. *Gels* **2023**, *9*, 340. [CrossRef]
35. Zampouni, K.; Mouzakitis, C.K.; Lazaridou, A.; Moschakis, T.; Katsanidis, E. Physicochemical properties and microstructure of bigels formed with gelatin and κ-carrageenan hydrogels and monoglycerides in olive oil oleogels. *Food Hydrocoll.* **2023**, *140*, 108636. [CrossRef]
36. Zampouni, K.; Filippou, A.; Papadimitriou, K.; Katsanidis, E. Evaluation of bigel systems as potential substitutes to partially replace pork backfat in semi-dry sausages. *Meat Sci.* **2024**, *208*, 109392. [CrossRef]

37. Han, L.; Chen, F.; Qiu, Y.; Gao, J.; Zhu, Q.; Wu, T.; Wang, P.; Zhang, M. Development and characterization of hydrogel-in-oleogel (bigel) system and their application in replacement of butter for bread making. *J. Sci. Food Agric.* **2023**, *104*, 1920–1927. [CrossRef]
38. Nutter, J.; Shi, X.; Lamsal, B.; Acevedo, N.C. Plant-based bigels as a novel alternative to commercial solid fats in short dough products: Textural and structural properties of short dough and shortbread. *Food Biosci.* **2023**, *54*, 102865. [CrossRef]
39. Nutter, J.; Shi, X.; Lamsal, B.; Acevedo, N.C. Designing and characterizing multicomponent, plant-based bigels of rice bran wax, gums, and monoglycerides. *Food Hydrocoll.* **2023**, *138*, 108425. [CrossRef]
40. Guo, Z.; Chen, Z.; Meng, Z. Bigels constructed from hybrid gelator systems: Bulk phase-interface stability and 3D printing. *Food Funct.* **2023**, *14*, 5078–5089. [CrossRef]
41. Saffold, A.C.; Acevedo, N.C. Development of novel rice bran wax/gelatin-based biphasic edible gels and characterization of their microstructural, thermal, and mechanical properties. *Food Bioprocess Technol.* **2021**, *14*, 2219–2230. [CrossRef]
42. Bollom, M.A.; Clark, S.; Acevedo, N.C. Development and characterization of a novel soy lecithin-stearic acid and whey protein concentrate bigel system for potential edible applications. *Food Hydrocoll.* **2020**, *101*, 105570. [CrossRef]
43. Bollom, M.A.; Clark, S.; Acevedo, N.C. Edible lecithin, stearic acid, and whey protein bigels enhance survival of probiotics during in vitro digestion. *Food Biosci.* **2021**, *39*, 100813. [CrossRef]
44. Liu, L.; Tian, W.; Chen, M.; Huang, Y.; Xiao, J. Oral sensation and gastrointestinal digestive profiles of bigels tuned by the mass ratio of konjac glucomannan to gelatin in the binary hydrogel matrix. *Carbohydr. Polym.* **2023**, *312*, 120765. [CrossRef] [PubMed]
45. Tian, W.; Huang, Y.; Song, Z.; Yu, Y.; Liu, J.; Cao, Y.; Xiao, J. Flexible Control of Bigel Microstructure for Enhanced Stability and Flavor Release during Oral Consumption. *Food Res. Int.* **2023**, *174*, 113606. [CrossRef]
46. Quilaqueo, M.; Iturra, N.; Contardo, I.; Millao, S.; Morales, E.; Rubilar, M. Food-grade bigels with potential to replace saturated and trans fats in cookies. *Gels* **2022**, *8*, 445. [CrossRef]
47. Vershkov, B.; Davidovich-Pinhas, M. The effect of preparation temperature and composition on bigel performance as fat replacers. *Food Funct.* **2023**, *14*, 3838–3848. [CrossRef]
48. De Los Santos-Trinidad, J.D.L.; Pérez-Alonso, C.; Cruz-Sosa, F.; Román-Guerrero, A. Effect of the use of sweet potato or chayote tuber starch hydrogels on the physical properties of oleogel-in-hydrogel (bigels) systems. *Int. J. Food Sci. Technol.* **2023**, *58*, 6871–6880. [CrossRef]
49. Qiu, R.; Wang, K.; Tian, H.; Liu, X.; Liu, G.; Hu, Z.; Zhao, L. Analysis on the printability and rheological characteristics of bigel inks: Potential in 3D food printing. *Food Hydrocoll.* **2022**, *129*, 107675. [CrossRef]
50. Martins, A.J.; Silva, P.; Maciel, F.; Pastrana, L.M.; Cunha, R.L.; Cerqueira, M.A.; Vicente, A.A. Hybrid gels: Influence of oleogel/hydrogel ratio on rheological and textural properties. *Food Res. Int.* **2019**, *116*, 1298–1305. [CrossRef]
51. Fernandes, A.S.; Neves, B.V.; Mazzo, T.M.; Longo, E.; Jacob-Lopez, E.; Zepka, L.Q.; de Rosso, V.V. Bigels as potential inks for extrusion-based 3d food printing: Effect of oleogel fraction on physical characterization and printability. *Food Hydrocoll.* **2023**, *144*, 108986. [CrossRef]
52. Li, J.; Han, J.; Xiao, Y.; Guo, R.; Liu, X.; Zhang, H.; Bi, Y.; Xu, X. Fabrication and characterization of novel food-grade bigels based on interfacial and bulk Stabilization. *Foods* **2023**, *12*, 2546. [CrossRef] [PubMed]
53. Hashim, A.F.; El-Sayed, S.M.; El-Sayed, H.S. Bigel formulations based on sesame oleogel with probiotics alginate hydrogel: A novel structure for nutritious spreadable butter. *Int. J. Biol. Macromol.* **2023**, *242*, 124782. [CrossRef] [PubMed]
54. Ghorghi, Z.B.; Yeganehzad, S.; Hesarinejad, M.A.; Faezian, A.; Kutsenkova, V.; Gao, Z.; Nishinari, K.; Nepovinnykh, N. Fabrication of novel hybrid gel based on beeswax oleogel: Application in the compound chocolate formulation. *Food Hydrocoll.* **2023**, *140*, 108599. [CrossRef]
55. Cho, K.; Tarté, R.; Acevedo, N.C. Development and characterization of the freeze-thaw and oxidative stability of edible rice bran wax-gelatin biphasic gels. *LWT* **2023**, *174*, 114330. [CrossRef]
56. Kibler, N.D.; Acevedo, N.C.; Cho, K.; Zuber-McQuillen, E.A.; Carvajal, Y.B.; Tarté, R. Novel biphasic gels can mimic and replace animal fat in fully-cooked coarse-ground sausage. *Meat Sci.* **2022**, *194*, 108984. [CrossRef]
57. Baltuonytė, G.; Eisinaitė, V.; Kazernavičiūtė, R.; Vinauskienė, R.; Jasutienė, I.; Leskauskaitė, D. Novel formulation of bigel-based vegetable oil spreads enriched with lingonberry pomace. *Foods* **2022**, *11*, 2213. [CrossRef]
58. Eisinaitė, V.; Jasutienė, I.; Vinauskienė, R.; Leskauskaitė, D. Development of bigel based dysphagia-oriented products, structured with collagen and carnauba wax: Characterisation and rheological behaviour. *Int. J. Food Sci. Technol.* **2023**, *58*, 145–153. [CrossRef]
59. Ghiasi, F.; Golmakani, M.T. Fabrication and characterization of a novel biphasic system based on starch and ethylcellulose as an alternative fat replacer in a model food system. *Innov. Food Sci. Emerg. Technol.* **2022**, *78*, 103028. [CrossRef]
60. Hashemi, B.; Varidi, M.; Jafari, S.M. Fabrication and characterization of novel whey protein-based bigels as structured materials with high-mechanical properties. *Food Hydrocoll.* **2023**, *145*, 109082. [CrossRef]
61. Scharfe, M.; Ahmane, Y.; Seilert, J.; Keim, J.; Flöter, E. On the effect of minor oil components on β-sitosterol/γ-oryzanol oleogels. *Eur. J. Lipid Sci. Technol.* **2019**, *121*, 1800487. [CrossRef]
62. Martins, A.J.; Vicente, A.A.; Cunha, R.L.; Cerqueira, M.A. Edible oleogels: An opportunity for fat replacement in foods. *Food Funct.* **2019**, *9*, 758–773. [CrossRef] [PubMed]
63. Chen, Z.; Bian, F.; Cao, X.; Shi, Z.; Meng, Z. Novel bigels constructed from oleogels and hydrogels with contrary thermal characteristics: Phase inversion and 3D printing applications. *Food Hydrocoll.* **2023**, *134*, 108063. [CrossRef]
64. Saffold, A.C.; Acevedo, N.C. The effect of mono-diglycerides on the mechanical properties, microstructure, and physical stability of an edible rice bran wax–gelatin biphasic gel system. *J. Am. Oil Chem. Soc.* **2022**, *99*, 1033–1043. [CrossRef]

65. Zhai, X.; Sun, Y.; Cen, S.; Wang, X.; Zhang, J.; Yang, Z.; Li, Y.; Wang, X.; Zhou, C.; Arslan, M.; et al. Anthocyanins-encapsulated 3D-printable bigels: A colorimetric and leaching-resistant volatile amines sensor for intelligent food packaging. *Food Hydrocoll.* **2022**, *133*, 107989. [CrossRef]
66. Machado, M.; Sousa, S.; Morais, P.; Miranda, A.; Rodriguez-Alcalá, L.M.; Gomes, A.M.; Pintado, M. Novel avocado oil-functionalized yogurt with anti-obesity potential: Technological and nutraceutical perspectives. *Food Biosci.* **2022**, *50*, 101983. [CrossRef]
67. Machado, M.; Sousa, S.C.; Rodríguez-Alcalá, L.M.; Pintado, M.; Gomes, A.M. Bigels as delivery systems of bioactive fatty acids present in functional edible oils: Coconut, avocado, and pomegranate. *Gels* **2023**, *9*, 349. [CrossRef]
68. Sennakesavan, G.; Mostakhdemin, M.; Dkhar, L.K.; Seyfoddin, A.; Fatihhi, S.J. Acrylic acid/acrylamide based hydrogels and its properties—A review. *Polym. Degrad. Stab.* **2020**, *180*, 109308. [CrossRef]
69. Ahmed, E.M. Hydrogel: Preparation, characterization, and applications: A review. *J. Adv. Res.* **2015**, *6*, 105–121. [CrossRef]
70. Himashree, P.; Sengar, A.S.; Sunil, C.K. Food thickening agents: Sources, chemistry, properties and applications—A review. *Int. J. Gastron. Food Sci.* **2022**, *27*, 100468. [CrossRef]
71. Nath, P.C.; Debnath, S.; Sridhar, K.; Inbaraj, B.S.; Nayak, P.K.; Sharma, M.A. Comprehensive review of food hydrogels: Principles, formation mechanisms, microstructure, and its applications. *Gels* **2022**, *9*, 1. [CrossRef]
72. Zheng, R.; Chen, Y.; Wang, Y.; Rogers, M.A.; Cao, Y.; Lan, Y. Microstructure and physical properties of novel bigel-based foamed emulsions. *Food Hydrocoll.* **2023**, *134*, 108097. [CrossRef]
73. Guo, Z.; Li, Y.; Song, X.; Meng, Z. Stabilization models of foams prepared from whippable bigels: Crystal absorption and droplet stability. *Food Hydrocoll.* **2023**, *147*, 109383. [CrossRef]
74. Cui, H.; Tang, C.; Wu, S.; McClements, D.J.; Liu, S.; Li, B.; Li, Y. Fabrication of chitosan-cinnamaldehyde-glycerol monolaurate bigels with dual gelling effects and application as cream analogs. *Food Chem.* **2022**, *384*, 132589. [CrossRef] [PubMed]
75. Xie, D.; Hu, H.; Huang, Q.; Lu, X. Influence of oleogel/hydrogel ratios and emulsifiers on structural and digestion properties of food-grade 3D printed bigels as carriers for quercetin and catechin. *Food Hydrocoll.* **2023**, *144*, 108948. [CrossRef]
76. Xie, D.; Hu, H.; Huang, Q.; Lu, X. Development and characterization of food-grade bigel system for 3D printing applications: Role of oleogel/hydrogel ratios and emulsifiers. *Food Hydrocoll.* **2023**, *139*, 108565. [CrossRef]
77. Zampouni, K.; Sideris, N.; Tsavdaris, E.; Katsanidis, E. On the structural and mechanical properties of mixed coconut and olive oil oleogels and bigels. *Int. J. Biol. Macromol.* **2024**, *268*, 131942. [CrossRef]
78. Kaimal, A.M.; Singhal, R.S. A bigel based formulation protects lutein better in the gastric environment with controlled release and antioxidant profile than other gel based systems. *Food Chem.* **2023**, *423*, 136304. [CrossRef]
79. Shakeel, A.; Farooq, U.; Iqbal, T.; Yasin, S.; Lupi, F.R.; Gabriele, D. Key characteristics and modelling of bigels systems: A review. *Mat. Sci. Eng. C* **2019**, *97*, 932–953. [CrossRef]
80. Tejada-Ortigoza, V.; Cuan-Urquizo, E. Towards the development of 3D-printed food: A rheological and mechanical approach. *Foods* **2022**, *11*, 1191. [CrossRef]

gels

MDPI

Review

Exploring the Feasibility of Direct-Dispersion Oleogels in Healthier Sausage Formulations

Niaz Mahmud, Md. Jannatul Ferdaus and Roberta Claro da Silva *

Food and Nutritional Sciences Program, North Carolina A&T State University, Greensboro, NC 27411, USA; nmahmud@aggies.ncat.edu (N.M.); mferdaus@aggies.ncat.edu (M.J.F.)
* Correspondence: rcsilva@ncat.edu; Tel.: +1-336-285-3641

Abstract: Oleogels developed through the direct-dispersion method offer an innovative, scalable, and efficient alternative to traditional fats in sausage production, providing a solution to health concerns associated with the high saturated fat content of conventional formulations. By closely mimicking the texture, stability, and mouthfeel of animal fats, these oleogels provide a novel approach to improving the nutritional profile of sausages while maintaining desirable sensory characteristics. This review critically evaluates cutting-edge research on oleogels, emphasizing innovations in their ability to enhance emulsion stability, increase cooking yield, reduce processing weight loss, and optimize fatty acid composition by reducing overall fat and saturated fat levels. Despite their potential, sausage formulations with oleogel still face challenges in achieving consistent sensory properties, texture, and oxidative stability, often failing to fully replicate the sensory qualities and shelf-life of animal fats. To push the boundaries of oleogel technology and meet the increasing demand for healthier, high-quality sausage products, we propose focused innovations in refining oil-to-gelator ratios, exploring a wider range of novel gelators, optimizing production methods, and developing cost-effective, scalable strategies. These advancements hold significant potential for revolutionizing the sausage industry by improving both the technological and nutritional qualities of oleogels.

Keywords: oleogel; direct-dispersion; fat replacement; sausage; nutritional enhancement; sensory; texture; oxidative stability

1. Introduction

With growing emphasis on nutritious eating habits, the food industry is challenged to create nutritious products that satisfy both health needs and consumer expectations for taste and texture [1,2]. Consumers today are more informed about the health risks associated with processed foods, particularly those high in saturated fats, sodium, and nitrates, which are commonly used in sausage production to enhance flavor, texture, and shelf-life [3,4]. Sausages, historically valued for their flavor and versatility, are traditionally produced with high levels of these ingredients, posing significant health risks such as cardiovascular disease and certain types of cancer [5]. In response, there has been a surge in innovation aimed at developing healthier alternatives that maintain product quality while meeting modern health standards. This shift has led to the exploration of various fat substitutes, among which oleogels have emerged as a promising option [6]. Due to their ability to mimic the texture and mouthfeel of animal fats while reducing saturated fat and cholesterol, oleogels offer unique advantages in sausage production by enabling a significant reduction in harmful fats without compromising taste and palate experience [7]. However, replicating the sensory qualities of traditional fats while reducing saturated fat content remains a formidable challenge, as achieving a comparable mouthfeel, emulsion stability, and flavor profile is critical to consumer acceptance.

This review begins by providing a brief overview of the historical development of sausages and an in-depth exploration of current market dynamics in sausage production, offering insights into the health implications associated with conventional ingredients and highlighting the growing trend toward healthier alternatives within the industry. With a particular emphasis on oleogels as an innovative solution, this analysis delves into their unique structuring mechanisms and evaluates their impacts on emulsion stability, cooking yield, and nutritional composition. Furthermore, it examines oleogels' potential to enhance the textural and sensory qualities of sausage products while improving oxidative stability, thus offering a viable pathway toward reducing harmful fats without compromising consumer expectations. By synthesizing the latest research findings and identifying persistent challenges, this review aspires to inform future advancements in oleogel applications, ultimately contributing to the creation of nutritionally enhanced sausage products that align with contemporary consumer demands for health and quality.

2. Sausages—Brief History and Current Market Size

Sausages have long been a staple in global cuisine, known for their versatility and flavor in preserving and consuming meat. They consist of comminuted meat mixed with fat, salt, and seasonings, encased in natural or synthetic casings [5]. Dating back to around 1500 BC, sausages were consumed by Babylonians and ancient Chinese, with the term "sausage" derived from the Latin word salsus, meaning salted or preserved [8,9]. By 449 BC, the Greeks frequently mentioned salami, and early sausage-making focused on preserving meat through salting and drying. Dry sausages were common in warm Southern Europe due to the lack of refrigeration, while fresh sausages thrived in cooler Northern Europe [9]. By the mid-19th century, during the U.S. Civil War, the sausage industry expanded with advances in refrigeration and meatpacking. Immigrants brought their traditional recipes, contributing to a wide range of sausage types. In the early 1900s, unsafe and unsanitary practices in meat production, including sausages, were exposed by Upton Sinclair's 1906 novel *The Jungle*. This public outcry led to the Federal Meat Inspection Act (FMIA), which required inspections to ensure cleanliness, accurate labeling, and transparency in meat products. The FMIA helped improve the safety and quality of sausages, ensuring they were made from safe ingredients, and laid the foundation for modern food safety standards. Today, stringent regulations continue to safeguard the production of diverse and safe sausage products worldwide [10].

The preparation methods, ingredients, and consumption of sausages vary greatly across regions, reflecting cultural preferences, local resources, and numerous variations in production styles. Sausages can be broadly classified into categories based on their processing methods—fresh, cured, cooked, smoked, and fermented—and this diversity makes them a versatile product suitable for everyday meals to fine dining creations [9]. The global sausage market is projected to generate \$110.70 billion in revenue in 2024, with an annual growth rate of 5.28%. Germany leads in revenue generation, contributing \$10.82 billion. By 2029, the global market volume is expected to reach 15.53 billion kg [11]. In the United States, the sausage market is expected to reach \$6.47 billion in revenue in 2024, growing by 5.48% annually. The U.S. market volume is projected to reach 730 million kg by 2029, with a 3.9% growth in 2025 [12]. These figures indicate strong and steady demand for sausages globally driven by consistent consumption patterns and volume expansion.

3. Health Risks of Sausage Consumption from Saturated Fats, Sodium, and Nitrates

As sausage consumption increases, so do the associated health risks, largely driven by traditional production methods that depend heavily on saturated fats, high sodium levels, and nitrate preservatives. Sausages are traditionally formulated with fats such as lard and tallow, both of which contain significant proportions of saturated fats—approximately 40% in lard and 50% in tallow [13]. Scientific research consistently shows a strong correlation between high saturated fat intake and elevated Low-Density Lipoprotein (LDL) cholesterol, a leading contributor to atherosclerosis [14,15]. This condition results in fatty deposits building up in the arteries, elevating the risk of cardiovascular diseases, including heart attacks and strokes. Saturated fats have also been linked to chronic inflammation, which can lead to insulin resistance—a key factor in the development of type 2 diabetes [16].

In addition to their high saturated fat content, traditional sausages are typically rich in salt and nitrates. Salt enhances flavor and extends the product's shelf-life; however, excessive consumption presents significant health risks. Research has consistently demonstrated that elevated salt intake is a major contributing factor to hypertension and is directly associated with an increased risk of cardiovascular diseases, including heart failure and stroke [17–19]. The American Heart Association recommends keeping sodium intake below 2300 milligrams daily, yet a single serving of sausage can contain up to half of this recommended daily limit, putting consumers at higher risk for sodium-related health issues [17–20]. Nitrates, commonly used to preserve and color sausages, can pose additional health risks. When exposed to high temperatures during cooking, nitrates can transform into nitrosamines, compounds classified as carcinogenic, particularly linked to an increased risk of colorectal cancer [21]. Research by the World Health Organization and various other studies in epidemiology has emphasized the cancer risk associated with high consumption of processed meats containing nitrates [22,23]. Frequent consumption of such processed meats, especially those rich in nitrosamines, has been shown to correlate with a higher incidence of digestive tract cancers, with colorectal cancer being the most prevalent among them [22,23]. With growing awareness of health risks, there is a clear need for changes in sausage production to promote healthier options. Reducing traditional saturated fats and minimizing sodium and nitrate-based preservatives can lower risks associated with heart disease, cancer, and hypertension [24]. Shifting to these healthier practices aligns with public health goals and meets consumer demand for better nutritional choices.

4. Current Market Trends in Healthier Sausage Production

Over recent years, sausage production has evolved with a growing emphasis on health, sustainability, and innovation. Consumers now have access to a variety of options, including low-fat, plant-based, and hybrid sausages, aimed at reducing fat content and catering to health-conscious individuals. Some of these alternatives are fortified with fiber and omega-3 fatty acids to boost their nutritional profiles [25–31]. In parallel, ethical sourcing and eco-friendly packaging have addressed sustainability concerns, while clean labels with simple, recognizable ingredients and global flavors have expanded the variety of offerings. Specialty options, such as gluten-free and keto-friendly sausages, further accommodate specific dietary needs.

The most significant trend in sausage products now is perhaps the growing demand for low-fat options, as consumers prioritize healthier choices in today's world [32]. The low-fat sausage market is projected to grow at a 7.93% annual growth rate from 2023 to 2031, reaching a value of $102.37 billion. By 2024, it is expected to hit $60 billion, driven by technological advancements, rising consumer demand, and strategic investments by industries, setting the stage for sustained market growth [33]. Key players in the low-fat sausage industry are actively innovating to meet the growing consumer demand for healthier options. Atria Oyj, a Finnish food company, is focusing on creating healthier, leaner sausage products as part of its broader strategy to cater to health-conscious consumers. Beyond Meat (El Segundo, CA, USA) is leveraging its expertise in plant-based alternatives, offering low-fat sausages to meet the increasing demand for meat substitutes. Heck Food (Kirklington, Bedale, UK) specializes in high-protein, low-fat sausages made with chicken and plant-based ingredients. Hormel Foods (Austin, MN, USA) has also expanded its portfolio to include leaner sausage options, aligning with consumer health trends. Meanwhile, JBS S.A., (São Paulo, Brazil) one of the world's largest meat processors, has introduced low-fat beef and pork sausages to reduce calorie content while maintaining flavor. Johnsonville (Sheboygan Falls, WI, USA) is focusing on reducing fat content in its traditional sausages to cater to health-conscious customers. Nestlé (Vevey, Switzerland), through its Garden Gourmet brand, is exploring plant-based sausage options that are low in fat. Premium Brands Holding Corporation (Richmond, ON, Canada) and Tyson Foods (Kernersville, NC, USA) are also innovating by using leaner cuts of meat and plant-based alternatives.

Smithfield Foods (Smithfield, VA, USA) produces leaner pork sausages as part of its healthier product line [34,35].

While these healthier endeavors in sausage production have gained considerable traction, offering low-fat options and replacing saturated fats with healthier lipid sources remain challenging. The exploration of alternative lipids and structuring them to mimic traditional fat sources is still in its early stages. Research into these substitutes aims to maintain the texture, flavor, and mouthfeel that consumers expect from conventional sausages while reducing fat content. As such, advancements in this area are ongoing, with many producers awaiting scalable and economically viable solutions to incorporate healthier lipid replacements effectively.

5. Challenges in Low-Fat and Reduced-Saturated-Fat Sausage: Oleogel as a Promising Solution

Despite advancements in low-fat sausage production, the primary challenge remains replicating the flavor, texture, and emulsion stability provided by traditional fats [6,36–39]. Traditional fats are integral to the sensory qualities that define sausages, making it difficult to replace them without compromising taste and mouthfeel. While plant-based and modified products offer lower-fat alternatives, they often fall short of matching the rich texture and binding properties consumers expect [40]. A recent survey indicated that while 36% of consumers actively seek low-fat meat options, nearly 38% would still opt for higher-fat versions when given the choice, reflecting a notable resistance to low-fat products. This reluctance is primarily driven by concerns over taste, texture, and cultural attachment to traditional meat products [41]. In response, research efforts have increasingly focused on identifying lipid sources that reduce both saturated and overall fat content in food products without compromising quality. Among these, oleogels have emerged as a leading innovation [42]. Oleogels are lipid-based materials that convert liquid oils into gel-like structures. They are composed of three key components: the oil phase, a gelator, and optional additives [7]. The oil phase, typically making up 80–99% of the oleogel, includes vegetable oils (e.g., sunflower oil, canola oil, soybean oil, olive oil), providing the liquid base for gelation. The gelator, which constitutes 1–20%, is the essential component that forms the gel network [43]. Gelators can be low molecular weight compounds (LMWGs) or polymers. Low molecular weight gelators include waxes (e.g., beeswax, carnauba wax), fatty acids (e.g., stearic acid), and mono- and diglycerides, which form a gel by crystallizing or self-assembling in the oil. High-molecular-weight gelators (HMWGs) (e.g., ethyl cellulose, proteins, polysaccharides, starches) form gels through polymer entanglement, cross-linking, microscopic phase separation, conformational transition, or mesophase formation [6,44]. Optional additives, such as emulsifiers (e.g., lecithin), antioxidants (e.g., tocopherols), and flavors, can be incorporated to enhance the texture, shelf-life, or sensory qualities of the oleogel emulsifiers stabilize the system by reducing interfacial tension between components and improving the dispersion of gelators within the oil phase. Additionally, they reinforce the gel network, preventing phase separation and enhancing the structural integrity and thermal stability of the oleogel [45,46]. Antioxidants, on the other hand, protect the oil from oxidation by neutralizing free radicals and reactive oxygen species, which can initiate or propagate oxidative reactions [47–49]. By preventing the breakdown of unsaturated lipids, antioxidants further contribute to the stabilization and preservation of the oleogel [50,51].

Oleogels stabilize liquid oils within a gel matrix, effectively mimicking the properties of solid fats. This allows manufacturers to significantly reduce saturated fat and cholesterol levels while preserving essential sensory qualities, such as mouthfeel and texture, to meet consumer expectations [52,53]. The reduction in saturated fat is essential for mitigating the risks of cardiovascular diseases and metabolic disorders, both of which are closely linked to high-fat diets [43]. Countless research efforts have demonstrated that incorporating oleogels into meat-based products can address these health concerns without compromising sensory qualities, which remain critical to consumer acceptance [37,39,43,52–55]. This balance between enhancing health benefits and maintaining product quality positions oleogels as a key innovation in the development of healthier meat products [6].

6. Optimized Method of Oleogel Preparation for Sausage Product Application

The distinct chemical composition and structure of various gelators require different strategies for forming oleogels, classified into direct and indirect technology. In direct methods, gelators are dispersed into heated oil, mixed, and cooled to form a network that solidifies the oil into a gel [56]. In contrast, indirect methods involve multi-step processes where the gelator is first dissolved in a secondary solvent, such as water or alcohol, before the solvent is removed through evaporation, drying, or other techniques. This leaves the gelator in the oil phase to form the oleogel. Indirect methods, including solvent exchange, emulsion-templating, and foam-templating, are particularly useful for hydrophilic polymers or gelators that do not dissolve easily in oil [57]. While these techniques provide better control over the microstructure of oleogels, they often require additional processing steps and specialized equipment, increasing complexity and costs. On the other hand, the direct-dispersion method is the most suitable for oleogel production in sausage applications due to its simplicity, scalability, and ease of integration into existing food manufacturing processes. It is especially advantageous for large-scale sausage production, where high-throughput processes demand a straightforward, cost-effective approach. Although high temperatures during processing may cause oil oxidation, this can be mitigated by using antioxidants, lowering processing temperatures, or reducing processing time. The simplicity, speed, and efficiency of the direct-dispersion method make it the superior choice for sausage production [58]. Table 1 highlighted the versatility in formulating direct-dispersion oleogels for sausage applications, showcasing a range of oils combined with specific gelators. These combinations yielded varied oil-to-gelator ratios, from 72:28 to 98:2, allowing for targeted adjustments in texture and fat content to meet desired nutritional and sensory qualities. Processing temperatures varied from 65 °C to 140 °C, indicating that certain gelators and oils required higher heat for full dissolution and stability. After processing, cooling and setting conditions were tailored to each oleogel, with some requiring room temperature while others necessitated precise cooling (e.g., 4 °C) for periods ranging from overnight to several days. These storage durations enhanced gel firmness and stability, ensuring that oleogels maintained a texture and mouthfeel comparable to traditional fats in sausages.

Table 1. Development of various direct-dispersion oleogels for sausage applications.

Oil ∣ Gelator	Oil:Gelator	Processing Temp. ∣ Time	Setting Conditions	Ref.
Conventional soybean oil, High oleic soybean oil ∣ Rice Bran Wax	90:10 and 97.5:2.5	90 °C ∣ 2 h	Set for 7 days at ambient temperature	[59]
Olive oil ∣ MG95 (Monoglycerides)	85:15	90–95 °C ∣ 60 min	Cooled for 40 min at ambient temperature, stored at 5 °C for 3 days	[60]
Sunflower oil, Peanut oil, Corn oil, Flaxseed oil ∣ Ethyl Cellulose	94:6, 92:8, 90:10, and 88:12	140 °C ∣ Until dissolved	Cooled to room temperature, stored at 4 °C	[61]
Sunflower oil ∣ Monoglycerides and Phytosterols	80:20	90 °C ∣ Up to 60 min	Cooled and stored at 18–20 °C	[62]
Conventional soybean oil ∣ Rice Bran Wax	90:10 and 97.5:2.5	90 °C ∣ 2 h	Stored at 2.7 °C for 5–7 days	[63]
High oleic sunflower oil, Sunflower oil ∣ Glyceryl Monostearate	95:5	90 °C ∣ 15 min	Cooled to room temperature, stored for 48 h at 25 °C, then 24 h at 4 °C	[64]
Flaxseed oil ∣ Sunflower Wax, Beeswax	90:10	80 °C ∣ Until dissolved	Cooled to ambient temperature (23 ± 2 °C) overnight	[65]
Canola oil ∣ Candelilla Wax	97.5:2.5	100 °C ∣ 20 min	Stored for 5 days at 4 °C	[66]
Sunflower oil ∣ Monoglycerides and Beeswax	80:20	90–95 °C ∣ 60 min	Cooled and stored at 4 °C for 24 h	[67]
Linseed oil ∣ Beeswax	92:8	80 °C ∣ 30 min	Cooled at ambient temperature until gel formation	[52]
Olive oil-chia oil ∣ Beeswax	90:10	65 °C ∣ Until dissolved	Cooled for 60 min at room temperature, then stored at 3 °C	[68]
Linseed oil ∣ Beeswax, γ-Oryzanol, and β-Sitosterol	92:8	80 °C ∣ Until dissolved	Stored at ambient temperature	[69]
Sunflower oil ∣ Candelilla Wax, Glyceryl Monostearate	90:10	80 °C ∣ Until dissolved	Cooled at 4 °C for 24 h before use	[70]
Rice bran oil ∣ Rice Bran Wax	98:2	80 °C ∣ Until dissolved	Cooled at room temperature and left for overnight	[71]
Canola oil ∣ Lecithin, Sorbitan Monostearate	72:28	80 °C ∣ 30 min	Cooled at 25 °C for 24 h, stored at 4 °C for 2 days	[72]

7. Incorporation of Direct-Dispersion Oleogels in Sausage Manufacturing

Sausage processing is a complex, multi-stage procedure designed to yield a product with desirable texture, flavor, and preservation qualities. Initially, ingredients are carefully prepared and measured to achieve a balanced composition. The primary components—meat and fat—are ground to achieve a specific particle size, facilitating a uniform texture and an ideal binding capacity, which are crucial for sausage structure. Seasonings, preservatives, and functional additives are then blended into the mixture to enhance flavor, prolong shelf-life, and stabilize the fat-protein matrix. This preparation creates a stable, uniform mix that can be further processed through cooking, smoking, curing, or fermentation, depending on the type of sausage being produced [9]. A critical point in sausage processing, particularly for modified formulations, is the mixing stage [58]. Here, oleogels are incorporated with other ingredients. These oleogels serve as a functional alternative to traditional fats, aiming to reduce saturated fat content while retaining essential textural and sensory qualities. When introduced at the mixing stage, oleogels are dispersed uniformly throughout the meat matrix. This uniform dispersion is essential not only for consistent fat replacement but also for ensuring that the oleogels contribute to a texture and mouthfeel similar to those provided by conventional animal fats. Figure 1A shows the incorporation of oleogels during the mixing steps in sausage production, while Figure 1B illustrates their preparation, demonstrating how oleogels can be integrated with other ingredients for uniform dispersion in the meat matrix, ensuring consistent fat replacement and desired product qualities.

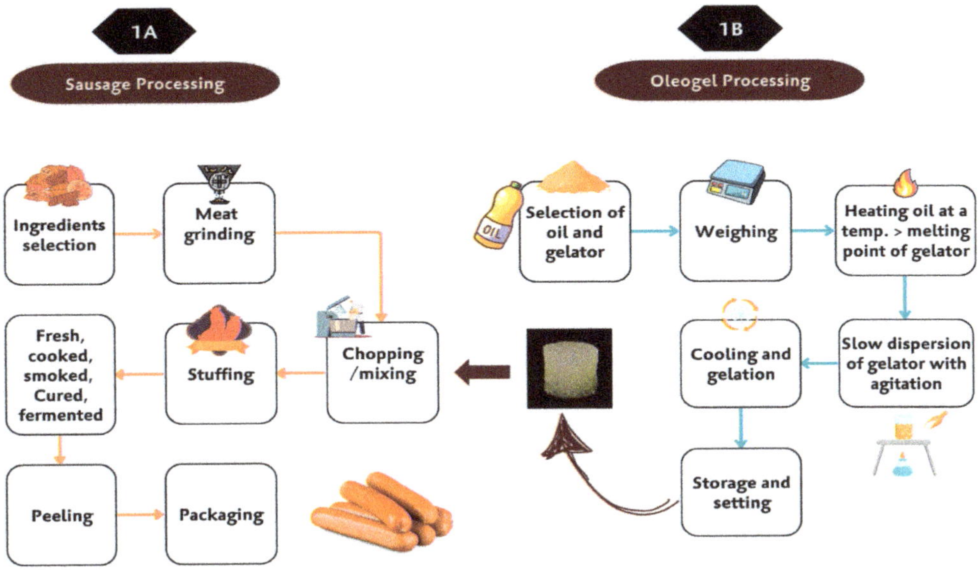

Figure 1. (**A**) Incorporation of oleogels during the mixing stage in sausage processing, (**B**) Preparation of oleogel for sausage application.

8. Structuring Mechanisms of Oleogels and Their Roles in Sausage Applications

Oleogels contribute to sausage production through three primary structuring mechanisms: crystal particle systems, self-assembly systems, and polymeric network systems [36]. Each mechanism leverages specific gelators and structuring methods, creating diverse textures and qualities optimized for various sausage types. In crystal particle systems, low molecular weight gelators (LMWGs), such as beeswax, candelilla wax, and rice bran wax, dissolve in oil at high temperatures and crystallize as the mixture cools [73]. This crystallization process traps the oil within a structured network, emulating the texture and stability of animal fats like lard or tallow.

This method is especially effective for imparting a firm texture, bite, and mouthfeel in emulsified meat products. The crystallized structure supports moisture retention, enhancing juiciness [74].

Self-assembly systems employ gelators such as phytosterols, sterols, and certain fatty acids, forming a three-dimensional network via weak, non-covalent interactions, such as hydrogen bonding and van der Waals forces [75]. Unlike crystal particle systems, these gelators create a softer, more flexible structure suited to sausages requiring a creamier texture, such as spreadable varieties. The network's softness improves moisture retention and provides a smoother mouthfeel, making it ideal for fresh sausages with a softer bite [76].

Polymeric network systems use high molecular weight gelators (HMWGs), including ethyl cellulose, proteins, and polysaccharides. These gelators create stable networks through physical entanglement or chemical cross-linking, offering significant heat resistance [77]. This stability is ideal for cooked sausages, as it ensures the product retains its structure and consistency during cooking. The result is a firm bite and even fat distribution, essential for products like frankfurters and hot dogs, where fat reduction is achieved without compromising texture [78].

Each oleogel structuring mechanism enhances sausage quality by replicating the texture and mouthfeel typically provided by animal fats while also improving moisture retention and thermal stability. These properties enable healthier sausage alternatives that preserve the essential sensory attributes of traditional formulations, adaptable to various sausage types and consumer preferences. Figure 2 illustrates the various oleogel mechanisms, accompanied by a word cloud highlighting oleogel formation and functions in sausage production.

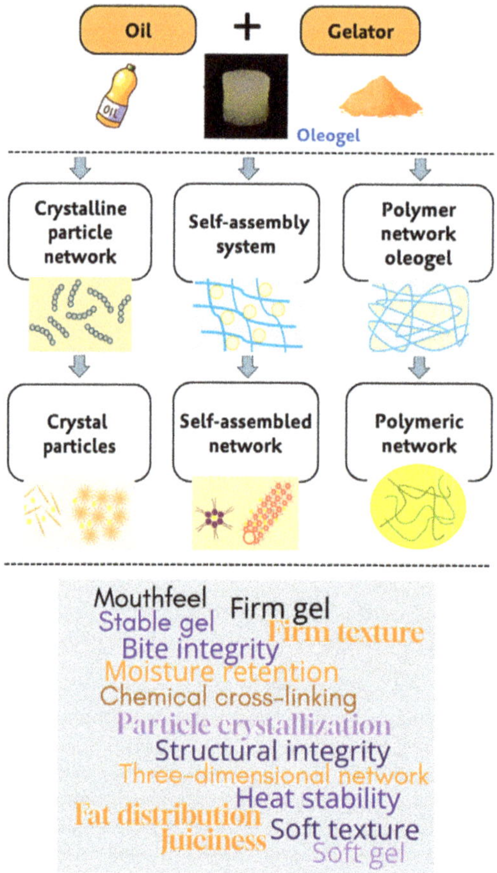

Figure 2. Oleogel mechanisms and word cloud of their formation and functions in sausage production.

9. Stability and Processing Dynamics in Oleogel-Enhanced Sausages

The incorporation of oleogels in sausage formulations presents a transformative approach to enhancing product stability and processing dynamics, crucial for maintaining structural integrity and optimizing production metrics. Oleogels, by their nature, could impact sausage emulsion stability, a critical factor in achieving a cohesive texture necessary for high-quality sausages [59]. This stability directly influences slice-ability and structural resilience, ensuring that sausages retain their shape and texture during handling, processing, and consumer use [64]. In terms of production outcomes, analyzing yield and loss parameters—specifically, total yield, fat retention, cooking losses, and overall weight stability—provides vital insights into oleogel's role in operational efficiency and product uniformity [59,65,70]. These metrics indicate how well oleogels support the functional demands of sausage manufacturing. Ultimately, understanding and optimizing these parameters are essential tasks for creating oleogel-enhanced sausages that meet industry quality standards, offering a synergy of robust processing performance and product consistency.

9.1. Emulsion/Batter and Structural Stability

Incorporating oleogels into sausage formulations significantly enhances emulsion stability and slice-ability, offering structural advantages over traditional formulations without oleogels. This approach allows for the replacement of animal fats, maintaining product quality and texture. Tarté et al. (2020) demonstrated the effectiveness of high oleic and conventional soybean oil oleogels structured with rice bran wax in bologna sausages, noting that both 10% and 2.5% rice bran wax concentrations improved emulsion stability compared to samples with only liquid soybean oil. Notably, samples with 10% rice bran wax showed significantly less water and lipid separation, creating a stable matrix that reduced fluid release and reinforced cohesion, closely approximating the stability of pork backfat [59]. Similarly, Wolfer et al. (2018) observed that in frankfurter sausages, 2.5% and 10% rice bran wax oleogels with soybean oil improved emulsion stability, with 10% concentrations achieving superior moisture retention and structural integrity. This highlights the significance of utilizing higher gelator concentrations, such as 10%, to achieve stability, thereby establishing structured oleogels as viable alternatives to traditional animal fats in emulsified products [63].

Further, Ferro et al. (2021) explored 5% glyceryl monostearate-based oleogels with sunflower and high oleic sunflower oils in bologna sausages. Despite a lower gelator concentration, these oleogels achieved high emulsion stability and minimal liquid separation, proving effective as fat substitutes. Additionally, oleogels improved slice-ability; for example, sausages with 10% pork fat replaced by oleogel exhibited a higher percentage of intact slices than those with pork fat, indicating better structural integrity. Sausages with 20% pork fat replaced by oleogel showed even greater consistency, yielding more cohesive slices [64]. The enhanced emulsion stability and slice-ability are attributed to the compact network formed by oleogels, which strengthens product cohesion and stability, highlighting the potential of oleogels to serve as a balanced, functional fat replacement in sausage products [74].

9.2. Cooking Yield and Loss Analysis

Oleogels structured with rice bran wax concentrations in soybean oil, particularly at the 10% level, have demonstrated remarkable efficacy in preserving cooking yields in Bologna sausages. This effectiveness is largely due to their role in stabilizing both moisture and fat retention. The concentration of rice bran wax within these oleogels is pivotal; at the 10% gelator level, a robust three-dimensional network forms, featuring a higher melting point that withstands thermal breakdown during cooking [59,63]. This network serves as a physical matrix within the meat structure, efficiently entrapping fat and moisture. Consequently, oleogels mitigate the release of fat and curtail moisture loss, leading to a substantial reduction in cooking loss—a crucial determinant in yield retention for processed meats such as frankfurters and Bologna sausages. Unlike traditional fats and liquid oils, which lack the structural cohesion to prevent "cook out", oleogels establish a semi-solid scaffold that anchors lipids [74].

Oleogels, crafted from a blend of flaxseed oil, sunflower wax, and beeswax at a 90:10 (*w*/*w*) ratio, have demonstrated considerable efficacy in reducing cooking loss in sucuk sausages compared to traditional formulations utilizing tallow fat [65]. This reduction is noteworthy as it suggests a higher retention of moisture and weight during cooking, thereby enhancing the juiciness and succulence of the final sausage product. The study's findings reveal a distinct pattern in cooking loss across different fat formulations: sucuk sausages made with tallow fat exhibit the greatest cooking loss, while those incorporating beeswax and sunflower wax oleogels show a marked decrease in cooking loss. The mechanisms underlying this reduction in cooking loss with oleogels, as opposed to traditional fats, can be largely attributed to differences in melting points. Traditional fats, with relatively low melting points, tend to liquefy rapidly when exposed to heat, leading to significant fat drainage and a corresponding increase in cooking loss. Conversely, oleogels exhibit higher melting points, allowing them to maintain a semi-solid state for longer under heat, which aids in moisture preservation by reducing fat migration and leakage [65]. Further enhancing their effectiveness, the structural properties of oleogels play a key role in minimizing cooking loss. The oleogel matrix forms a stable network that effectively encapsulates moisture, limiting the evaporation of free water during cooking. This structural stability results in reduced cooking and weight loss, yielding a final product with improved moisture retention, better texture, and reduced fat loss throughout the cooking process [36]. Table 2 lists recent findings on the yield and processing losses associated with oleogel incorporation in sausage products.

Table 2. Effects of direct-dispersion oleogels on sausage properties with varying levels of pork backfat replacement.

Oleogel Matrix	Sausage Type	%Fat Replacement	Sausage Properties						Ref.
			Yield And Losses	Fat Content	Fatty Acid Profile	Texture	Sensory	Oxidative Stability	
Conventional soybean oil, High oleic soybean oil | Rice Bran Wax	Bologna	41.9% of pork backfat	No yield differences in oleogel-treated and control samples	Slightly lower fat content in CSO-oleogel-treated samples than control	Linoleic & α-linolenic acids were highest in CSO-oleogel-treated samples, lower in HOSO-oleogel-treated samples, lowest in control. Oleic ↓ in CSO, ↑ in HOSO vs. Control	Lipid source and storage time had no significant effects on texture profile or incisor peak force	No differences were found in aroma, flavor, texture, or moistness between control and oleogel-treated samples	TBARS values stayed consistently low over the 98-day storage period	[59]
Olive oil | MG95 (Monoglycerides)	Fermented	50% pork backfat	Lower weight loss after cooking in oleogel-treated samples than control	Fat content reduced by 6.72% in oleogel-treated samples	SFA reduced by 17.4%, MUFA increased by 9.4%, and cholesterol reduced by 18.8% in oleogel-treated samples compared to control	ND	Oleogel samples ranked second after control samples	ND	[60]
Peanut oil | Ethyl Cellulose	Harbin red sausage	10–50% pork backfat	ND	All oleogel-treated samples showed a significant ↓ in total fat, with the lowest fat content in 50% treatment	In oleogel-treated samples, SFAs dropped from 39.9% to 28.4%, while UFAs rose from 55.7% to 68.4%. ω-6 FAs increased by 23–122%, but ω-3 FAs showed a slight ↓ vs. pork fat control	No significant influence in oleogel-treated samples compared to control	No significant influence in oleogel-treated samples compared to control	Higher lipid oxidation in oleogel-treated samples due to greater degree of unsaturation of peanut oil	[61]
Sunflower oil | Monoglycerides and Phytosterols	Frankfurter	50% pork backfat	ND	ND	ND	The control samples showed higher hardness, brittleness, gumminess, and chewiness than oleogel-treated samples, while cohesiveness and elasticity showed no difference	No differences were found	No differences were found	[62]
Conventional soybean oil | Rice Bran Wax	Frankfurter	80.65% pork backfat	No differences were found in cooking yield	No differences were found	Oleogel-treated samples had higher PUFAs (18:2n6, 18:3n3) and lower SFAs (18:0, 16:0) and MUFA (18:1c9) than control samples	Oleogel-treated samples were similar to pork fat treatment in firmness, chewiness, and springiness	Pork fat replacement did not affect aroma but significantly reduced flavor	Higher lipid oxidation in oleogel-treated samples due to higher degree of unsaturation in soybean oil	[63]
High oleic sunflower oil, Sunflower oil | Glyceryl Monostearate	Bologna	25, 50, 75 and 100% of pork backfat	ND	50% replacement samples showed a significant fat reduction (~25%) compared to the control	SFA was reduced by up to 67% in oleogel-treated samples, while MUFA (high oleic oleogels) and PUFA (sunflower oleogels) increased	Hardness increased in the oleogel-treated samples compared to control	Both oleogel-treated and control samples had good consumer acceptance, with no significant differences	ND	[64]

Table 2. *Cont.*

Oleogel Matrix	Sausage Type	%Fat Replacement	Sausage Properties						Ref.
Canola oil | Candelilla Wax	Frankfurter	50–100% of pork backfat	ND	Higher contents of fat in oleogel-treated samples	ND	Sausages with 50% fat replacement had lower hardness, while 100% oleogel samples showed higher hardness. Adhesiveness, cohesiveness, elasticity, and resilience remained unchanged	ND	ND	[66]
Sunflower oil | Monoglycerides and Beeswax	Semi-Smoked	7% and 10% of pork backfat	ND	No differences were found	SFA decreased by 35–38%, while MUFA and PUFA increased	ND	No differences were found	ND	[67]
Linseed oil | Beeswax	Frankfurter	25–10% of pork backfat	ND	Higher contents of fat in oleogel-treated samples	SFA reduced from 35.15 g/100 g to 33.95 g and 32.34 g/100 g in 25% and 50% replacement samples, achieving more balanced n-6/n-3 ratios, compared to control	Cohesiveness, gumminess, and chewiness slightly increased in oleogel-treated samples compared to control sausages	Control had better appearance, lower yellowness (less rancidity), and higher scores in odor, taste, and juiciness	ND	[52]
Linseed oil | Beeswax, γ-Oryzanol, and β-Sitosterol	Fermented	20% and 40% of pork backfat	ND	Fat content ranged from 33.01% to 39.19%, lower than in traditional sausages	The fatty acid profile improved nutritionally, with PUFA/SFA and n-6/n-3 ratios of 1.41 and 0.93 in oleogel-treated samples	Beeswax oleogel resulted in notable changes in hardness, cohesiveness, gumminess, and chewiness, with sausages containing beeswax oleogel being less hard than control	Control was most accepted, while 20% replacements scored positively on the hedonic scale, and 40% replacements scored negatively	ND	[69]
Sunflower oil | CW, Glyceryl Monostearate	Bologna	100% of pork backfat	100% fat replacement raised fat loss, highest in GMS (7.75%) and CW (5.81%) oleogels	ND	ND	Control samples had higher hardness compared to oleogel-treated samples. Adhesiveness, cohesiveness, springiness, and gumminess were similar across samples	ND	ND	[70]

ND; Not determined, ↓; Decrease, ↑; Increase.

10. Nutritional Modifications in Sausage Products with Oleogel

Exploring the potential of oleogels in sausage formulations opens up a promising approach for enhancing nutritional value while maintaining the desired sensory and texture qualities of sausage products. Oleogels, with their ability to modulate both fat and moisture levels, can significantly influence the texture and health profile of sausages. By replacing traditional fats with oleogels, formulations achieve a balance that helps retain moisture and, importantly, reduces overall fat content. This substitution offers a path toward healthier meat products without sacrificing quality [6]. The integration of oleogels as partial replacements for pork backfat in sausage production impacts moisture retention and fat content due to the unique structural and compositional properties of oleogels. By carefully controlling gelator-to-oil ratios, oleogels form a stable, semi-solid matrix that both immobilizes oil and retains water within the sausage, enhancing moisture levels [69]. A study on fermented sausages demonstrated that an olive oil-based oleogel with a monoglyceride ratio of 85:15 (*w*/*w*) as a fat substitute significantly affected moisture and fat retention. During fermentation, this oleogel matrix served as a moisture barrier, effectively slowing dehydration and leading to higher moisture content, which ranged from 41.49% to 47.28% depending on the specific formulation [60]. This retention likely results from the oleogel's capacity to coat meat particles and create a barrier, preserving moisture during processing [55]. Furthermore, substituting pork backfat with olive oil oleogel was associated with a modest reduction in total fat content, particularly in formulations where 50% of the pork fat was replaced. This fat reduction demonstrates the oleogel's potential for lowering lipid content while maintaining essential structural integrity in the sausage. Similarly, an investigation into Harbin red sausage formulations using peanut oil oleogels structured with ethyl cellulose revealed how varying ethyl cellulose concentrations, especially at 10%, enhanced stability and reduced oil loss. Replacing pork fat at levels from 10% to 50% with the peanut oil oleogels resulted in a significantly lower total fat content compared to the control. This reduction is attributed to the oleogel's lower lipid density and high oil-binding capacity, which limit fat release during cooking [61]. The maintained moisture content, comparable to traditional sausage formulations, further underscores the oleogel's ability to retain water, likely due to the stable three-dimensional network formed by ethyl cellulose. Oleogels also facilitate fat reduction by substituting part of the saturated animal fat with plant-derived oils, inherently lower in saturated fats and free from cholesterol [79]. In semi-smoked and red sausage formulations, replacing pork backfat with oleogels—ranging from 50% to 100%—led to significant decreases in total fat content, as plant oils in the oleogels contribute fewer lipids than animal fat [67]. This balance between moisture retention and fat reduction can be attributed to the oleogel's gelled network, which not only limits fat release during thermal processes but also supports a more stable moisture environment. Consequently, oleogels provide a scientifically backed strategy for enhancing the nutritional profile of meat products by reducing fat content while preserving essential moisture, thereby improving both the health aspects and textural qualities of the final product.

Moreover, oleogels offer the possibility of a more beneficial fatty acid profile, favoring unsaturated fats that are associated with improved cardiovascular health [80]. The composition of the gelators used in oleogels—such as ethylcellulose, rice bran wax, and glyceryl monostearate—plays a key role in enhancing the fatty acid profile of meat products without contributing to higher levels of saturated fatty acids (SFAs) [81]. Unlike traditional animal fats, which are high in SFAs, these gelators themselves do not introduce additional SFAs into the formulation. Instead, they structure unsaturated vegetable oils, which are typically rich in monounsaturated (MUFAs) and polyunsaturated fatty acids (PUFAs), into a gel matrix, thereby preserving and enhancing the product's nutritional profile [82,83]. For instance, in the reformulated Harbin red sausage, the replacement of pork back fat with peanut oil oleogels structured with ethylcellulose at 6–12% concentrations resulted in a reduction in SFA while increasing MUFA and PUFA. The ethylcellulose-based oleogel effectively gelled the peanut oil, which is naturally high in MUFAs, thereby reducing the overall SFA content of the sausage without compromising its texture and sensory qualities [61]. In frankfurters, the use of rice bran wax at 2.5% and 10% concentrations with soybean oil, replacing up to 50% of pork back fat, showed a significant decrease in SFA and a corresponding increase in PUFA and MUFA content. The rice bran wax oleogel improved the PUFA/SFA ratio of the sausage without introducing additional SFAs, as the rice bran wax serves as a structuring agent rather than a fat source. Data from the study indicated that PUFA content increased substantially in frankfurters with 10% rice bran wax oleogel, and the reduction in SFA contributed to a more heart-healthy fatty acid profile [63]. In Bologna sausages, oleogels using glyceryl monostearate (5%) with sunflower oil or high oleic sunflower oil replaced pork fat by up to 100%, further enhancing the PUFA and MUFA levels while reducing SFAs. Sunflower oil, particularly high oleic variants, is rich in MUFAs and PUFAs, which are retained within the gel matrix formed by glyceryl monostearate. As a result, the Bologna sausage exhibited a PUFA/SFA ratio that was markedly improved, with the data showing higher MUFA and PUFA levels than the control samples. For example, sausages with 100% replacement showed an increase in PUFAs while maintaining low SFA levels, thanks to the properties of glyceryl monostearate, which does not add saturated fats but enables the structuring of healthier oils. The low SFA contribution of these gelators and their ability to stabilize unsaturated oils in a gel matrix ensure that the oleogels contribute to a more favorable fatty acid profile [64]. By preventing the oils from undergoing rapid oxidation and by encapsulating them effectively, the gelators preserve the nutritional benefits of high-MUFA and high-PUFA oils, making them valuable alternatives for producing meat products with healthier lipid profiles. Table 2 summarizes recent studies on the effects of oleogel on fat content and the enhancement of the fatty acid profile in sausage formulations.

Additionally, by integrating oleogels, there is potential to lower cholesterol content in sausages, aligning these products with current nutritional guidelines. The influence of oleogels on cholesterol reduction in meat products is closely linked to their ability to replace traditional animal fats. For instance, in Thai sweet sausage, rice bran wax/rice bran oil (2:98) oleogels (RBOG) were used as partial fat replacers, substituting pork backfat at levels of 0%, 25%, 50%, and 75%. By replacing animal fats with RBOG, which inherently lacks cholesterol, these formulations reduce overall cholesterol levels, providing a healthier alternative without compromising the fat-binding properties essential for texture and moisture retention [71]. The rice bran wax and oil structure also promote a stable gel matrix that supports these benefits, showing that oleogels are not only functional fat substitutes but also contribute positively to the nutritional profile by lowering cholesterol content in final products.

11. Effects of Oleogel on Texture and Sensory Properties in Sausages

For the successful adoption of oleogel-enriched sausages, it is imperative to achieve specific quality attributes, including texture and sensory properties. Texture and microstructure play a foundational role in emulating the firm, succulent bite characteristic of traditional sausages, with oleogels providing structural integrity and moisture retention necessary for consumer satisfaction. Equally vital are the sensory properties—taste, aroma, and mouthfeel—that oleogels must effectively replicate to deliver an experience akin to that of conventional animal fats, ensuring a rich texture and pleasant flavor release. These integrated properties collectively enhance the likelihood of oleogel-enriched sausages meeting both consumer acceptance criteria and rigorous industry standards, positioning them as a viable, health-conscious alternative in the market.

11.1. Instrumental Texture Profile Analysis and Microstructure

Incorporating oleogels into sausage formulations has yielded varied effects on textural properties, influenced by factors such as the meat matrix, type of oil, and selected gelator. While some studies indicate minimal impact on texture, others reveal that oleogels affect textural values. Importantly, none of the research reviewed reports significant or adverse deviations from control samples, underscoring the compatibility of oleogels with traditional sausage textures and highlighting their potential as adaptable components in healthier sausage products without compromising sensory quality. For instance, Tarté and colleagues (2020) demonstrated that oleogels made with 90.0% or 97.5% of either conventional or high oleic soybean oil and rice bran wax as a gelator could replace up to 41.9% of pork backfat in bologna sausage without significantly affecting texture. The rice bran wax gelator, applied at 2.5% and 10% of the oil content, effectively replicated the firmness and chewiness of pork fat by forming a stable matrix that maintained key texture profile parameters and incisor peak force, thus mimicking the solid fat structure [59]. Similar results were observed in frankfurter-type sausages, where rice bran wax/soybean oil oleogels (2.5:97.5 and 10:90) replicated the firmness, chewiness, and springiness of traditional pork fat [63]. In contrast, Ferro (2021) reported that monoglyceride-based oleogels (5%) made from conventional and high-oleic sunflower oils, used to replace varying amounts of pork fat (25%, 50%, 75%, and 100%) in Bologna sausages, influenced texture by increasing hardness as the proportion of pork fat replacement increased. This effect was attributed to the compact structure formed with higher oleogel content, affecting the sausage's slice-ability and resulting in a higher slice count compared to the control. These textural changes are likely due to the stable gel-like properties ($G' > G''$) of monostearate-based oleogels, which partially mimic pork fat's structure while enhancing unsaturated fatty acid content [64]. Similar findings by Shao (2023) noted that replacing 10–30% of pork fat with peanut oil oleogels structured with 6–12% ethylcellulose led to increased hardness and structural stability in Harbin red sausages, indicated by higher storage (G') and loss (G'') moduli [61]. On the other hand, a different study structured sunflower oil with a 15:5 weight ratio of monoglycerides to phytosterols, replacing 50% of pork backfat in frankfurter sausages. The oleogel-substituted sausages showed lower values in hardness, brittleness, gumminess, and chewiness compared to controls with full pork backfat, likely due to structural differences in crystalline properties between the oleogel and pork fat [62]. Table 2 presents recent research on the influence of oleogel on texture characteristics in sausage formulations.

Regarding microstructure analysis, sausages containing oleogels revealed distinct differences compared to traditional pork fat formulations. In a recent study, control samples with pork fat exhibited the largest fat globules, with 49.9% of globules exceeding 100 μm^2, significantly more than in samples containing rice bran wax oleogels. In oleogel-treated samples, the structure appeared more compact, with fewer visible fat globules compared to pork fat and soybean oil samples. Specifically, image analysis indicated that 14.46% of the pork fat sample area consisted of lipids, a marked contrast to 2.65% in 2.5% rice bran wax and 6.88% in 10% rice bran wax oleogel treatments. This suggests that the oleogel structure may have fragmented into smaller particles, undetectable by imaging software, or dissolved into the protein phase due to the hydrophilic properties of certain rice bran wax components [63]. Scanning electron microscopy in another study further supported these findings, showing that reduced pork fat levels led to a denser, more compact sausage structure. Control formulations with 20% and 10% pork fat displayed a uniform and porous structure typical of stable meat emulsions, with more pores observed in 10% pork fat, likely due to the higher water content from lower fat. In contrast, formulations with 100% oleogel replacement exhibited a significantly denser network, correlating with increased hardness and higher G' values in the meat batters, indicating a firmer structure likely due to the oleogel's integration into the protein-fat matrix [64]. These findings imply that partial oleogel replacement effectively preserves traditional emulsion structures, while full replacement might alter microstructural characteristics, increasing density and firmness. Figure 3 illustrates the microstructural properties of sunflower oil oleogels structured with a 2:1 monoglyceride-beeswax blend at 7% and 10% concentrations, and their effects on semi-smoked sausages.

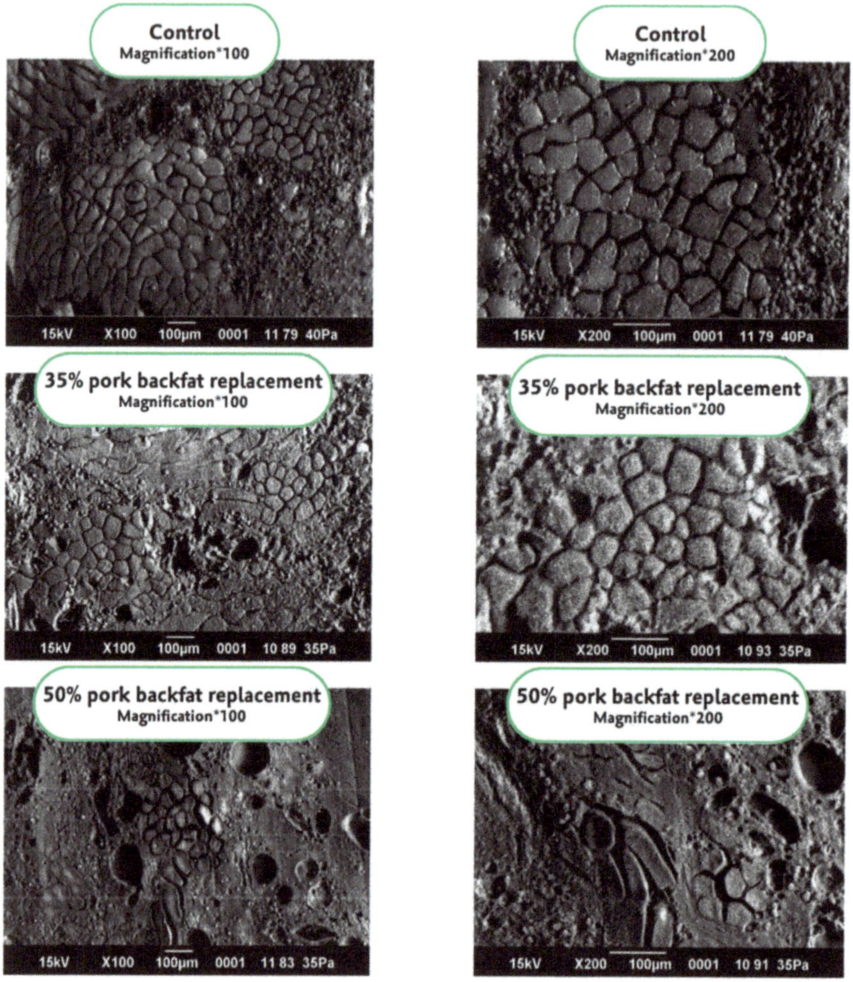

Figure 3. Microstructure images of semi-smoked sausages with sunflower oil oleogels structured using a 2:1 monoglyceride-beeswax blend at 7% and 10% concentrations, replacing 35% and 50% pork backfat, respectively, showing the control sample with notable porosity and large lipid granules, while increased oleogel content resulted in a more homogeneous structure with reduced granules, with the 7% oleogel at 35% replacement closely resembling the control structure and exhibiting improved uniformity. Reproduced from [67] with permission.

The textural effects of oleogel incorporation in sausage formulations depend on key variables: the oil-to-gelator ratio, fat replacement level, oleogel type, and processing method. Higher oil-to-gelator ratios generally yield a more structured texture, while lower ratios contribute to softer textures. Partial fat replacements often preserve traditional textures more effectively, whereas complete replacements may lead to a firmer product due to increased compactness. The oleogel type influences the replication of conventional fat's structural attributes, particularly affecting chewiness and cohesiveness. Additionally, the processing method, including heat application and mechanical handling, modifies oleogel behavior within the sausage matrix, necessitating tailored adjustments for optimal textural outcomes. Exploring each of these factors relative to specific sausage types will be crucial in developing oleogel-based formulations that align with targeted textural expectations.

11.2. Sensory Performance

Sensory analysis plays a pivotal role in food product development and optimization, providing insights into consumer acceptance through attributes such as taste, texture, aroma, color, overall mouthfeel, and acceptability. In the context of sausage formulations utilizing oleogels as a fat substitute, sensory evaluations yield mixed results, underscoring the delicate balance between health benefits and desirable sensory qualities. While some formulations have successfully retained textural properties similar to traditional fat-based sausages, achieving a familiar mouthfeel and taste profile, others faced rejection due to noticeable deviations in texture or flavor from conventional sausage expectations. For instance, Igenbayev (2023) explored sausage formulations where pork fat was partially replaced with beeswax-structured oleogels. Two experimental samples were developed, containing 7% and 10% oleogel, respectively, using a formulation of sunflower oil structured with monoglyceride and beeswax in a 2:1 ratio to ensure texture and stability. Sensory analysis revealed that both oleogel samples exhibited a uniform texture, appealing red-pink color, and the absence of gray spots, along with a pleasant aroma and semi-smoked sausage taste. However, while the 7% oleogel sample maintained favorable sensory qualities, the 10% inclusion resulted in significant declines in texture and juiciness. Sensory panelists rated the 10% sample lowest for appearance and cross-sectional quality, suggesting that higher oleogel levels might negatively impact consumer acceptability due to textural and juiciness losses [67].

In another investigation, sensory analysis of sucuk sausages incorporating oleogels (Flaxseed oil-sunflower wax/beeswax) compared to a traditional tallow-based control provided valuable insights. Panelists evaluated samples across seven sensory descriptors: appearance, hardness, chewiness, fattiness, juiciness, aroma, and flavor. The control sample scored highest in appearance, followed by samples containing sunflower wax oleogel and beeswax oleogel. The control also ranked significantly firmer in sensory hardness, aligning with measurements that confirmed oleogels' limitations in replicating tallow fat's hardness. Conversely, oleogel samples were perceived as chewier due to oleogels' softer consistency at mouth temperature, producing a tender, chewable texture. Interestingly, although all samples had the same fat content, oleogel-enriched samples scored highest in fattiness perception, possibly enhanced by beeswax oleogel. Differences in juiciness were also notable, with beeswax oleogel samples ranking highest, followed by sunflower wax and the control. This variation may result from higher melting points of beeswax and sunflower wax oleogels, which retain moisture during cooking. However, aroma and flavor ratings favored the control sample, highlighting that while oleogel replacements provide nutritional benefits, they fall short of delivering the rich sensory qualities of traditional fat-based sucuk [65].

Franco (2019) assessed sensory preferences when pork backfat was partially substituted with linseed oleogel, composed of linseed oil and beeswax (8% *w*/*w*). The control sample was favored in appearance, particularly regarding color, as it exhibited lower yellowness, which often correlates with better consumer acceptance [84–86]. Higher yellowness in meat products is generally linked to rancidity from lipid oxidation, reducing consumer appeal. Significant differences were also observed in odor and taste, with the control sample rated most favorably. These findings indicate that while linseed oleogel can replace pork fat, higher substitution levels may compromise consumer acceptability due to altered appearance, odor, taste, and juiciness [52]. These differences often stem from the specific types and concentrations of oleogels used, as well as their interactions with other ingredients, which influence characteristics such as firmness, juiciness, and cohesiveness. This feedback is essential for refining oleogel formulations, highlighting the need to optimize oleogel attributes to closely mimic traditional pork fat's sensory qualities while offering a healthier alternative. Table 2 lists recent findings on the sensory properties of sausages with oleogel incorporation.

12. Effect of Oleogels on Oxidative Stability in Sausage Products

The application of oleogels as partial fat replacements in sausage formulations has shown promising results in managing lipid oxidation while offering healthier alternatives to traditional animal fats. In Harbin red sausage, ethylcellulose oleogels with peanut oil at concentrations of 6–12% were used to replace pork fat at levels ranging from 10–50%. Results indicated that lower replacement levels (10–30%) maintained oxidative stability similar to the pork fat control, making these levels optimal for reducing saturated fat without compromising product oxidative stability. However, higher replacement levels (40–50%) led to increased lipid oxidation due to the unsaturated nature of peanut oil [61]. In frankfurter sausages, a sunflower oil oleogel structured with a 15:5 monoglycerides-to-phytosterols ratio replaced 50% of pork backfat, achieving comparable oxidative stability to the control over 40 days. The interaction of monoglycerides and phytosterols formed a robust network structure that limited oxidation effectively, supporting its suitability for fat replacement in processed meats [62]. Additionally, in another frankfurter study, rice bran wax oleogels with soybean oil at 2.5% and 10% concentrations were used to replace pork fat. While the 2.5% rice bran wax oleogel maintained oxidation levels close to the pork fat control, the 10% rice bran wax showed higher Thio-barbituric acid values, likely due to oxidation occurring from prolonged heating during oleogel formation. Nevertheless, both concentrations kept lipid oxidation within acceptable limits throughout storage, with lower concentrations proving particularly effective in balancing fat reduction and oxidation control [63]. Collectively, these studies suggest that oleogels structured with specific gelators and used in controlled concentrations—typically up to 30% replacement—can serve as viable fat substitutes in various sausage types, promoting healthier profiles while managing lipid oxidation effectively. Table 2 details recent studies on the oxidative stability of sausage products enhanced with oleogel incorporation.

13. Conclusions and Future Perspectives

This review highlights the potential of oleogels developed using the direct-dispersion approach as a compelling alternative to conventional fats in sausage production, emphasizing crucial quality considerations for integrating oleogels into sausage formulations. A diverse range of gelators, encompassing both low and high molecular weight compounds, can be utilized to structure various plant-based oils. The selection of the oil/gelator type and the corresponding ratios significantly influences the functional properties of sausages. Different gelators can elicit distinct effects on the texture, stability, and overall quality of the product. Furthermore, the choice of oil, particularly those rich in favorable fatty acids, can enhance the nutritional profile of the sausage, contributing to improved health outcomes by reducing saturated fat levels and increasing the content of unsaturated, health-promoting fatty acids. Oleogels have the potential to address the rising consumer demand for healthier meat products without compromising quality. Despite these advantages, a key challenge lies in refining oleogel formulations to closely replicate the sensory and texture characteristics of animal fats. Advancements in structuring techniques and sensory optimization could further enhance oleogels' suitability for commercial sausage production. Scaling the application of oleogels in industrial sausage production will require careful optimization of formulation, processing, and equipment to maintain consistent quality and functionality at larger scales. At the industrial level, precise blending is crucial to ensure uniform distribution of oleogels in the sausage matrix while preserving the gel's structure under varying temperature conditions. Additionally, rigorous quality control measures will be necessary to monitor texture, shelf-life, and sensory characteristics, ensuring the final product aligns with consumer expectations. Key considerations for commercial success include cost efficiency, regulatory compliance, and proper labeling, particularly as health-conscious consumers increasingly demand alternatives to saturated fats in processed foods. Future research should prioritize enhancing oleogel formulations for greater stability, scalability, and sensory appeal, thus supporting broader industry adoption and consumer acceptance.

This progress will not only help meet public health goals but also provide nutritionally improved meat products without compromising on quality or appeal.

Author Contributions: Conceptualization: N.M. and R.C.d.S.; writing—original draft preparation: N.M. and M.J.F.; writing—review and editing: R.C.d.S., N.M. and M.J.F.; supervision: R.C.d.S. All authors have read and agreed to the published version of the manuscript.

Funding: This research was funded by NC.X-267-5-12-170-1/National Institute of Food and Agriculture 482 (NIFA), the Department of Family and Consumer Sciences and the Agriculture Research 483 Station at North Carolina Agricultural and Technical State University (Greensboro, NC 27411, 484 USA).

Institutional Review Board Statement: Not applicable.

Informed Consent Statement: Not applicable.

Data Availability Statement: No new data were created or analyzed in this study.

Conflicts of Interest: The authors declare no conflicts of interest.

References

1. Mena, B.; Sîrbu, A.; Eze, C.C. Global Consumer Perception Towards Healthy Foods: Influencing Factors and Current Trends. In *Consumer Perceptions and Food*; Bogueva, D., Ed.; Springer Nature: Singapore, 2024; pp. 605–623, ISBN 978-981-9778-69-0.
2. Cena, H.; Calder, P.C. Defining a Healthy Diet: Evidence for the Role of Contemporary Dietary Patterns in Health and Disease. *Nutrients* **2020**, *12*, 334. [CrossRef] [PubMed]
3. Safa, H.; Portanguen, S.; Mirade, P.-S. Reducing the Levels of Sodium, Saturated Animal Fat, and Nitrite in Dry-Cured Pork Meat Products: A Major Challenge. *Food Nutr. Sci.* **2017**, *8*, 419–443. [CrossRef]
4. Avelar De Sousa, C.C.; De Jesus Pereira Dos Santos, L.; Da Silva, M.C.A.; Cavalheiro, C.P. A Comprehensive Overview of Sodium, Total and Saturated Fat Content in Meat Products Sold in Brazil. *Nutr. Food Sci.* **2024**. [CrossRef]
5. Carballo, J. Sausages: Nutrition, Safety, Processing and Quality Improvement. *Foods* **2021**, *10*, 890. [CrossRef]
6. Silva, R.C.D.; Ferdaus, M.J.; Foguel, A.; Da Silva, T.L.T. Oleogels as a Fat Substitute in Food: A Current Review. *Gels* **2023**, *9*, 180. [CrossRef]
7. Perța-Crișan, S.; Ursachi, C.-. Ștefan; Chereji, B.-D.; Munteanu, F.-D. Oleogels—Innovative Technological Solution for the Nutritional Improvement of Meat Products. *Foods* **2022**, *12*, 131. [CrossRef]
8. Perricone, M.; Arace, E.; Calò, G.; Sinigaglia, M. Ethnic Fermented Foods. In *Starter Cultures in Food Production*; Speranza, B., Bevilacqua, A., Corbo, M.R., Sinigaglia, M., Eds.; Wiley: Hoboken, NJ, USA, 2017; pp. 384–406, ISBN 978-1-118-93376-3.
9. Lonergan, S.M.; Topel, D.G.; Marple, D.N. Sausage Processing and Production. In *The Science of Animal Growth and Meat Technology*; Elsevier: Amsterdam, The Netherlands, 2019; pp. 229–253, ISBN 978-0-12-815277-5.
10. *Handbook of Animal-Based Fermented Food and Beverage Technology*; Hui, Y.H., Evranuz, E.Ö., Eds.; CRC Press: Boca Raton, FL, USA, 2016; ISBN 978-0-429-10752-8.
11. Statista Sausages—Worldwide. Available online: https://www.statista.com/outlook/cmo/food/meat/processed-meat/sausages/worldwide (accessed on 9 October 2024).
12. Statista Sausages—United States. Available online: https://www.statista.com/outlook/cmo/food/meat/processed-meat/sausages/united-states (accessed on 9 October 2024).
13. Hertrampf, J.W.; Piedad-Pascual, F. Animal Fats. In *Handbook on Ingredients for Aquaculture Feeds*; Springer: Dordrecht, The Netherlands, 2000; pp. 43–59, ISBN 978-1-4020-1527-4.
14. Chiu, S.; Williams, P.T.; Krauss, R.M. Effects of a Very High Saturated Fat Diet on LDL Particles in Adults with Atherogenic Dyslipidemia: A Randomized Controlled Trial. *PLoS ONE* **2017**, *12*, e0170664. [CrossRef] [PubMed]
15. Lawrence, G.D. Perspective: The Saturated Fat–Unsaturated Oil Dilemma: Relations of Dietary Fatty Acids and Serum Cholesterol, Atherosclerosis, Inflammation, Cancer, and All-Cause Mortality. *Adv. Nutr.* **2021**, *12*, 647–656. [CrossRef]
16. Ruiz-Núñez, B.; Dijck-Brouwer, D.A.J.; Muskiet, F.A.J. The Relation of Saturated Fatty Acids with low-Grade Inflammation and Cardiovascular Disease. *J. Nutr. Biochem.* **2016**, *36*, 1–20. [CrossRef]
17. Halagarda, M.; Wójciak, K.M. Health and Safety Aspects of Traditional European Meat Products. A Review. *Meat Sci.* **2022**, *184*, 108623. [CrossRef]
18. Zhou, T.; Sheng, B.; Gao, H.; Nie, X.; Sun, H.; Xing, B.; Wu, L.; Zhao, D.; Wu, J.; Li, C. Effect of Fat Concentration on Protein Digestibility of Chinese Sausage. *Food Res. Int.* **2024**, *177*, 113922. [CrossRef] [PubMed]
19. Whelton, P.K.; Appel, L.J.; Sacco, R.L.; Anderson, C.A.M.; Antman, E.M.; Campbell, N.; Dunbar, S.B.; Frohlich, E.D.; Hall, J.E.; Jessup, M.; et al. Sodium, Blood Pressure, and Cardiovascular Disease: Further Evidence Supporting the American Heart Association Sodium Reduction Recommendations. *Circulation* **2012**, *126*, 2880–2889. [CrossRef] [PubMed]
20. Food and Drug Administration Sodium in Your Diet. Available online: https://www.fda.gov/food/nutrition-education-resources-materials/sodium-your-diet (accessed on 12 March 2024).

21. Bastide, N.M.; Pierre, F.H.F.; Corpet, D.E. Heme Iron from Meat and Risk of Colorectal Cancer: A Meta-analysis and a Review of the Mechanisms Involved. *Cancer Prev. Res.* **2011**, *4*, 177–184. [CrossRef] [PubMed]
22. Kalaycıoğlu, Z.; Erim, F.B. Nitrate and Nitrites in Foods: Worldwide Regional Distribution in View of Their Risks and Benefits. *J. Agric. Food Chem.* **2019**, *67*, 7205–7222. [CrossRef]
23. Bouvard, V.; Loomis, D.; Guyton, K.Z.; Grosse, Y.; Ghissassi, F.E.; Benbrahim-Tallaa, L.; Guha, N.; Mattock, H.; Straif, K. Carcinogenicity of Consumption of Red and Processed Meat. *Lancet Oncol.* **2015**, *16*, 1599–1600. [CrossRef]
24. Wang, Y.; Beydoun, M.A. Meat Consumption is Associated with Obesity and Central Obesity Among US Adults. *Int. J. Obes.* **2009**, *33*, 621–628. [CrossRef]
25. Guedes-Oliveira, J.M.; Brad Kim, Y.H.; Conte-Junior, C.A. What are the Potential Strategies to Achieve Potentially More Healthful Meat Products? *Int. J. Food Sci. Technol.* **2021**, *56*, 6157–6170. [CrossRef]
26. Taladrid, D.; Laguna, L.; Bartolomé, B.; Moreno-Arribas, M.V. Plant-Derived Seasonings as Sodium Salt Replacers in Food. *Trends Food Sci. Technol.* **2020**, *99*, 194–202. [CrossRef]
27. Flores, M.; Piornos, J.A. Fermented Meat Sausages and the Challenge of Their Plant-Based Alternatives: A Comparative Review on Aroma-Related Aspects. *Meat Sci.* **2021**, *182*, 108636. [CrossRef]
28. Jankauskienė, A.; Kiselioviene, S.; Aleknavičius, D.; Miliūnaitė, I.; Kerzienė, S.; Gaižauskaitė, Ž.; Juknienė, I.; Zaviztanavičiūtė, P.; Kabašinskienė, A. Innovative Applications of Tenebrio Molitor Larvae in the Production of Sustainable Meat Sausages: Quality and Safety Aspects. *Foods* **2024**, *13*, 1451. [CrossRef]
29. Santos, J.M.D.; Ignácio, E.O.; Bis-Souza, C.V.; Silva-Barretto, A.C.D. Performance of Reduced Fat-Reduced Salt Fermented Sausage with Added *Microcrystalline cellulose*, Resistant Starch and Oat Fiber Using the Simplex Design. *Meat Sci.* **2021**, *175*, 108433. [CrossRef] [PubMed]
30. Stoica, M.; Antohi, V.M.; Alexe, P.; Ivan, A.S.; Stanciu, S.; Stoica, D.; Zlati, M.L.; Stuparu-Cretu, M. New Strategies for the Total/Partial Replacement of Conventional Sodium Nitrite in Meat Products: A Review. *Food Bioprocess Technol.* **2022**, *15*, 514–538. [CrossRef]
31. Mahmud, N.; Valizadeh, S.; Oyom, W.; Tahergorabi, R. Exploring functional plant-based seafood: Ingredients and health implications. *Trends Food Sci. Technol.* **2024**, *144*, 104346. [CrossRef]
32. James, N. Savoring Innovation: Trends Transforming the Sausages Market. Available online: https://www.verifiedmarketreports.com/blog/top-7-trends-in-sausages/ (accessed on 9 October 2024).
33. Allied Market Research Low Fat Sausage Market Size, Share, Competitive Landscape and Trend Analysis Report, by Type, by End User, by Distribution Channel: Global Opportunity Analysis and Industry Forecast, 2021–2031. Available online: https://www.alliedmarketresearch.com/low-fat-sausage-market-A31786#:~:text=What%20is%20the%20total%20market,Q2 (accessed on 3 December 2024).
34. Valuates Global Low Fat Sausage Market Research Report. 2023. Available online: https://reports.valuates.com/market-reports/QYRE-Auto-37B13560/global-low-fat-sausage (accessed on 9 October 2024).
35. Allied Market Research Low Fat Sausage Market Will Gain Momentum by 2031 to Surpass $4.8 Billion | New Highs—Current Trends And Growth. Available online: https://www.alliedmarketresearch.com/request-sample/32236 (accessed on 9 October 2024).
36. Ferdaus, M.J.; Barman, B.; Mahmud, N.; Da Silva, R.C. Oleogels as a Promising Alternative to Animal Fat in Saturated Fat-Reduced Meat Products: A Review. *Gels* **2024**, *10*, 92. [CrossRef] [PubMed]
37. Mahmud, N.; Islam, J.; Tahergorabi, R. Utilizing Foam-Structured Hydroxypropyl Methylcellulose Oleogels with Curcumin for Deep-Fried Chicken Nuggets in Carnauba Wax-Canola Oil Oleogel. *J. Agric. Food Res.* **2024**, *18*, 101364. [CrossRef]
38. Airoldi, R.; Da Silva, T.L.T.; Ract, J.N.R.; Foguel, A.; Colleran, H.L.; Ibrahim, S.A.; Da Silva, R.C. Potential Use of Carnauba Wax Oleogel to Replace Saturated Fat in Ice Cream. *J. Am. Oil Chem. Soc.* **2022**, *99*, 1085–1099. [CrossRef]
39. Islam, J.; Mahmud, N.; Oyom, W.; Aminzare, M.; Tahergorabi, R. Effects of Β-Sitosterol Oleogel on Fat Uptake Reduction, Physicochemical Properties, Oxidative Stability, and Sensory Characteristics of Fried Surimi Fish Balls. *J. Agric. Food Res.* **2024**, *17*, 101240. [CrossRef]
40. Lonkila, A.; Kaljonen, M. Ontological Struggle over New Product Category: Transition potential of Meat Alternatives. *Environ. Innov. Soc. Transit.* **2022**, *42*, 1–11. [CrossRef]
41. Food Insight Consumer Survey: Purchasing Behaviors, Eating Decisions and Health Perceptions of Dietary Fats and Oils. Available online: https://foodinsight.org/consumer-survey-purchasing-behaviors-eating-decisions-and-health-perceptions-of-dietary-fats-and-oils/ (accessed on 9 October 2024).
42. Ferdaus, M.J.; Blount, R.J.S.; Silva, R.C. da Assessment of Natural Waxes as Stabilizers in Peanut Butter. *Foods* **2022**, *11*, 3127. [CrossRef]
43. Mahmud, N.; Islam, J.; Oyom, W.; Adrah, K.; Adegoke, S.C.; Tahergorabi, R. A Review of Different Frying Oils and Oleogels as Alternative Frying Media for Fat-Uptake Reduction in Deep-Fat Fried Foods. *Heliyon* **2023**, *9*, e21500. [CrossRef]
44. Ilyin, S.O. Structural Rheology in the Development and Study of Complex Polymer Materials. *Polymers* **2024**, *16*, 2458. [CrossRef] [PubMed]
45. Perez-Santana, M.; Cedeno-Sanchez, V.; Carriglio, J.C.; MacIntosh, A.J. The Effects of Emulsifier Addition on the Functionalization of a High-Oleic Palm Oil-Based Oleogel. *Gels* **2023**, *9*, 522. [CrossRef] [PubMed]
46. Li, S.; Zhu, L.; Wu, G.; Jin, Q.; Wang, X.; Zhang, H. Relationship Between the Microstructure and Physical Properties of Emulsifier Based Oleogels and Cookies Quality. *Food Chem.* **2022**, *377*, 131966. [CrossRef]

47. Qiu, H.; Qiu, Z.; Chen, Z.; Liu, L.; Wang, J.; Jiang, H.; Zhang, H.; Liu, G. Antioxidant Properties of Blueberry Extract in Different Oleogel Systems. *LWT* **2021**, *137*, 110364. [CrossRef]
48. Chen, Z.; Shi, Z.; Meng, Z. Development and Characterization of Antioxidant-Fortified Oleogels by Encapsulating Hydrophilic tea Polyphenols. *Food Chem.* **2023**, *414*, 135664. [CrossRef]
49. Jeong, S.; Lee, S.; Oh, I. Development of Antioxidant-Fortified Oleogel and Its Application as a Solid Fat Replacer to Muffin. *Foods* **2021**, *10*, 3059. [CrossRef]
50. Domínguez, R.; Lorenzo, J.M.; Pateiro, M.; Munekata, P.E.S.; Alves Dos Santos, B.; Basso Pinton, M.; Cichoski, A.J.; Bastianello Campagnol, P.C. Main Animal Fat Replacers for the Manufacture of Healthy Processed Meat Products. *Crit. Rev. Food Sci. Nutr.* **2024**, *64*, 2513–2532. [CrossRef]
51. Wang, Z.; Chandrapala, J.; Truong, T.; Farahnaky, A. Oleogels Prepared with Low Molecular Weight Gelators: Texture, Rheology and Sensory Properties, a Review. *Crit. Rev. Food Sci. Nutr.* **2023**, *63*, 6069–6113. [CrossRef]
52. Franco, D.; Martins, A.; López-Pedrouso, M.; Purriños, L.; Cerqueira, M.; Vicente, A.; Pastrana, L.; Zapata, C.; Lorenzo, J. Strategy towards Replacing Pork Backfat with a Linseed Oleogel in Frankfurter Sausages and Its Evaluation on Physicochemical, Nutritional, and Sensory Characteristics. *Foods* **2019**, *8*, 366. [CrossRef]
53. Da Silva, S.L.; Amaral, J.T.; Ribeiro, M.; Sebastião, E.E.; Vargas, C.; De Lima Franzen, F.; Schneider, G.; Lorenzo, J.M.; Fries, L.L.M.; Cichoski, A.J.; et al. Fat Replacement by Oleogel Rich in Oleic Acid and Its Impact on the Technological, Nutritional, Oxidative, and Sensory Properties of Bologna-Type Sausages. *Meat Sci.* **2019**, *149*, 141–148. [CrossRef]
54. Mahmud, N.; Tahergorabi, R. Modification of Frying Oil and Batter for Fat Uptake Reduction in Deep-Fried Chicken Products: An Overview. In *Proceedings of the National Conference on Next-Generation Sustainable Technologies for Small-Scale Producers (NGST 2022)*; Advances in Biological Sciences Research; Uzochukwu, G., Niroj, A., Ejimakor, G., Tahergorabi, R., Uzoechi, S., Eds.; Atlantis Press International BV: Dordrecht, The Netherland, 2023; Volume 34, pp. 54–63, ISBN 978-94-6463-281-1.
55. Oyom, W.; Mahmud, N.; Islam, J.; Valizadeh, S.; Awuku, R.B.; Ibrahim, S.A.; Tahergorabi, R. Enhancing the Oxidative Stability, Physicochemical and Sensory Quality of Deep-Fat Fried Chicken Nuggets Using Thyme Essential Oil-Loaded Oleogel Coatings. *Prog. Org. Coat.* **2024**, *186*, 107977. [CrossRef]
56. Pinto, T.C.; Sabet, S.; Kazerani García, A.; Kirjoranta, S.; Valoppi, F. Oleogel Preparation Methods and Classification. In *Advances in Oleogel Development, Characterization, and Nutritional Aspects*; Palla, C., Valoppi, F., Eds.; Springer International Publishing: Cham, Swizerland, 2024; pp. 77–114, ISBN 978-3-031-46830-8.
57. Gravelle, A.J.; Bovi Karatay, G.G.; Hubinger, M.D. Oleogels Produced by Indirect Methods. In *Advances in Oleogel Development, Characterization, and Nutritional Aspects*; Palla, C., Valoppi, F., Eds.; Springer International Publishing: Cham, Swizerland, 2024; pp. 231–269, ISBN 978-3-031-46830-8.
58. De Souza Paglarini, C.; Câmara, A.K.F.I.; Da Silva, T.L.T.; Pollonio, M.A.R. Emulsified Meat Product with Oleogels for Reducing Saturated Fat Content. In *Functional Meat Products*; Verruck, S., Teixeira Marsico, E., Eds.; Methods and Protocols in Food Science; Springer US: New York, NY, USA, 2024; pp. 69–80, ISBN 978-1-07-163572-8.
59. Tarté, R.; Paulus, J.S.; Acevedo, N.C.; Prusa, K.J.; Lee, S.-L. High-Oleic and Conventional Soybean Oil Oleogels Structured with Rice Bran Wax as Alternatives to Pork Fat in Mechanically Separated Chicken-Based Bologna Sausage. *LWT* **2020**, *131*, 109659. [CrossRef]
60. Zampouni, K.; Soniadis, A.; Dimakopoulou-Papazoglou, D.; Moschakis, T.; Biliaderis, C.G.; Katsanidis, E. Modified Fermented Sausages with Olive Oil Oleogel and NaCl–KCl Substitution for Improved Nutritional Quality. *LWT* **2022**, *158*, 113172. [CrossRef]
61. Shao, L.; Bi, J.; Li, X.; Dai, R. Effects of Vegetable oil and Ethylcellulose on the Oleogel Properties and Its Application in Harbin Red Sausage. *Int. J. Biol. Macromol.* **2023**, *239*, 124299. [CrossRef] [PubMed]
62. Kouzounis, D.; Lazaridou, A.; Katsanidis, E. Partial Replacement of Animal Fat by Oleogels Structured with Monoglycerides and Phytosterols in Frankfurter Sausages. *Meat Sci.* **2017**, *130*, 38–46. [CrossRef]
63. Wolfer, T.L.; Acevedo, N.C.; Prusa, K.J.; Sebranek, J.G.; Tarté, R. Replacement of Pork fat in Frankfurter-Type Sausages by Soybean Oil Oleogels Structured with Rice Bran wax. *Meat Sci.* **2018**, *145*, 352–362. [CrossRef]
64. Ferro, A.C.; De Souza Paglarini, C.; Rodrigues Pollonio, M.A.; Lopes Cunha, R. Glyceryl Monostearate-Based Oleogels as a New Fat Substitute in Meat Emulsion. *Meat Sci.* **2021**, *174*, 108424. [CrossRef]
65. Yılmaz, E.; Toksöz, B. Flaxseed Oil-Wax Oleogels Replacement for Tallowfat in Sucuk Samples Provided Higher Concentrations of Polyunsaturated Fatty Acids and Aromatic Volatiles. *Meat Sci.* **2022**, *192*, 108875. [CrossRef]
66. Avendaño Vásquez, G.; Varela Santos, E.D.C.; Vargas García, K.A.; Chávez Alfaro, O.M.; Bustos Ramírez, K. Study of Textural Properties Frankfurter's Sausages Modified with Candelilla Wax Oleogel. *Agro Product.* **2024**, *17*, 181–187. [CrossRef]
67. Igenbayev, A.; Ospankulova, G.; Amirkhanov, S.; Aldiyeva, A.; Temirova, I.; Amirkhanov, K. Substitution of Pork Fat with Beeswax-Structured Oleogels in Semi-Smoked Sausages. *Appl. Sci.* **2023**, *13*, 5312. [CrossRef]
68. Pintado, T.; Cofrades, S. Quality Characteristics of Healthy Dry Fermented Sausages Formulated with a Mixture of Olive and Chia Oil Structured in Oleogel or Emulsion Gel as Animal Fat Replacer. *Foods* **2020**, *9*, 830. [CrossRef] [PubMed]
69. Franco, D.; Martins, A.J.; López-Pedrouso, M.; Cerqueira, M.A.; Purriños, L.; Pastrana, L.M.; Vicente, A.A.; Zapata, C.; Lorenzo, J.M. Evaluation of Linseed Oil Oleogels to Partially Replace Pork Backfat in Fermented Sausages. *J. Sci. Food Agric.* **2020**, *100*, 218–224. [CrossRef] [PubMed]
70. Tanislav, A.E.; Cornea, A.A.; Radu, E.D.; Țibulcă, D.; Mureșan, V.; Mudura, E. Candelilla Wax and Glycerol Monostearate-Based Oleogels as Animal Fat Substitutes in Bologna Sausages. *Gels* **2024**, *10*, 399. [CrossRef] [PubMed]

71. Issara, U. Improvement of Thai Sweet Sausage (Goon Chiang) Properties by Oleogel Made of Rice Bran Wax and Rice Bran Oil: A Textural, Sensorial, and Nutritional Aspect. *IOP Conf. Ser. Earth Environ. Sci.* **2022**, *995*, 012045. [CrossRef]
72. Kong, W.; Wang, L.; Xu, H.; Liu, D. Effects of Lecithin/Sorbitol Monostearate-Canola Oil Oleogel as Animal Fat Replacer on the Fatty Acid Composition and Physicochemical Properties of Lamb Sausage. *J. Food Process. Preserv.* **2023**, *2023*, 2567854. [CrossRef]
73. Patel, A.R. CHAPTER 1. Oil Structuring: Concepts, Overview and Future Perspectives. In *Food Chemistry, Function and Analysis*; Patel, A.R., Ed.; Royal Society of Chemistry: Cambridge, UK, 2017; pp. 1–22, ISBN 978-1-78262-829-3.
74. López-Pedrouso, M.; Lorenzo, J.M.; Gullón, B.; Campagnol, P.C.B.; Franco, D. Novel Strategy for Developing Healthy Meat Products Replacing Saturated Fat with Oleogels. *Curr. Opin. Food Sci.* **2021**, *40*, 40–45. [CrossRef]
75. Tan, T.H.; Chan, E.; Manja, M.; Tang, T.; Phuah, E.; Lee, Y. Production, Health Implications and Applications of Oleogels as Fat Replacer in Food System: A Review. *J. Am. Oil Chem. Soc.* **2023**, *100*, 681–697. [CrossRef]
76. Manzoor, S.; Masoodi, F.A.; Naqash, F.; Rashid, R. Oleogels: Promising Alternatives to Solid Fats for Food Applications. *Food Hydrocoll. Health* **2022**, *2*, 100058. [CrossRef]
77. Bascuas, S.; Morell, P.; Hernando, I.; Quiles, A. Recent Trends in Oil Structuring Using Hydrocolloids. *Food Hydrocoll.* **2021**, *118*, 106612. [CrossRef]
78. Li, Y.; Zou, Y.; Que, F.; Zhang, H. Recent Advances in Fabrication of Edible Polymer Oleogels for Food Applications. *Curr. Opin. Food Sci.* **2022**, *43*, 114–119. [CrossRef]
79. Dubois, V.; Breton, S.; Linder, M.; Fanni, J.; Parmentier, M. Fatty Acid Profiles of 80 Vegetable Oils with Regard to Their Nutritional Potential. *Eur. J. Lipid Sci. Technol.* **2007**, *109*, 710–732. [CrossRef]
80. Lunn, J.; Theobald, H.E. The Health Effects of Dietary Unsaturated Fatty Acids. *Nutr. Bull.* **2006**, *31*, 178–224. [CrossRef]
81. Davidovich-Pinhas, M. Oleogels. In *Polymeric Gels*; Elsevier: Amsterdam, The Netherlands, 2018; pp. 231–249, ISBN 978-0-08-102179-8.
82. Sagiri, S.S.; Rao, K.J. Natural and bioderived molecular gelator–based oleogels and Their Applications. In *Biopolymer-Based Formulations*; Elsevier: Amsterdam, The Netherlands, 2020; pp. 513–559, ISBN 978-0-12-816897-4.
83. Li, S.; Wu, G.; Li, X.; Jin, Q.; Wang, X.; Zhang, H. Roles of Gelator Type And Gelation Technology on Texture and Sensory Properties of Cookies Prepared with Oleogels. *Food Chem.* **2021**, *356*, 129667. [CrossRef]
84. Bozkurt, H.; Bayram, M. Colour and Textural Attributes of Sucuk During Ripening. *Meat Sci.* **2006**, *73*, 344–350. [CrossRef] [PubMed]
85. Savadkoohi, S.; Hoogenkamp, H.; Shamsi, K.; Farahnaky, A. Color, Sensory and Textural Attributes of Beef Frankfurter, Beef Ham and Meat-Free Sausage Containing Tomato Pomace. *Meat Sci.* **2014**, *97*, 410–418. [CrossRef]
86. Ruiz, J.; García, C.; Muriel, E.; Andrés, A.I.; Ventanas, J. Influence of Sensory Characteristics on the Acceptability of Dry-Cured Ham. *Meat Sci.* **2002**, *61*, 347–354. [CrossRef]

MDPI AG
Grosspeteranlage 5
4052 Basel
Switzerland
Tel.: +41 61 683 77 34

Gels Editorial Office
E-mail: gels@mdpi.com
www.mdpi.com/journal/gels

www.ingramcontent.com/pod-product-compliance
Lightning Source LLC
Chambersburg PA
CBHW061515210926
51850CB00008B/1929
9783725863747